AF614791

# Designing 2D and 3D Network-on-Chip Architectures

Konstantinos Tatas · Kostas Siozios
Dimitrios Soudris · Axel Jantsch

# Designing 2D and 3D Network-on-Chip Architectures

Springer

Konstantinos Tatas
Department of Computer Science
and Engineering, School of
Applied Sciences
Frederick University
Nicosia
Cyprus

Axel Jantsch
Department of Electronic Systems
Royal Institute of Technology
Kista
Sweden

Kostas Siozios
Dimitrios Soudris
Department of Computer Science, School
of Electrical and Computer Engineering
National Technical University of Athens
Athens
Greece

ISBN 978-1-4614-4273-8 ISBN 978-1-4614-4274-5 (eBook)
DOI 10.1007/978-1-4614-4274-5
Springer New York Heidelberg Dordrecht London

Library of Congress Control Number: 2013944755



Printed on acid-free paper

Springer is part of Springer Science+Business Media (www.springer.com)

# Preface

Moore's law continues unabated and new design challenges lead to new design methodologies and even paradigm shifts. One such recent development is the introduction of three-dimensional integration technology. Efficiently utilizing novel technologies poses new design challenges and therefore require new design methodologies and EDA tools. Training engineers in these methodologies and design techniques is essentially done at the graduate level and once these technologies become the established paradigm, at the undergraduate level.

Network-on-Chip technology has been a popular research topic for a while now, and is the current design paradigm for multi- and many-core architectures. It is also a natural complement for 3D integration technology. Its multifaceted and multidisciplinary nature imposes a number of challenges both in the industrial and academic environments. While at the graduate level it is common or even preferable to use papers, case studies and assignments as the main teaching tools, at the undergraduate level a suitable textbook is indispensable. Since there is an increased need to include an introduction to Networks-on-Chip in undergraduate curricula, such a textbook is required, and has been missing from the literature for too long. At the same time, the large body of research work in the field must be also made available in an organized way for graduate students, researchers, and professionals to use as reference.

Therefore, the purpose of this book is two-fold. First, to be used as an undergraduate or graduate level textbook for introduction to Network-on-Chip technology providing students and practising engineers with the fundamentals as well as details in the many facets of Network-on-Chip design, including recent work on 3D NoCs. Second, to be used as reference for researchers in the field.

For this purpose, each of the first seven chapters in the first part of the book begins with the introduction of the fundamental concepts assuming only little prior knowledge of the reader in the field of digital design, computer architecture, and VLSI design. Instead, relative information is often summarized here in order to make each chapter as self-contained as possible. At the same time, after the introduction to the fundamentals, following advances in the area are summarized in a survey manner with appropriate references, so that the student can immediately build upon the fundamentals while the practising researcher can easily find relative information. Furthermore, each chapter contains a number of questions, problems, and design assignments that can be used in the class or the laboratory.

These first seven chapters can be used as an introductory course in NoC design either at the undergraduate or graduate level.

The next four chapters in the first part of the book contain case studies from the academia and industry, which can be used as a second (graduate-level) course in NoC design.

The second part of the book contains proposed design projects covering all aspects of NoC design, starting with design space exploration and high-level simulation, down to the VLSI design of NoC components.

The authors would like to thank all the people who helped to make this book possible, by contributing, providing reviews, and experimental results.

Nicosia, June 2013 — Konstantinos Tatas
Athens — Kostas Siozios
Athens — Dimitrios Soudris
Stockholm — Axel Jantsch

# Contents

# Acronyms

| | |
|---|---|
| BE | Best effort |
| CA | Cycle accurate |
| DFS | Dynamic frequency scaling |
| DPM | Dynamic power management |
| DUV | Design under verification |
| DVS | Dynamic voltage scaling |
| FIFO | First input first output |
| HDL | Hardware description language |
| IC | Integrated circuit |
| NI | Network interface |
| NoC | Network-on-chip |
| OCP | Open core protocol |
| OSI | Open systems interconnection |
| PE | Processing element |
| QoS | Quality of service |
| RTL | Register transfer level |
| SAF | Store-and-forward |
| SoC | Systems-on-chip |
| TDMA | Time division multiple access |
| TSV | Through silicon via |
| VC | Virtual channel |
| VCT | Virtual cut-through |
| WS | Wormhole switching |

# Part I
# Network-on-Chip Design Methodology

# Chapter 1
# Network-on-Chip Technology: A Paradigm Shift

**Abstract** The first chapter introduces the fundamental Network-on-Chip (NoC) concepts starting with the motivation that caused the paradigm shift from bus-based to NoC-based architectures. Similarities and differences between NoCs and computer networks are discussed. Furthermore, the components of a NoC-based System-on-Chip (SoC) are introduced and the NoC functionality is outlined in terms of the OSI layer structure. A significant part of this chapter is spent explaining both the benefits and challenges of adopting NoC as the SoC communication infrastructure. Finally, current research topics in the area are classified.

## 1.1 Introduction

The driving force behind Integrated Circuit (IC) technology has been Moore's law for almost five decades [1]. Although this is projected to slow down to a doubling every 3 years in the next few years for fixed chip sizes [2], the exponential trend is still in force. Though the evolution is continuous, the system level focus, or system scope, moves in steps. When a technology matures for a given implementation style, it leads to a paradigm shift. Past examples of such shifts were moving from room- to rack-level systems (LSI-1970s) and later from rack- to board-level systems (VLSI-1980s). This trend allowed in the 1990's the introduction of Systems-on-Chip (SoC), the integration of many components such as microprocessors, custom IP, and even analog in a single die. One of the implications of Moore's law is the increase in IC operating frequency. However, in the past few years integrated circuits have not been keeping up with the operating frequencies predicted by Moore's law mainly due to power consumption and thermal density constraints, as shown in Fig. 1.1. Since, however, transistor density has continued to increase unencumbered, designers worked around this problem by introducing multicore and many-core architectures [3], increasing performance without a proportional increase in operating frequency. This trend continues today with the introduction of heterogeneous many-core architectures [4].

K. Tatas et al., *Designing 2D and 3D Network-on-Chip Architectures*,
DOI: 10.1007/978-1-4614-4274-5_1, © Springer Science+Business Media New York 2014

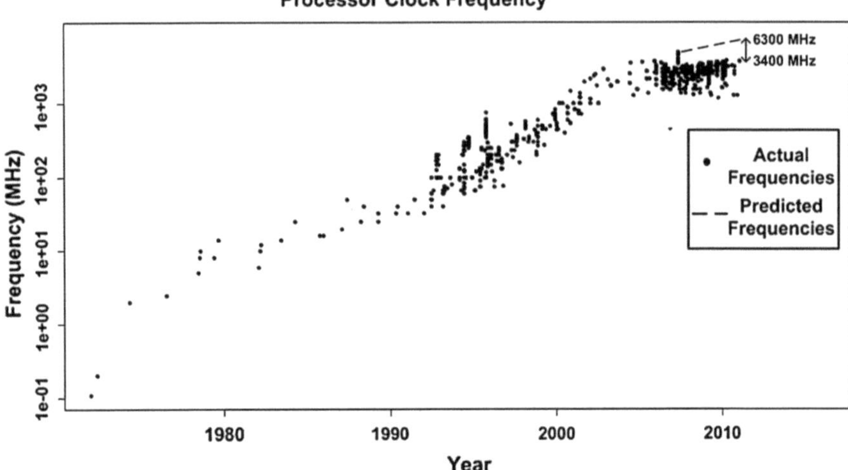


**Fig. 1.1** Processor actual operating frequencies (Data from Stanford CPU-DB [5]) vs predicted operating frequencies (from ITRS roadmap [2])

Historically, computation has been expensive and communication cheap. However, with scaling microchip technologies, this changed. More specifically, last years computation is becoming ever cheaper, while communication encounters fundamental physical limitations such as time-of-flight of electrical signals, power use in driving long wires/cables, etc. In comparison with off-chip, on-chip communication is significantly cheaper. There is room for lots of wires on a chip. Thus, the shift to single-chip systems has relaxed system communication problems. However on-chip wires do not scale in the same manner as transistors do, and, as we shall see in the following, the cost gap between computation and communication is widening, as it is depicted in Fig. 1.2. Meanwhile the differences between on- and off-chip wires make the direct scaling down of traditional multi-computer networks suboptimal for on-chip use.

In addition to that, silicon technologies face other challenges. Synchronization of future chips with a single clock source and negligible skew will be extremely difficult, if not impossible. The most likely synchronization paradigm for future chips, globally asynchronous and locally synchronous, involves using many different clocks. In the absence of a single timing reference, SoC chips become distributed systems on a single silicon substrate. Global control of the information traffic is unlikely to succeed because the system needs to keep track of each component's states. Thus, components will initiate data transfers autonomously, according to their needs. The global communication pattern will be fully distributed, with little or no global coordination.

As SoC complexity scales, capturing the system's functionality with fully deterministic operation models will become increasingly difficult. As global wires span multiple clock domains, synchronization failures in communicating between differ-

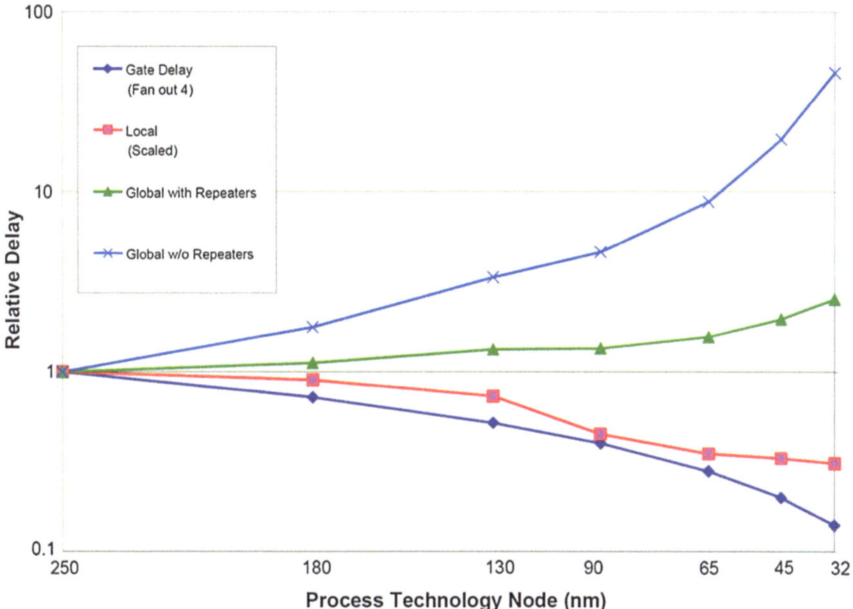


**Fig. 1.2** Projected relative delay for local and global wires and for logic gates in technologies of the near future [2]

ent domains will be rare but unavoidable events [6]. Moreover, energy and device reliability concerns will impose low levels of logic swing and power supply voltage, most likely less than one volt [2]. Electrical noise due to crosstalk, electromagnetic interference, and radiation-induced charge injection will likely produce data errors, also called upsets. Thus, transmitting digital values on wires will be inherently unreliable and non-deterministic. Other causes of non-determinism include design components with a high level of abstraction and coarse granularity and distributed communication control.

The integration of many processing and memory cores in a single chip introduced in turn a communication overhead that traditional bus-based architectures cannot handle for a number of reasons [7]. More specifically, the interconnection infrastructure has a significant impact on SoC costs because among others it influences four key factors of SoC design: die size, power consumption, design time, and performance.

- *Die size* can increase if conventional interconnect routing wire and gate area requirements explode due to the increasing number of IP blocks in a SoC.
- *Power consumption* can mushroom if an SoC's interconnect cannot easily be configured for advanced power management schemes like dynamic frequency and voltage scaling (DVFS).
- *Project design time* can extend if the SoC's interconnect becomes difficult to configure and verify. This can slow downstream designs if a platform-based design methodology is used as a basis for derivative SoC designs.

| BUS Pros & Cons | | | NoC Pros & Cons |
|---|---|---|---|
| Every unit attached adds parasitic capacitance, therefore electrical performance degrades with growth. | - | + | Only point-to-point one-way wires are used, for all network sizes, thus local performance is not degraded when scaling. |
| Bus timing is difficult in a deep submicron process. | - | + | Network wires can be pipelined because links are point-to-point. |
| Bus arbitration can become a bottleneck. The arbitration delay grows with the number of masters. | - | + | Routing decisions are distributed, if the network protocol is made non-central. |
| The bus arbiter is instance-specific. | - | + | The same router may be re-instantiated, for all network sizes. |
| Bus testability is problematic and slow. | - | + | Locally placed dedicated BIST is fast and offers good test coverage. |
| Bandwidth is limited and shared by all units attached. | - | + | Aggregated bandwidth scales with the network size. |
| Bus latency is wire-speed once arbiter has granted control. | + | - | Internal network contention may cause a latency. |
| Any bus is almost directly compatible with most available IPs, including software running on CPUs. | + | - | Bus-oriented IPs need smart wrappers. Software needs clean synchronization in multiprocessor systems. |
| The concepts are simple and well understood. | + | - | System designers need reeducation for new concepts. |

**Fig. 1.3** Qualitative comparison between NoC and bus interconnections [8]

- *Performance* can suffer if an interconnect approach cannot adapt to changing requirements over the SoC's design cycle and product life for changing SoC IP blocks, quality of service (QoS), bandwidth, latency, security, and clock frequency.

Whereas the shared bus is a simple interface, since it is built on well-understood concepts and it is easy to model, however, it introduces a mentionable drawback especially in a highly interconnected (multicore) system. This mainly occurs because the buses do not scale well for a large number of components, while the long global wires and the system behavior becomes unpredictable from the component's point of view. More specifically, as the number of cores integrated to a system increases, the power usage per communication event grows as well due to more attached units leading to higher capacitive load. Apart from power dissipation, the arbitration problem especially for multi-master busses is also crucial. A qualitative comparison between Network-on-Chip (NoC) and bus-based approaches can be found in Fig. 1.3.

Based on this analysis, we can conclude that NoCs overcome some of the limitations found in buses; however even this interconnection paradigm cannot be thought of as ultimately scalable and, as such, it is an intermediate solution. Specifically, the dedicated point-to-point links are optimal in terms of bandwidth availability, latency, and power usage, since they usually are designed with a full-custom approach (application-specific interconnection solution). Also, it is much simpler to model, design, and verify the proper functionality of point-to-point links, as compared to the corresponding network. On the other hand, as the number of cores increases, there is an exponential increase to the number of links. Thus an area and possibly a routing problem arises.

From the point of view of design-effort, we can claim that in small size systems (consisting of a few cores), it might be efficient to provide the desired communication through an ad hoc communication structure. But, as the system size grows and the design cycle time requirements decrease, there is an increasing demand for proposing more generalized solutions. For maximum flexibility and scalability, it is generally accepted that a move toward a shared, segmented global communication structure is needed. The segmented long wires in this approach avoid signal degradation, and busses are implemented as multiplexed structures in order to reduce power and increase responsiveness. This notion translates into a data-routing network consisting of communication links and routing nodes that are implemented on the chip. In contrast to traditional SoC communication methods outlined previously, such a distributed communication media scales well with chip size and complexity. Additional advantages include increased aggregated performance by exploiting parallel operation. Furthermore, the recent introduction of 3-D integration technologies [9] literally added a new dimension to the bus-based core interconnection problem.

A solution that proved popular was adopting a well-established paradigm, that of computer networks. Thus, Networks-on-Chip (NoC) were introduced. NoC technology is often called "a front-end solution to a back-end problem". As semiconductor transistor dimensions shrink and increasing amounts of IP block functions are added to a chip, the physical infrastructure that carries data on the chip and guarantees QoS begins to crumble. Many of today's SoC are too complex to utilize a traditional hierarchal bus or crossbar interconnect approach. Yesterday's village traffic has turned into today's congested freeways.

## 1.2 Network-on-Chip

NoC technology is a relatively new approach to signaling that enables not only more efficient interconnects but also more efficient design and verification processes for modern SoCs. NoC is an approach to signaling that matches the needs of the signal to various communications protocols in a way that reduces the complexity of the chip's interconnect. Slow or low bandwidth signals can be multiplexed onto a single line with other signals, while only the highest speed, highest bandwidth signals communicate directly over space-consuming parallel paths.

A typical NoC-based MPSoC is shown in Fig. 1.4. It is composed of a number of components, called *nodes*, including Processing Elements (PEs), such as CPUs, custom IPs, DSPs, etc., and storage elements (embedded memory blocks). The PEs are attached to *network adapters* through the *Network Interfaces* (NIs), while their functionality is to decouple computation (the cores) from communication (the network).

The communication through the NoC is performed by enabling PEs to send and receive packets through the network fabric composed of *switches/routers* connected together through physical *links*, or *channels*. Typically, each link is a pair of opposite, unidirectional, point-to-point buses, possibly including appropriate repeaters. Since the desired properties for NoC usually is application-oriented, there are different approaches for designing these links (e.g., they might consist of one or more logical or physical channels). Often a switch together with its host PE, or memory, is referred to as a *tile*.

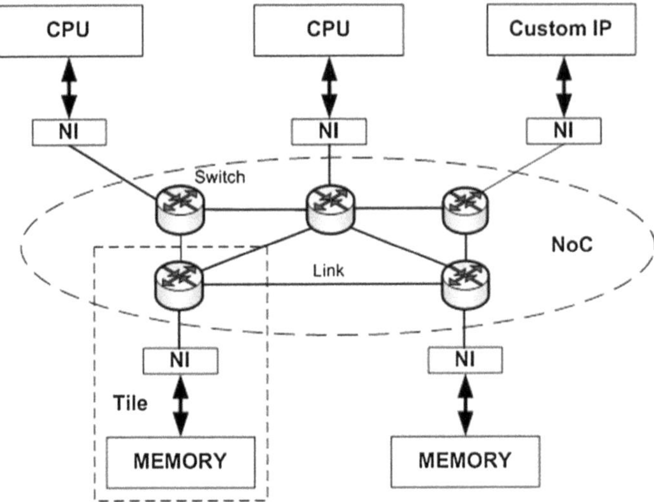


**Fig. 1.4** NoC-based MPSoC

A critical parameter at NoCs affects the decision about how the packets have to cross the network. Since NoC is a distributed network, these decisions are taken according to the employed *routing strategy* usually at each router. Since this parameter highly affects the performance of NoC-based platform, up to now different routing algorithms have been proposed in relevant literature [10].

For many designers the use of a NoC results in more flexibility to optimize their SoCs. Prior to using a NoC approach these designers only had time to change their interconnect about 5–6× per project. Today these same designers are able to refine their interconnect 9–10× per project. This gives them greater flexibility to accommodate changes coming from both engineering and customers while also providing more time for system-level power and performance optimization [11].

The NoC paradigm, of course, was not established overnight. Initially, it was not obvious that NoC was the best candidate architecture for alleviating the productivity gap problem. Jantch and Tenhunen [12] introduced two properties required of a design process in order to become a true paradigm shift:

1. *Arbitrary Composability*: A system composed of components and combinators that allow the components to be connected and integrated into larger component assemblages is arbitrarily composable if a given component assemblage $A$ can be extended with any component by using the existing combinators without changing the relative behavior of $A$.
2. *Linear Effort*: A design process that builds a system from the set of components and combinators features this property if the design effort of integrating a given set of $n$ assemblages by means of the combinators is dependent on $n$ but not the size of the assemblages.

Even though the above properties are meant as heuristics and not rigid mathematical properties, they are important criteria that should be met by a design process in order to be considered as a new paradigm. Consequently, the design of NoCs have to be appropriately performed in order to fulfill them, if not guarantee them in all cases. Some of the features that allow NoC-based platforms to adhere to the above fundamental properties are summarized as follows:

- Allows design reuse;
- separates communication from computation;
- avoids global, centralized controller for communication;
- allows arbitrary number of terminals;
- allows the addition of links as the system size grows (scalability);
- does not utilize long, global wires spanning the whole chip;
- facilitates customization (e.g., link width, buffer sizes, topology, etc.);
- allows multiple voltage and frequency domains;
- delivers data in order either naturally or via layered protocol;
- offers varying guarantees for transfers;
- offers support for system testing;
- adapts easily to 3-D integration platforms.

Next, we describe in more detail some of the most important challenges for NoC architectures [13]. Of course, we have to mention that whenever an architect has to design a new NoC architecture, these features have to be appropriately tuned in order to better address the product's specifications.

## 1.3 OSI Layer Roles in a NoC

NoCs are packet-switched communication networks derived from the parallel computing domain. By exploiting the lessons learned by the telecommunications community, the global on-chip communication is decomposed into layers similar to the ISO/OSI Reference Model seen in the network on computers (see Fig. 1.5).

The ISO/OSI model does not define exactly how a system should be build, but acts as a conceptual guideline. More specifically, by allowing each layer to hide its own complexity from the layers above and only communicates with adjacent layers, it is possible to enable different services, providing to the programmer an abstraction of the communication framework [14]. A series of these layers is known as a "protocol stack" and can be designed in hardware or software.

In the OSI model, network functionality is divided into seven different layers, whereas the goals of this classification can be summarized as follows:

- The layers should exchange a minimum of information through the interfaces;
- there should be sufficiently many layers so that unrelated functions are not put in the same layer;
- each layer should have boundaries with its upper and lower layer only.

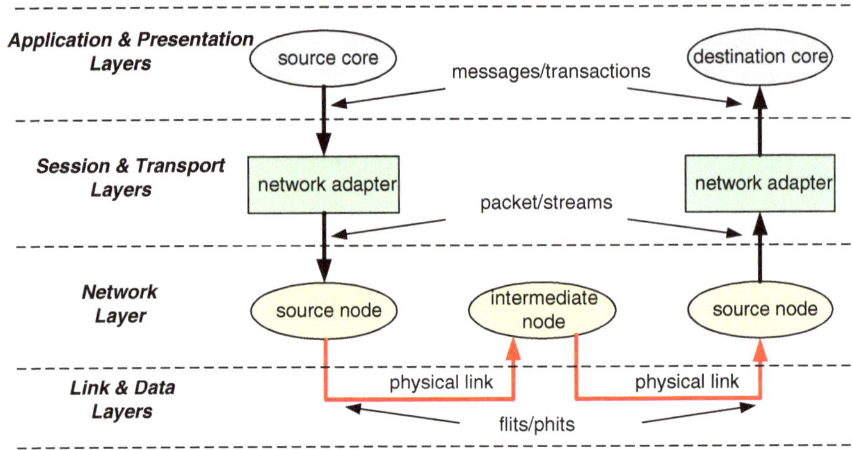


**Fig. 1.5** ISO/OSI reference model for NoCs

## 1.3.1 Application Layer

This layer is where the resources are located. Notice that a resource may be performing many different tasks (e.g., a microprocessor can execute several processes simultaneously).

## 1.3.2 Presentation Layer

As resources may have different representations for numbers, there must be some conversion between them. This layer performs the conversion among the different representations used from hardware resources. For instance, typical examples are representations with different resolution (i.e., integers in big-endian and little-endian format). Another example could be if a microprocessor uses a signal processor to speed up calculations. In such a case, if the microprocessor uses a floating point representation, whereas the signal processor uses a fixed point format, then conversion between these formats is absolutely necessary. Such a conversion is actually realized inside the presentation layer.

## 1.3.3 Session Layer

Using the transport layer, the session layer will be responsible for establishing connections between resources. Two different protocols are used for this scope: (i) a connection-oriented protocol, which is more energy efficient under heavy traffic due to retransmissions, and (ii) a connectionless protocol, where additional work at receiver is necessary in order to support out-of-order delivery of data.

### *1.3.4 Transport Layer*

This layer incorporates all the necessary mechanisms for checking and verifying that no packages are lost in the lower layers. The transport layer deals with the message segmentation into packages, as well as their reassembling. Another critical task that is handled inside the transport layer affects the flow control, which highly affects the performance of the underlying NoC. More specifically, network congestion increases cost per transmitted bit due to contention resolution overhead, whereas the amount of data that enters the network, can be regulated, at the price of throughput.

### *1.3.5 Network Layer*

The network layer handles issues related to the topology and the consequent routing scheme. More specifically, both hierarchical and heterogeneous architectures can be supported from this layer in order to better address the communication requirements of modern platforms. For instance, an optimum solution usually involves the clustering of nodes with high-bandwidth requirements and their connection through dedicated short length channels. This selection leads to mentionable energy and performance gains, instead of using the rest of the network, since the intracluster communication is by far more efficient compared to the corresponding intercluster links. In addition to this, the network layer is also responsible for employed routing algorithm. The two most widely accepted techniques for NoCs are the circuit and packet switching. The advantages and disadvantages of employing these algorithms are summarized as follows:

- Circuit switching:
  - network control overhead incurs only once;
  - best in case of persistent communication.
- Packet switching:
  - distributed network control overhead;
  - more energy-efficient for irregular communication.

### *1.3.6 Data Link Layer*

The data link layer ensures a reliable transfer despite the physical unreliability. For this purpose, it deals with the tasks of detecting and correcting errors occurred at the physical layer. In addition to that, this layer is also responsible for medium access control for sharing a common channel resource, with contention-based access. A typical approach for correcting an error is through the retransmission of data in case of error. However, this approach can be costly in terms of energy and performance degradation. Of course, the optimal choice for this parameter is derived after taking into consideration also the constraints posed by the target system, as well as the characteristics of physical channels.

### 1.3.7 Physical Layer

The physical layer refers to all that concerns the electrical characteristics of wires, the circuits, and techniques to drive information (drivers, repeaters, layout, etc). Different techniques can be employed during the design of physical layer targeting to optimize the performance and reduce power/energy dissipation. Among others are the usage of low-swing signaling at transmitter through the reduction in $V_{dd}$, the usage of differential receivers, as well as the employ of techniques that provide less reliable data reception (if this is affordable from the specifications of target system). In addition to that, the design of the physical layer includes tasks that affect the pseudo-differential signaling at receiver (e.g., reference signal sharing, less signal transitions, and reduced noise margin). Finally, since the clocks are extremely energy-inefficient, whereas the global synchronization of interconnection architecture is not optimal, the usage of Global Asynchronous Local Synchronous (GALS) units might be a viable solution.

Although such a structured hierarchy in network design is similar to the one found in LAN networks, the usage of existing algorithms and/or topologies is not suitable for realizing on-chip communication due to the numerous fundamental differences.

- *Dynamic insertion and removal of nodes* which is essential in computer networks, since they are designed for perpetual expansion is not required in NoC-based MPSoC since the number of nodes is a known priori. The only special consideration required is when a node or link fails.
- *Cost of buffers and links*: While buffers are cheap and links costly in computer networks, the opposite is true for the NoC environment, where it is common to have 128-bit wide links and even bufferless routers have been proposed.
- *Reliability and non-determinism*: Designing a reliable NoC architecture is especially challenging since the interconnection infrastructure is potentially affected by failure of any node that a flit visits, or failure of any communication link that it needs to traverse.
- *Performance*: Despite the common goal for all networks to achieve as much as possible energy savings, smaller delay and design-time specialization, the usage of traditional network technologies (such as TCP/IP) to NoCs, entails too much latency. In addition to that, NoCs exhibit local proximity and less non-determinism. The first attempts at NoC design were perhaps too computer network-influenced, often underestimating or even ignoring the above fundamental differences.

## 1.4 Benefits and Challenges of Adopting NoCs

Traditionally, ICs have been designed with dedicated point-to-point connections, with one wire dedicated to each signal. For large designs, in particular, this has several limitations from a physical design viewpoint. The wires occupy much of the area of the chip, and in nanometer CMOS technology, interconnects dominate both performance and dynamic power dissipation, as signal propagation in wires across the chip requires multiple clock cycles. (See Rent's rule for a discussion of wiring requirements for point-to-point connections).

NoC links can reduce the complexity of designing wires for predictable speed, power, noise, reliability, etc., thanks to their regular, well-controlled structure. From a system design viewpoint, with the advent of multicore processor systems, a network is a natural architectural choice. A NoC can provide separation between computation and communication, support modularity and IP reuse via standard interfaces, handle synchronization issues, serve as a platform for system test, and, hence, increase engineering productivity.

### *1.4.1 IP Reuse*

Initially, the massive reuse of existing IP blocks is an absolute necessity. The shapes the IPs to be-reused can take is quite variable: hard (layout-ready) or soft, in house or third-party, simple or complex (e.g., complete kernels). The NoC interconnect has to convey transactions between sockets of different protocols (OCP, AXI, proprietary, etc.), each of which is highly parameterized.

This variety has major impact from a performance perspective. For instance, the IPs may have to be clocked at different frequencies, whereas regarding their sockets they may present different data widths, leading to a wide range of peak throughputs which must be adapted between initiators and targets.

In addition to that, initiators usually exhibit different traffic patterns, in terms of transaction lengths, address alignments, as well as regularity. Consequently, their reaction and resistance to latency, or transient back-pressure, might differ as well. Also, meeting the QoS required by each initiator despite the presence of the other traffic becomes more challenging as the number of IPs increases.

### *1.4.2 Power Management*

In most SoCs, aggressive power management is becoming a requirement, not only for handheld devices where it is critical for the system autonomy, but also for desktop or automotive applications as well, where it has direct consequences on the chip packaging and system cooling. This power management includes local clock gating on groups of several flip-flops, global clock gating saving the energy dissipated in an idle block, and active power supply of some IPs or subsystems. As the NoC crosses power domains, it must collaborate with IPs and the SoC power controller, so as to guarantee the to-be-shutdown domain is empty of transactions, or request the power supply restoration.

In addition to that, dynamic voltage and frequency scaling (DVFS) is becoming a common practice, again as the NoC is likely to cross voltage domains, the places to insert required level-shifters must be cleanly identified. The sensitivity to power consumption also calls for a tight control of the gate and wire count.

### *1.4.3 Physical Constraints and Timing Closure*

Most IPs shrink with the CMOS process, the NoC does not: it is by definition spread across the die, making the back-end phase a difficult challenge. On the one hand, depending on clock frequencies, pipeline stages may have to be inserted, just to leave margin for spatial propagation delay. On the other hand, long wires are an expensive resource; the ability to locally adjust the number of wires in the communication bundles to back-end constraints and performance requirements is key.

### *1.4.4 Software Observability and Security*

If the hardware complexity rises, the software complexity explodes, with consequences on the hardware. Toward this direction, many different software layers collaborate on a SoC: Firmware, operating system, drivers, middleware, applications, sensitive data subsystems, etc., all distributed on the different IPs. The interconnect infrastructure has to provide mechanisms in order to grant, or deny, the right for transactions to reach certain targets, depending on initiators and in-band security qualifiers.

For software debugging reasons, the NoC must be observable to a certain extent. Error detection and reporting is a minimum, but transaction probing and sometimes on chip performance measurements are required as well. As on-chip traffic becomes more and more distributed, observation collection becomes more expensive. Since the whole traffic on the NoC cannot be observed in real-time, and observation must focus on strategic traffic such as DRAM accesses. In the end, the cost versus observability tradeoffs must be left in the hands of the designer.

### *1.4.5 Verification*

In general, IP quality verification is a well-known problem that designers face during the development of new products. Because of its wide configuration space (thousands of parameters is a good order of magnitude) verifying the implementation of a particular NoC is a much more complex problem. To make matters worse, checking the conformance of the sockets to their respective protocol is not enough: the correct translation of transactions between the initiator side and the target side must be verified as well. In order to address this requirement, depending on the initiator and target socket configurations, transactions may have to be split, realigned, unwrapped, etc. On top of this functional verification, the performance of the NoC also has to be validated.

### *1.4.6 EDA Design Flow*

NoC development overlaps the whole SoC design project lifetime: From early SoC architecture phase, to final place and route (P&R), passing through performance qualification, RTL design, IP integration, verification, software integration. One must pay special attention that a decision taken during an early phase, such as design architecture, does not jeopardize the feasibility of a later phase, such as back-end. Moreover, it is not unusual that marketing requirements for a given SoC change while the chip is being designed, adding or removing some IPs, and thus impacting the NoC specification.

In addition to that, as the complexity and cost of brand new architectures increases, organizing projects around reusable platforms becomes more efficient. Small variations of the platform, or derivatives, may then be brought to the market in shorter cycles. The interconnect system design methodology must make sure that those late specification changes are smoothly integrated in the design and verified.

## 1.5 Research on On-Chip Networks

There are a couple of requirements that every NoC implementation has to meet. Among others, performance requirements include small latency, guaranteed throughput, path diversity, sufficient transfer capacity, and low power consumption. Similarly, regarding the architectural requirements, there is a need for scalability, generality, and programmability. Apart from them, and due to technology scaling and different operating conditions, NoCs have also to address issues related to fault tolerance, as well as valid operation under different QoS demands.

In this section, we provide a review of the approaches of various research groups. Fig. 1.6 illustrates a simplified classification of this research.

Although NoCs can borrow concepts and techniques from the well-established domain of computer networking, it is impractical to blindly reuse features of "classical" computer networks and symmetric multiprocessors. In particular, NoC switches should be small, energy-efficient, and fast. Neglecting these aspects along with proper, quantitative comparison was typical for early NoC research but nowadays they are considered in more detail. The routing algorithms should be implemented by simple logic, and the number of data buffers should be minimal. Network topology and properties may be application-specific.

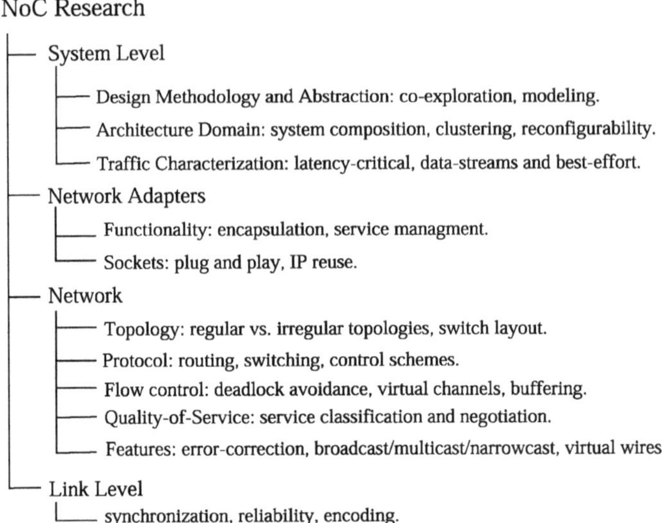


**Fig. 1.6** Typical NoC research area classification [8]

Some researchers think that NoCs need to support QoS, namely achieve the various requirements in terms of throughput, end-to-end delays, and deadlines. Real-time computation, including audio and video playback, is one reason for providing QoS support. However, current system implementations like VxWorks, RTLinux, or QNX are able to achieve sub-millisecond real-time computing without special hardware. This may indicate that for many real-time applications the service quality of existing on-chip interconnect infrastructure is sufficient, and dedicated hardware logic would be necessary to achieve microsecond precision, a degree that is rarely needed in practice for end users (sound or video jitter need only tenth of milliseconds latency guarantee). Another motivation for NoC-level quality-of-service is to support multiple concurrent users sharing resources of a single chip multiprocessor in a public cloud computing infrastructure. In such instances, hardware QoS logic enables the service provider to make contractual guarantees on the level of service that a user receives, a feature that may be deemed desirable by some corporate or government clients.

To date, several prototype NoCs have been designed and analyzed in both industry and academia but only few have been implemented on silicon. However, many challenging research problems remain to be solved at all levels, from the physical link level through the network level, and all the way up to the system architecture and application software. Additionally, a number of scientific publications onto peer-reviewed transactions, journals, conferences, and workshops can be found in the past few years. These publications study various hot topics of NoC, spanning from software level up to architecture, and hardware implementation.

## Questions

**1.1.** Why interconnection delay is by far more important as compared to logic delay?
**1.2.** Summarize the advantages and disadvantages of employing buses and NoCs.
**1.3.** What are the main reasons for adopting the NoC design paradigm?
**1.4.** List the main components of a NoC-based MPSoC, providing appropriate definitions.
**1.5.** What are the two fundamental properties for an arbitrarily scalable architecture?
**1.6.** Summarize the main features of NoC-based architecture.
**1.7.** What are the main challenges in NoC design?
**1.8.** Can you describe the OSI layers for a NoC architecture? What is the role for each of them?
**1.9.** What are the differences between NoC infrastructure, as compared to the conventional LAN networks?
**1.10.** Why GALS (Global Asynchronous Local Synchronous) is considered a viable solution for on-chip interconnection?
**1.11.** Describe briefly the advantages and disadvantages of using circuit switching as compared to packet switching.
**1.12.** Can you describe some open research issues at the NoC domain?

## References

1. G. Moore, No Exponential is Forever: But "Forever" Can Be Delayed! [semiconductor industry], Solid-State Circuits Conference (ISSCC), pp. 20–23, 2003
2. ITRS 2011.http://www.itrs.net
3. V. Lo, S. Rajopadhye, S. Gupta, D. Keldsen, M. Mohamed, B. Nitzberg, J. Telle, X. Zhong, OREGAMI: tools for mapping parallel computations to parallel architectures. Int. J. Parallel Programming **20**(3), 237–270 (1991)
4. S. Murali, G. De Micheli, SUNMAP: a tool for automatic topology selection and generation for NoCs, Design Automation Conference (DAC), pp. 914–919, 2004
5. Stanford CPU database. http://cpudb.stanford.edu
6. B. Ackland, A. Anesko, D. Brinthaupt, S. Daubert, A. Kalavade, J. Knobloch, E. Micca, M. Moturi, C. Nicol, J. O'Neill, J. Othmer, E. Sackinger, K. Singh, J. Sweet, C. Terman, J. Williams, A single chip, 1.6-billion, 16-b MAC/s multiprocessor DSP. IEEE J. Solid-State Circuits **35**(3), 412–424 (2000)
7. A comparison of Network-on-Chip and Busses, white paper by Arteris Corp. http://www.design-reuse.com/articles/10496/a-comparison-of-network-on-chip-and-busses.html
8. T. Bjerregaard, S. Mahadevan, A survey of research and practices of Network-on-chip. ACM Comput. Surv. **38**(1), Article 1 (2006)
9. V. Pavlidis, E. Friedman, *Three-Dimensional Integrated Circuit Design* (Morgan Kaufmann, San Francisco, 2010)
10. V. Rantala, T. Lehtonen, J. Plosila, Network on Chip Routing Algorithms, TUCS Technical Report, Turku Center for Computer Science, No. 779, August 2006

11. NoC Interconnect Improves SoC Economics, Objective Analysis - Semiconductor Market Research, 2011. http://www.objective-analysis.com/uploads/NoC_Interconnect_Improves_SoC_Economics_-_Objective_Analysis.pdf
12. A. Jantch, H. Tenhunen, *Networks on Chip* (Springer, Berlin, 2003)
13. J.-J. Lecler, G. Baillieu, Application driven network-on-chip architecture exploration & refinement for a complex SoC. J. Des. Autom. Embed. Syst. **15**(2), 133–158 (2011)
14. G. De Micheli, L. Benini, *Networks on Chips: Technology and Tools (Systems on Silicon)* (Morgan Kaufmann, San Francisco, 2006)

# Chapter 2
# NoC Modeling and Topology Exploration

**Abstract** This chapter describes two of the most important tasks for designing NoC-based systems dealing with NoC modeling, as well as the topology exploration. For this purpose, state-of-the-art architectural solutions are discussed and open research topics are highlighted. Additionally, this chapter provides a description of alternative traffic models used as input to the NoC domain for evaluating the efficiency of various architectural parameters. The last topics discussed in this chapter are topology synthesis and application mapping onto the derived NoC architecture under various constraints.

## 2.1 Introduction

With the advent of chip multi-processors (CMP) and multi-core systems-on-chip (SoC), the network-on-chip (NoC) paradigm has been proposed as a viable solution to the problem of connecting the continuously increasing number of processing cores that are integrated on a single die. The communication problem is expected to become far more important with the demand for higher processing power posed by the majority of application domains and the technology scaling.

Even though the NoC-based communication scheme introduces architectural scalability and performance enhancement, as already discussed in Chap. 1, however, careful design is absolutely required in order to achieve these gains. In this chapter, we discuss a number of architectural issues related to widely accepted NoC topologies. Additionally, the available academic and commercial solutions for topology modeling and synthesis are presented, whereas the main features/limitations for each of them are highlighted. In order to quantify the efficiency for each architectural selection, appropriate benchmarks are also required. For this purpose, throughout the second chapter we also introduce a number of typical traffic models employed for the scope of evaluating NoC architectures. Since the design, as well as the quantification of these NoC parameters are rather complex tasks, which cannot be easily performed without the usage of dedicated software tools, a number of frameworks are also introduced.

K. Tatas et al., *Designing 2D and 3D Network-on-Chip Architectures*,

DOI: 10.1007/978-1-4614-4274-5_2, © Springer Science+Business Media New York 2014

## 2.2 Topology Exploration

Topology refers to the structure of the network and its organization, since it determines the connections between nodes on chip. More specifically, it has to do with the number of nodes (either processing cores or storage elements), the routers, the communication links, as well as their interconnection. Similarly, topology exploration deals with the evaluation of alternative topologies in order to quantify the efficiency for each of them under various design criteria. Based on the outcome from topology exploration, device architects perform topology selection. During this task, the most suitable NoC topology, which satisfies the application's communication requirements and imposes the minimum cost (in terms of power/energy consumption, silicon area, etc), is selected. Note that during topology selection, physical-level constraints are also taken into consideration. Finally, there is an evaluation step, where the efficiency of previous selections is quantified.

Up to now, researchers propose the usage of various basic topologies such as a bus, star, mesh, point to point, as well as hierarchical ones, which can have the same or different topologies locally and globally (for instance, a locally bus globally mesh topology).

### *2.2.1 Regular Versus Irregular Topologies*

A first classification of NoC topologies concerns their regularity. More specifically, depending on the regularity structure of network's layout, the NoC is characterized either as regular, or irregular. Both of these topologies exhibit advantages that should be exploited by the target application domain. More specifically, a regular NoC topology assumes a homogeneous distribution of routers, which leads among others to lower design time and cost. Even though it is possible to incorporate such a topology to general-purpose SoC designs, it is mostly suitable for mesh- and torus-based architectures. Additionally, regular topologies are highly reusable and impose the minimum re-design effort, in case they are employed to different applications/architectures.

Despite the previously mentioned advantages, the usage of regular topologies is not widely accepted for commercial products because they impose a number of shortcomings. These limitations mainly occur due to the nonoptimal utilization of interconnection network, which in turn results to increased delay and power consumption. In order to alleviate these limitations, the adoption of irregular, or custom, topologies is also proposed. Since these topologies are mostly application oriented, they are suitable for SoCs consisted from heterogeneous cores. Additionally, the design of irregular topologies requires more advanced routing algorithms that take into consideration the nonuniformity of the underlying network.

For instance, assume a NoC architecture where each of the routers has to be attached to a different number of nodes. In case a regular topology is employed,

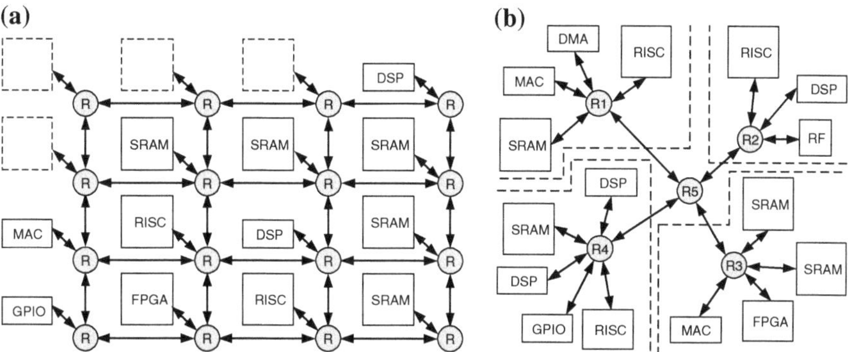


Fig. 2.1 **a** regular and **b** irregular NoC topologies

some of the routers will remain nonfully utilized, which in turn leads to significant delay, power, and area overheads. On the other hand, it is possible to design a heterogeneous NoC architecture, where each router has a different number of ports. Such a customization of hardware components results to higher performance, as compared to the case where a regular topology is employed. However, the introduced heterogeneity imposes additional design effort both for the routers, as well as for the entire NoC.

Figure 2.1 gives an example of these two topologies.

As a conclusion, we can mention that it is not possible to design a single NoC topology suitable for all the application domains, since each of them introduce inherent features and constraints that have to be exploited as much as possible in order to maximize the efficiency of underlying communication network. For this purpose, apart from the two previously mentioned extreme solutions, architects also incorporate to their designs heterogeneous, or piece-wise homogeneous, topologies. These topologies consist of highly configurable network building blocks (e.g., routers), which can be customized for a specific application domain (e.g. multimedia, telecom, etc).

## *2.2.2 Direct Versus Indirect Topologies*

Similarly, it is feasible to classify the network's topology either as direct, or indirect. At a direct topology, each node has a direct point-to-point link to a subset of other nodes in the system, called neighboring nodes. Direct topologies usually lead to a higher availability of communication bandwidth, but they impose an increase of nodes in the system. Hence, there is a fundamental trade-off between connectivity and cost. A number of existing NoC products incorporate a direct topology, among others are the Nostrum [1], SoCBus [2], Proteo [3], Octagon [4], etc.

The majority of directed network topologies have orthogonal implementation (layout) and the incorporated nodes usually are arranged in an $n$-dimensional orthogonal space. Such an architectural arrangement exhibits the competitive advantage of simplicity. Hence, it is feasible to employ light and fast enough routing algorithms that achieve packet routing with comparable performance, but with significant lower complexity. Typical instantiations of this topology are the $n$-dimension mesh, the torus, the folded torus, the hypercube, and the octagon.

Section 2.2.3 highlights the most important of these topologies.

### 2.2.3 2-D NoC Topologies

A 2D mesh, depicted schematically in Fig. 2.2, is the simplest and most popular topology for NoCs. It consists of an $M \times N$ mesh of switches interconnecting nodes (e.g., processing cores, memories, etc) placed along with the switches. Every switch, except those at the edges, is connected to four neighboring switches and one node. In this case, the number of switches is equal to the number of nodes. In order to realize the physical communication path between nodes and switches (or between switches), we employ communication channels, each of which consists of two unidirectional links between either a switch and a node, or two switches.

The 2D mesh topology assumes that all the links have the same length, and hence it imposes an inherent regularity which simplifies the physical design considerably. Also, it is easier to predict the area requirements for mesh topologies, because it grows almost linearly with the increase in number of nodes. Apart from these advantages, the usage of mesh topology has also some drawbacks. The plethora of routers incorporated in this topology usually leads to congested regions over the device. For this purpose, careful design and application mapping has to be performed to avoid traffic accumulating, especially in the center of the mesh. Note that 2D mesh topolo-

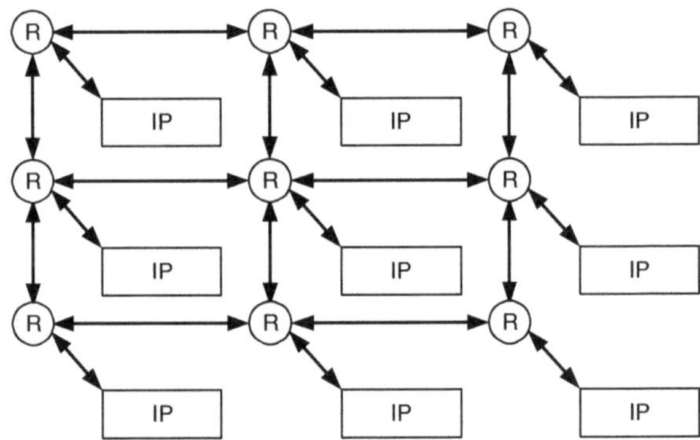


**Fig. 2.2** 2-D mesh topology

gies are most likely to exhibit congestion problems at the center of the architecture, rather than in the periphery, due to the employed routing algorithms.

An example of this topology can be found in the Chip-Level Integration of Communicating Heterogeneous Elements (CLICHE) NoC [5].

Torus is another direct topology, which is formed by an $n$-dimensional grid with $k$ nodes in each dimension. Different instantiations of this topology have been proposed. Next, we describe in more detail two approaches widely accepted in designs. Fig. 2.3 depicts a $k$-array 1-cube, also referred as 1-D torus, which is essentially a ring network composed by $k$-nodes. Even though this topology is simply enough to be incorporated into designs; however, it leads to a limited scalability and performance with the increase of nodes.

Similarly, the $k$-array 2-cube (2-D torus) topology, depicted in Fig. 2.4, exhibits the layout of a regular mesh except that nodes at the edges are connected to switches at the opposite edge via wrap-around routing channels. Every switch has five ports, one connected to the local resource and the others connected to the closest neighboring switches. The limitation of this topology affects the long end-around connections, since their delay might be significantly increased, depending on the size of the architecture. Consequently, in order to avoid any timing violation problems, the length, delay, and power consumption of these links should be carefully taken into consideration during the design phase.

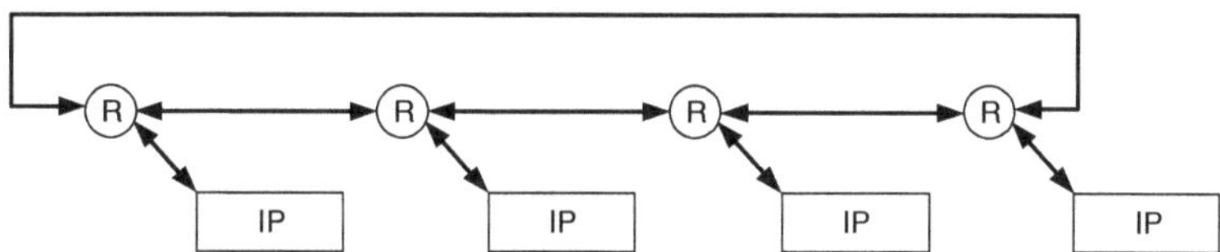


**Fig. 2.3** 1-D torus

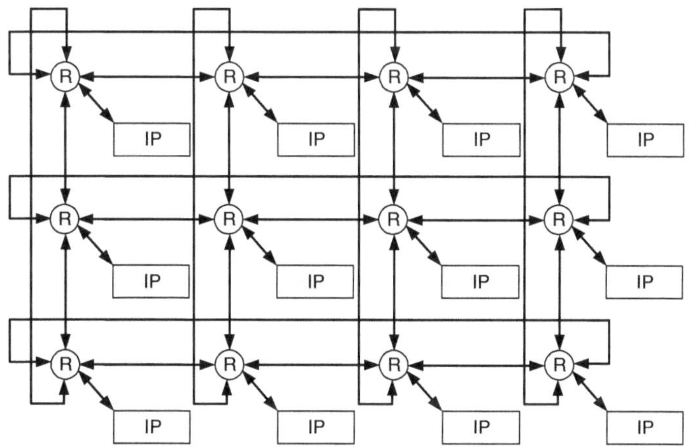


**Fig. 2.4** 2-D torus

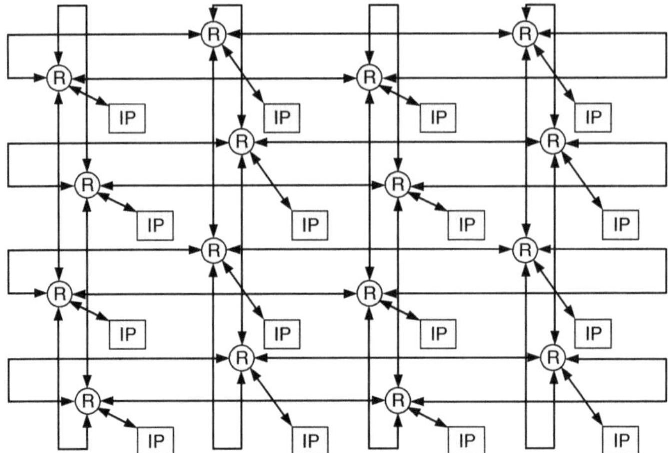


**Fig. 2.5** Folded torus

The folded torus, depicted schematically at Fig. 2.5, is an extension of torus topology. More specifically, the different pattern of connections found in folded torus topology, alleviates the limitation of excessive delays for long end-around connections. Additionally, the inherent regularity of this topology imposes that all the links exhibit the same physical length and hence delay. Consequently, it is much easier for the router to determine a shortest path with the same effort. Since the NoCs usually are employed for application-specific designs, usually architects prefer to modify (or extend) mesh and tori topologies by adding bypass links in order to achieve even higher performance at the cost of higher silicon area.

We have to mention that both mesh and torus topologies are possible to be implemented either as direct, or indirect networks.

Octagon is another well-established direct topology found in NoCs. Its simplest configuration, as it is depicted in Fig. 2.6, consists of a ring of eight nodes, which are connected by 12 bi-directional links. These links provide two-hop communication between any pair of nodes in the ring. This allows the usage of simple algorithms for performing fast yet efficient shortest-path routing. This topology assumes that each node is associated with a node and a switch, whereas in order to accomplish the necessary communication between any pair of nodes, at the most two hops are required within the basic octagonal unit. In case, a platform consists of more than eight nodes, the octagon is extended to multidimensional space. More specifically, at this approach one of the nodes per octagons is employed as a bridge between different octagons. Of course, such type of interconnection mechanism usually imposes significant increase in wiring complexity.

Apart from the previously mentioned direct topologies, a number of indirect topologies are also widely accepted for the design of NoC architectures. Figures 2.7 and 2.8 give two similar topologies named fat-tree and butterfly fat-tree, respectively. In both topologies, the nodes are connected to an architecture's external

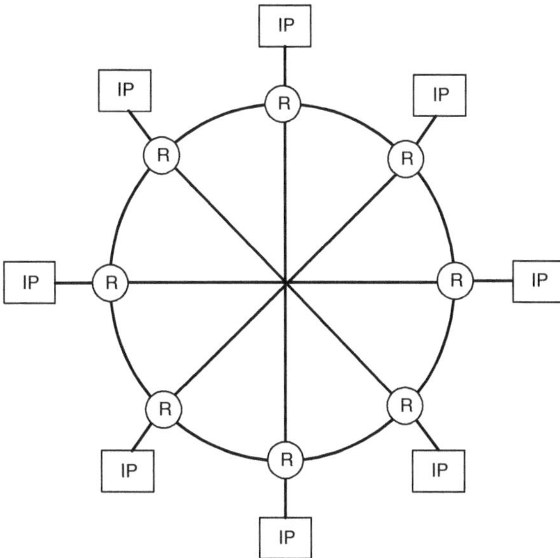


**Fig. 2.6** Octagon NoC

switch, whereas the switches have point-to-point links to other switches. More specifically, whenever an architecture incorporates this type of interconnection network, the processing units and memory modules are assigned to the leafs of the trees, the switches are placed at the vertices, whereas the communication path involves climbing up and down some part of the tree.

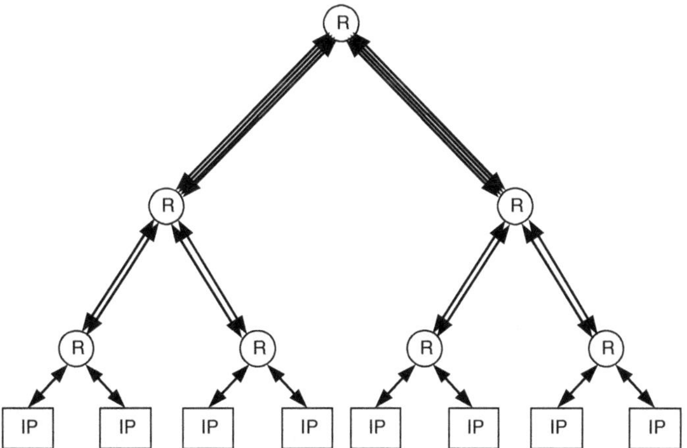


**Fig. 2.7** Fat-tree topology

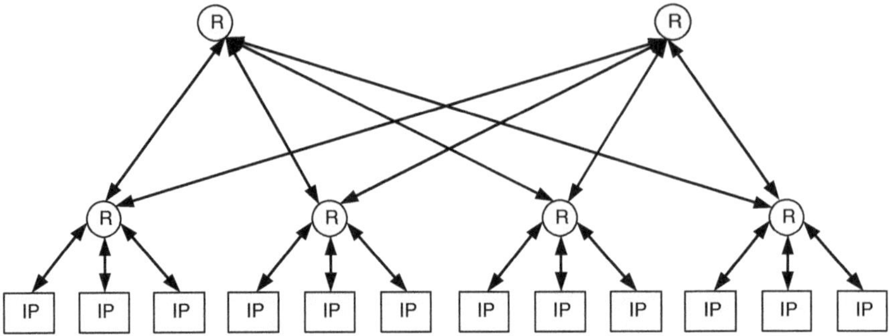


**Fig. 2.8** Butterfly fat-tree topology

A pair of coordinates is used to label each node, $(l, p)$, where $l$ denotes a node's level and $p$ gives its position within this level. Based on this assumption, the addresses for the $N$ nodes found at the lowest level range from 0 to $(N - 1)$. The pair $(0, N)$ denotes the locations of nodes at that lowest level. In order to accomplish the requirement for higher communication demand closer to the root of the tree, as compared to its leafs, the fat-tree-based topologies incorporate more links near the root of the tree.

A butterfly network, depicted in Fig. 2.9 can be either uni-, or bidirectional. For instance, a simple unidirectional butterfly network consists of eight input ports, eight output ports and three router levels, each of which contains four routers. Packets arriving to the inputs on the left side of the network are routed to the correct output on the right side of the network. In a bidirectional butterfly network, all the inputs and outputs are on the same side of the network. At this approach, all the packets coming to inputs are first routed to the other side of the network, then turned around and routed back to the correct output [6].

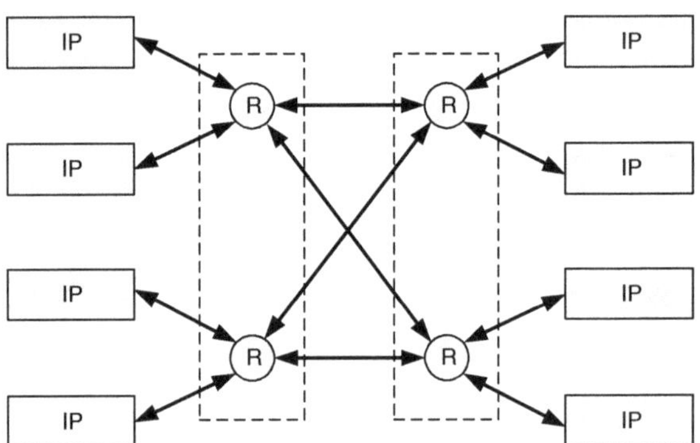


**Fig. 2.9** Butterfly topology with four inputs, four outputs, and two router stages each containing two routers

Polygon, shown in Fig. 2.10, is another widely accepted topology. The simplest polygon network corresponds to a circular network, where packets travel in a loop from one router to the next. In case we add chords to the circle, the network become more diverse. However, if chords are inserted only between opposite routers, the topology is called a spidergon.

A star network consists of a central router in the middle of the star, and computational resources, or subnetworks, in the spikes of the star. The capacity requirements of the central router are quite large, because all traffic between spikes goes through the central router. That causes a significant possibility of congestion in the middle of the star. Figure 2.11 shows a typical instance of a star topology.

Finally, there are designs where the usage of NoC is fully customized, or ad-hoc, topologies, are suitable. Figure 2.12 shows a typical example of two custom NoC topologies. These topologies are composed of a mix of shared bus, direct, and indirect NoC topologies. The competitive advantage of such application-specific (custom) topologies is the significant improvement to the overall network performance. On the other hand, the shortcoming of this selection concerns the penalty in the structured wiring, which is nonetheless one of the main advantages offered by regu-

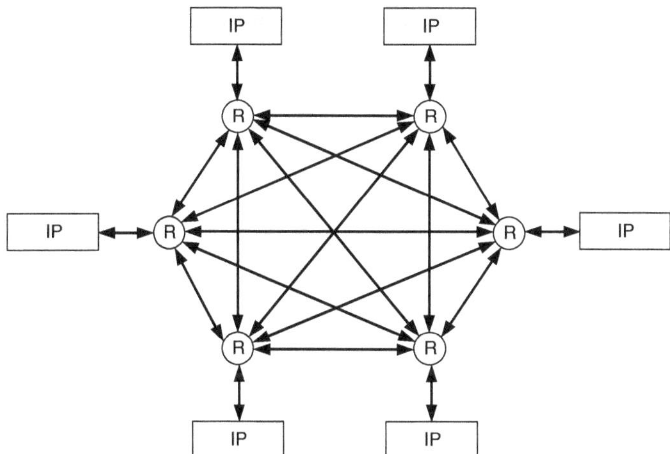


**Fig. 2.10** Polygon (hexagon) topology with all the potential links

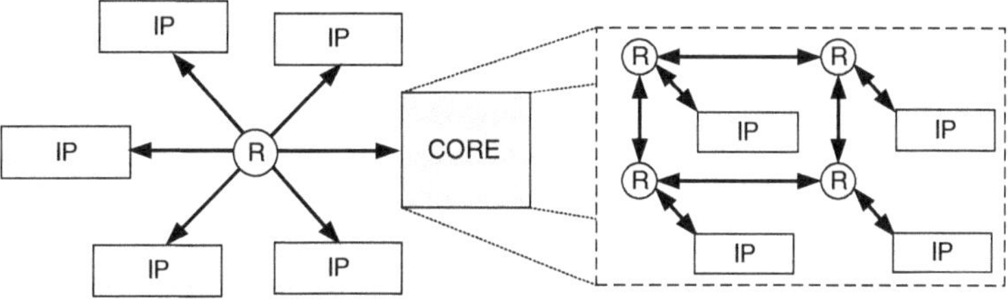


**Fig. 2.11** Star topology

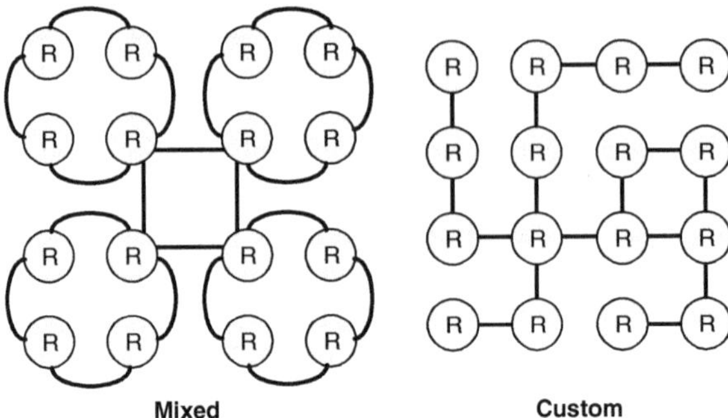


**Fig. 2.12** Examples of customized NoC topologies

lar on-chip networks. Designers pay effort to provide an acceptable trade-off between these two approaches through the appropriate insertion of application-specific long-range links to regular architectures. Even though such a selection boosts the network performance, without mentionable impact on the network's regularity and its area requirements, the long-range paths at regular topologies usually impose links with varying lengths and performance metrics (which are not desirable especially for sufficient packet routing).

### 2.2.4 3D Topologies

Three dimensional (3D) chip stacking is touted as the silver bullet technology that can keep Moore's momentum and fuel the next wave of consumer electronics products. Apart from the flexibility imposed by this new design paradigm, one of the major challenges that designers face today in 3D integration is how to achieve both the interconnection across the components within a layer, as well as across the layers in a scalable and efficient manner. A viable solution to this problem is the usage of NoCs.

The previously mentioned topologies are applied to provide traffic communication for planar (2D) architectures. However, in the last few years there is an effort from academia, research centers and industry on designing 3D architectures, where a number of nodes are assigned to different layers, whereas the connectivity between adjacent layers is provided through vertically aligned Through-Silicon Vias (TSVs). By combining the advantages provided by 3D integration with the increased scalability found in NoCs, this new design paradigm seems promising to provide the communication infrastructure for the next generation of complex systems. More specifically, the locality along the $z$-axis leads among others to significantly reduced interconnect delay, canonical interconnect structure, increased flexibility, as well as integration of dissimilar systems and technologies. However, in order for these architectures to become widely accepted, novel design methodologies and tools, which

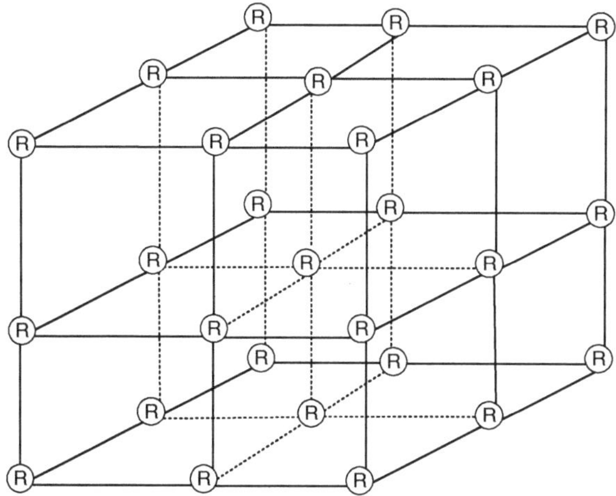


**Fig. 2.13** 3D mesh topology

take into consideration the inherent features provided by this new process technology, are absolutely critical.

As we will discuss in upcoming chapters, almost all of the previously mentioned topologies are also applicable to the 3D domain, if they are appropriately extended. For instance, a typical extension affects the increase to the number of router's ports. Assuming we have a mesh topology, then it is possible to make the transition from 2D to 3D NoC architectures by equally distributing tiles onto the device layers of the 3D architecture. However, as we depict in Fig. 2.13, the 3D mesh topology imposes the usage of 7-port switches rather than the 5-port employed at the corresponding 2D architecture.

The diameter of 3D mesh can be defined as $D = d(k-1)$, where $d$ represents the dimension and $k$ is the number of nodes in plane. A Torus network is same as mesh network with boundary nodes connected by wrap-around edges. These wrap-around edges significantly reduce the overall diameter of the network and thus improving the throughput and latency. Figure 2.14 shows 3D torus architecture and partitioning approach into quadrants. The diameter of torus can be defined as follows: $D = d\left(\lfloor\frac{nx}{2}\rfloor + \lfloor\frac{ny}{2}\rfloor + \lfloor\frac{nz}{2}\rfloor\right)$, where $nx$, $ny$ and $nz$ is the number of nodes in plane $x$, $y$, and $z$, respectively.

An alternative approach for designing 3D NoCs with improved area footprint per tile assumes the implementation of the processing element and the router in a distributed fashion across layers. Additionally, a low-diameter 3D NoC is studied in [7], where device architects leverage long wires to connect remote intralayer nodes. A novel layer-multiplexed 3D network architecture with vertical multiplexing and de-multiplexing links is proposed in [8], whereas a survey that summarizes the pros and cost of using various topologies for 3D NoCs can be found in [69, 70].

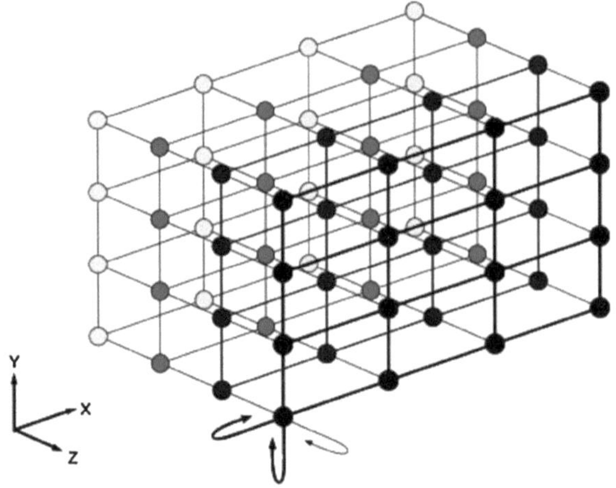


**Fig. 2.14** 3-D torus topology

As a conclusion we can mention that it is not possible to find an optimal NoC topology, since its efficiency depends highly on the inherent constraints posed by the target application domain. We have already highlighted that the mesh topology is the one easiest to implement because it introduces the minimum design complexity. More specifically, the mesh topology exhibits some desirable properties of its own, such as a very simple addressing scheme and multiple source-destination routes, which give robustness against to network disturbances. The low-dimension topologies are favored in case there is an increased demand for bandwidth between switches and the delay caused by switches is comparable to the inter-switch delay. On the other hand, if the target application exhibits a high degree of locality at the communication pattern, then there is a demand for higher order dimension topologies.

## 2.3 Traffic Modeling

Usually, the formal description of NoC architectures is based on states and transitions among states that are caused by various actions in the system. Such a model represents an abstraction of the system's behavior, which is sufficiently accurate, yet tractable, for analysis, and verification purposes. Measuring and comparing performance, cost, and other architecture parameters, is an important challenge that has been hardly addressed so far. Unfortunately, conventional benchmarks for multiprocessor systems [9, 10] are usually application oriented, and hence they cannot be used directly for quantifying communication-intensive architectures, such as NoCs. Moreover, the current SoC benchmark circuits (like ITC 2002 [11]), consist only of a very limited number of blocks, which is not the case for scalable NoC-based architectures. Finally, since the traffic at NoCs varies considerably during the application execution, more advanced traffic models have to be incorporated in order to study in detail the behavior of target system. For this purpose, throughout this section we describe a number of well-established traffic models employed for evaluation purposes at the NoC domain.

### 2.3.1 Available Models for Traffic

The design of NoC architectures requires mechanisms for accurately predicting the network usage in advance, instead of designing, or fabricating, the NoC and afterwards evaluating its performance. In order to address this requirement, architects employ various traffic models for determining and quantifying the impact of critical parameters for the underlying network. More specifically, a traffic pattern [6] is a graph that describes both the spatial, as well as the temporal distribution of network traffic (data and signal communication among nodes). In general, this graph focuses on the amount of traffic that has to be transmitted to the NoC and not on the dependencies between nodes. Even though this implies that the conclusions derived by employing such graphs might not correspond to the actual system requirements; however, their usage is desirable especially during the initial design phase in order to quantify specific characteristics of the underlying architecture. Additionally, with these traffic models it is feasible to have an abstract overview about the impact of various traffic patterns at the performance of NoC architecture.

Traffic models are classified either as realistic or synthetic. Specifically, realistic traffic models are traces of application execution onto NoCs. On the other hand, the synthetic traffic patterns correspond to abstract models of packets exchanged between nodes of the NoCs. In contrast to realistic traffic, which is representative of a more specific class of applications, synthetic traffic is generated based on mathematical models. Hence, it covers a broad class of applications executed onto NoC platforms. The competitive advantage of incorporating synthetic models is that it allows a network to be stressed with a regular and predictable pattern. However, since they do not represent traffic from real-life applications, they cannot be employed for accurate design-space exploration, whenever an application-specific NoC platform has to be designed.

Apart from the traffic modeling, tools that enable traffic generation are also important during the exploration phase. Due to the importance of this task, up to now numerous traffic generators have been proposed [12–16]. In Brebner and Levi [17], a traffic generator is introduced that supports both uniform and burst mode distributions. Additionally, this solution is able to generate stochastic traffic based on input traces generated by real-life applications. A similar work that explores and analyzes numerous architectural characteristics of NoCs, by generating different types of traffic, can be found in [18].

Next, we summarize the main features provided for a number of well-established traffic models.

### 2.3.2 Synthetic Traffic: Temporal Distribution

A temporal distribution determines how an individual node generates traffic over time and how this traffic is propagated to the NoC architecture. This distribution includes a list of traffic properties, such as the period (or rate) of generating messages, while typical values for this parameter are constant (periodic), random, normal, etc. Similarly, it is possible to perform another classification based on their timing information. Next, we describe in more detail the properties for three different temporal distribution models.

**Perfect synchronous model**: The perfect synchronous model is based on the "perfect synchrony hypothesis" [19]. This approach assumes that neither communication nor computation tasks take time. In other words, whenever we employ the perfect synchronous model, the result of a computation is available at the same time that the input is applied. Additionally, if several processes are interconnected, then the result of computations will ripple instantly through the system. Even though the perfect synchronous model is easy to understand, it leads to nonrealistic results, since it does not take into consideration any timing information. Hence, the conclusions derived by incorporating this approach might differ a lot compared to a more accurate approach, or the actual NoC implementation.

**Clocked synchronous model**: This approach is based on "synchronous hypothesis assumption" [19], which assumes that a global clock signal is responsible to control the start for each computation in the system. More specifically, in the clocked synchronous model, the communication takes no time, while the computation is performed at only one clock cycle. Consequently, this approach is suitable for being used at cycle-true models. On the other hand, the limitation of this model affects the lack of awareness to take into account physical time, since the time period is expressed in clock cycles.

**Discrete event model**: In contrast to the previous two time-driven models, the functions that take place at the discrete event model are evaluated only on certain events. This enables time to be expressed in floating point format, which in turn allows the simulation of physical time. This feature simplifies considerably the simulation procedure, whereas the corresponding flexibility becomes far more important whenever the model consists of multiple time domains. On the other hand, the usage of an event queue might lead to a performance degradation. Even though such a penalty might be negligible for systems that are frequently idle; however, it becomes especially critical for architectures with increased switching activity.

Additionally, there are available various traffic models for NoC topologies, which support the packet generation for different rates. More specifically, these models are summarized as follows:

**Uniform**: This is the simplest traffic model. Assuming a network with $N$ nodes, the probability for a node to send a packet to another node at this model is equals to $1/(N-1)$. A restriction to this approach is that a node does not send data to itself. Additionally, for simplicity reasons, uniform distribution assumes that the length of a packet, the latency between packets, as well as the destination node for packets are randomly chosen using utilization ratios, which are defined either by the designer or through a random process.

**Uniform Random**: A node that is supposed to use this traffic pattern sends packets randomly to other nodes with an equal probability $\lambda = 1/N$, where $N$ is the number of nodes in the network. A typical solution for providing such kind of traffic is with the usage of TGFF tool [6], which allows to define both the number of computational tasks, as well as the communication transactions.

**Locality**: At this model, the probability $P$ for a node to send a packet to a destination node depends on the source-destination distance (i.e., the number of links from a source router to a destination router). This probability is computed as follows: $P(d) = \frac{1}{A(D)\times 2^d}$, where $D$ corresponds to the maximum distance in the network and $A(D) = \sum_{d=1}^{D}\left(\frac{1}{2^d}\right)$ is a normalizing factor guaranteeing that the sum of all probabilities is equals to 1. Note that within a set of nodes with the same distance, each node is selected with uniform probability.

**Nearest Neighbor Traffic**: At this model, a constant percentage of traffic goes to the nearest neighbors (with a predefine radius $r$) and the rest of the traffic is distributed with uniform and random models. Such an approach is commonly used to evaluate the impact of communication locality on the performance and power consumption of the NoC [6].

**HotSpot**: This scenario selects $\lfloor N/M \rfloor^2$ of the nodes ($N$ is the total number of nodes) as traffic hotspots, where $M \in \{2, 4, 8, \ldots, N\}$. Certain fraction $p$ (usually $p \in \{0.5, 0.7\}$) of traffic is targeted to these hotspots (one is selected at a time by uniform random selection). The other traffic is sent uniformly to all other nodes. Both number of hotspot nodes $M$, as well as their fraction $\rho$ are user-defined parameters. A variant of this scheme selects different hotspots for each source.

**Burst-Mode**: This traffic distribution assumes that all the packets are sent according to a fixed packet generation rate, whereas at the stable state, there is no traffic between nodes for a predefined amount of time. Both of these time periods are tunable by the designer in order to study different utilization of network architecture. Such a type of traffic is suitable for emulating typical burst modes generated from real cores.

### *2.3.3 Synthetic Traffic: Spatial Distribution*

**Uniform**: At this approach, each node sends packets to randomly chosen destination node. Even though this is one of the simplest traffic models, its simplicity might lead to nonacceptable results for evaluation of NoC architecture.

**Bit Permutation**: At the bit permutation traffic, each source sends all of its traffic to a single destination. The destination node in this approach is performed based on a function of the source address (this is also known as bit permutation pattern). On other words, a given source node sends only to one destination node, whereas the address of destination node is a function of the source node address. Because this type of traffic concentrates load on individual source-destination pairs, they tend to stress the load balance of a topology and routing algorithm. Details on how to generate these traffic patterns are given in [6].

**Transpose**: Transpose: In transpose traffic, the destination coordinates are the transpose of the source coordinates. Under this load, the network's diagonal bisection is a bottleneck as all packets must cross it. By incorporating transpose traffic in conjunction to with appropriate NoC routing algorithms, it can produce a highly imbalanced network load, since the links counter-clockwise about the center of the mesh are utilized while the clockwise links are unused.

**Bit Complement**: Another commonly used load traffic is bit complement in which each node exchanges packets with a node on the opposite side of the network at a uniform random distributed offered rate. Since not all nodes are equal senders or receivers of traffic, usually bit complement model does not closely match the actual traffic of the NoC architecture. In order to compute the coordinates of destination node at this traffic model, we perform a bit-wise inversion of the source coordinates. This load stresses the horizontal and vertical network bisections. Under this load, a NoC statically spreads traffic across all of the bisection links, providing a perfectly balanced network load.

**Bit Reverse**: In the traffic pattern generated according to the bit-reversal permutation, a message generated in the source node $X = X_1 X_2 \ldots X_n$ is destined for node $B(x) = X_n X_{n-1} \ldots X_2 X_1$; that is, the reversed order of the bit pattern $x$.

**Bit Rotation**: The bit rotation pattern is similar to the bit permutation and assumes that the destination address is obtained by rotating the bit string representation of the source node address to the right by one. For instance, suppose that the number of nodes $N$ is a power of 2, and $m$ is the number of bits used to express the addresses of source and destination nodes. Then, bit rotations are those in which each bit $d_i$ of the $m$-bit destination address is a function of one bit of the source address, $s_j$, where $j$ is a function of $i$. For the bit rotation pattern, $d_i = s_{(i+1) \bmod m}$.

**Shuffle**: This traffic pattern is similar to bit rotation, but the destination node is retrieved as follows: $d_i = s_{(i-1) \bmod m}$.

*N* **Complement**: Similarly to *Bit Rotation*, this scenario creates load on source-destination pairs. Suppose that nodes are numbered as naturals 0, 1, 2, ..., $N-1$. In case the address of a source node is $n_s$, then the corresponding destination address at this traffic model is node $n_d$, such that $n_s + n_d = N$. For example, if a network has seven nodes, the nodes are numbered as 0, 1, ..., 6. Source nodes 0, 1, 2, 3, 4, 5, and 6 will select destination nodes 6, 5, 4, 3, 2, 1, respectively. We have to notice that for both permutations traffics (*Bit Rotation* and *N Complement*), the numbering of nodes determines the actual spatial distribution of traffic. It is advised that nodes should

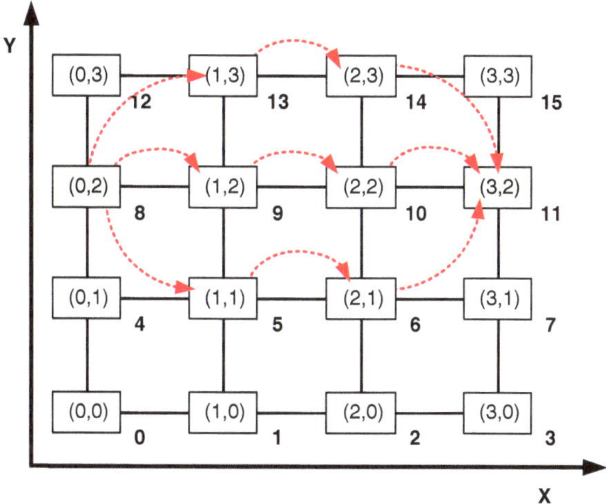


**Fig. 2.15** A fork-join pipeline with three parallel branches and a depth of 3, i.e., $c = 3$ and $e = 3$

be regularly numbered according to their topology. This also implies that these two traffic patterns are most beneficial to regular topologies.

**Tornado**: At this traffic model, each node sends packets halfway around the mesh along the $X$-dimension. This workload is a challenge especially for rings and meshes.

**Fork-Join Pipeline**: A fork-join pipeline [20] is a traffic pattern where a fork node feeds $c$ nodes that are the starting point of $c$ parallel pipelines. Each pipeline has a depth of $e$ nodes. At the end of the pipelines (after $e$ stages), the data are merged into a join node. An example of a fork-join pipeline in a 4×4 mesh network with $c = 3$ and $e = 3$ is shown in Fig. 2.15. In general for 2D meshes, $c$ is set to $c = h = \lfloor\sqrt{N} - 2\rfloor$. Hence, for a 4×4 network, we have $c = e = 3$.

At the example depicted in Fig. 2.15, we map the nodes' coordinates to integers in order to compare different topologies, whereas this approach is applicable both for 2D (mesh, torus, fat tree, irregular, etc) and 3D (cube, irregular etc) topologies. For instance, as illustrated in Fig. 2.15, a node $(x, y)$ maps to an integer $I$ by $I = x + (y \times N_y)$, where $N_y$ is the maximum number of nodes along the Y axis for the 2D mesh. Apart from this example, the user may define his/her own mapping from 2D/3D coordinates to integers.

## *2.3.4 Realistic Traffic*

Apart from the previously mentioned synthetic traffic models, it is common to evaluate NoC architectures with the usage of traffic captured from real applications. This traffic, also known as realistic traffic, allows the designer to analyze with acceptable precision a number of architecture-oriented parameters such as delay and power consumption.

It should be noted, however, that the traffic patterns generated by different modules in a NoC strongly depends on the application for which the NoC is designed. Since the performance of the NoC is a function of the traffic profile, the most accurate way to assess the characteristics of the NoC would be to generate application-dependent traffic profiles.

Even though this is the optimum scenario, usually the design of systems affects devices that can perform sufficiently for a (wide) range of applications. In these cases, the employed traffic profiles have to stress the underlying NoC with a way that (almost) corresponds to the the actual application execution. Various approaches for realistic traffic have been proposed the last years. Among others are the GSM voice CODEC [21], SPLASH-2 [22], MediaBench [23], and SPEC [24] traffic profiles.

Unfortunately, such a selection might lead to excessive time period for profiling, even if all the applications are known beforehand. In order to overcome this limitation, designers usually employ synthetic traffic profiles (as they were discussed in previous section), which can represent the traffic for a class of applications. This suggests that the use of both realistic and synthetic traffic profiles forms a complete set for the evaluation of the techniques proposed for NoC systems.

## 2.4 Topology Modeling

This chapter discusses the modeling of NoC topology, which is used for the purposes of encoding the application implementation onto the underlying architecture. Conceptually, the purposes of NoC modeling are to explore the vast design and feature space, as well as to evaluate trade-offs between various design parameters (e.g., power, area, design-time, etc); while adhering to application requirements on one side and technology constraints on the other side. As it is discussed in survey [25], NoC modeling has three intertwined aspects: *modeling environment*, *abstraction levels*, and *result analysis*.

### *2.4.1 Modeling Environment*

A key aspect in NoC design is model creation, since it is a concrete representation of functionality for a target NoC. Hence, whenever a NoC is going to be incorporated at a SoC, proper modeling approaches both for the NoC's topology, as well as the NoC stack protocol, are absolutely required. This enables the flexible architecture specification and the considerable speed up of exploration procedure for different interconnection solutions.

The NoC models found in relevant literature can be classified either as analytical, or simulation based. More specifically, an analytical model relies on mathematic equations that describe the functionality of underlying network. Even though this approach might lead to the most accurate results (in case the NoC is properly modeled), it imposes the maximum computational effort. This limitation makes the usage of analytical models to be applied only to simple and small NoCs, whereas the architectural parameters that are studied are usually very limited. On the other hand, recently there is an increased interest for NoC simulation environments. These solutions provide a number of competitive advantages which enable designers to study the NoC under different topologies and operating conditions. Among others, by using simulation-based approaches it is feasible to model the entire NoC protocol stack, which allows designers to guarantee proper functionality for the communication infrastructure. Additionally, the opportunity to perform co-simulation between the network and the rest of the chip is very important in order to test the proper functionality of complex systems.

Madsen et al. [26] and Mahadevan et al. [27] introduce an abstract framework for NoC modeling using allocators, scheduler, and synchronizer. The allocator translates the path traversal requirements of the message in terms of its resource requirements (e.g., bandwidth, buffers, etc) and attempts to minimize resource conflicts. The scheduler executes the message transfer according to the particular network service requirements having as goal to minimize resource occupancy. Finally, the synchronizer models the dependencies among communicating messages allowing concurrency.

A commercially available network simulator, named OPNET, is discussed in [28, 29]. This software supports also hierarchical modeling of networks, including processes (state machines), various topologies, and simulation of different traffic scenarios. The work described at [29] uses OPNET to model a QoS-based NoC architecture and design with irregular network topology.

Another framework for NoC modeling, named OCCN, is discussed in [30]. This framework is based on open source C++ code build on top of SystemC. In order to simplify the task of implementing communication drivers at different levels of abstraction, it uses a number of generic message passing APIs.

A cycle accurate RTL model for quantifying the performance and power/energy consumption of NoCs is discussed in [31]. Both delay and power/energy consumptions are quantified at fine-grain level (e.g., building components of routers and links), with the usage of SPICE simulations for a 0.18 um CMOS technology. Power analysis for dynamic power consumption can also be evaluated with the usage of Orion simulator proposed by [32].

Even though simulation-based NoC models are able to simulate complex interconnection architectures, they exhibit the limitation of increased computational complexity. In order to alleviate this problem, researchers have proposed the usage of NoC emulation. Such a selection reduces the simulation period from hours (required to process through many millions of cycles as would be necessary in any thorough communication co-exploration) to a few minutes. Additional speedup at this procedure can be achieved with the usage of hardware accelerators, such as the FPGA-based emulation discussed in [19].

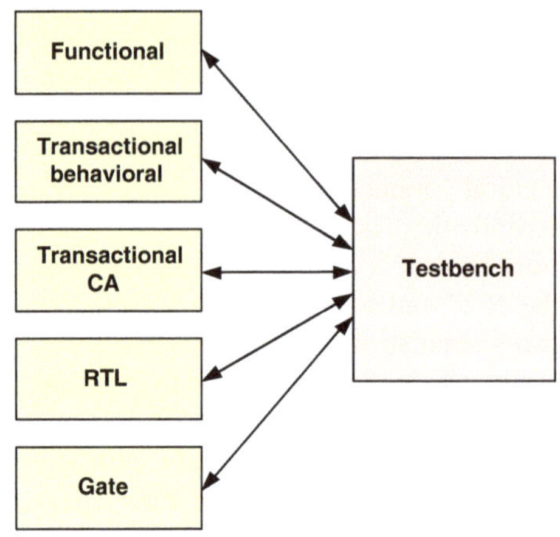


**Fig. 2.16** Modeling of different abstraction levels

## *2.4.2 Different Abstraction Levels for Noc Modeling*

Another critical issue for NoC modeling is the usage of different abstraction levels. This task is software supported with a number of hardware description languages, like SystemC (a library of C++) [33] and SystemVerilog [34]. A summary of available communication primitives at different levels of abstractions can be found in [25, 71]. Based on analysis discussed in [35], there are five abstraction levels, as they are depicted in Fig. 2.16.

**Functional models** usually have no notion of resource sharing or time. This imposes that the functionality is executed instantaneously, or as an ordered sequence of events. The functional model may or may not be bit-accurate. This layer is suitable among others for system concept validation, functional partitioning between control and data, including abstract data type definition, hardware or software communication, and synchronization mechanisms. Models are usually based on core functionality written in ANSI C and a SystemC-based wrapper.

**Transactional behavioral models**, denoted also as transactional, are functional models mapped to a discrete time domain. The transactions at this model are assumed to be atomic operation, while the duration for each of them is stochastically determined. Even though it is not always possible to model general transactions on bus protocols, transactional models are particularly important for protocol design and analysis, as well as communication model support. Below this level of abstraction, NoCs impose additional identifiers (i.e., addressing), in order to uniquely define the traversal path, or to provide services for end-to-end communication.

**Transactional Cycle-Accurate (CA) models** map transactions to a clock cycle, in contrast to the previously mentioned asynchronous models. This allows synchronous protocols, wire delays, and device access times to be accurately modeled. The usage of transactional CA model is very important especially for functional and cycle-accurate performance modeling of abstract processor core wrappers (called bus functional models), bus protocols, signal interfaces and testbenches, in a simple, generic and efficient way using discrete-event systems. We have to notice that transactional CA models are similar to corresponding RTL models, except that they are not synthesizable.

**Register-Transfer Level (RTL) models** correspond to the abstraction level from which synthesis tools can generate gate-level, or netlist, descriptions. Systems that are modeled with such an approach are usually visualized as having two components: data and control. More specifically, the data part is composed of registers, operators, and data paths, whereas the control part provides the time sequence of signals that evoke activities in the data part. Data types are bit-accurate, interfaces are pin-accurate, and register transfer is accurate. Similarly, propagation delay is usually back annotated from gate models.

**Gate models** are the last of the available abstraction levels, aiming to provide a system description in terms of primitives, such as logic with timing data and layout configuration. For simulation purposes, gate models may be internally mapped to a continuous time domain, including currents, voltages, noise, clock rise, and fall times.

A comparative study for the three communication models used in parallel computation, namely the synchronous communication which rewards burst-mode message transfers, the asynchronous communication with fixed message size, and the asynchronous with variable message size communication while also accounting for network load, can be found in [36]. A similar study of parallel computation applications, but with a more detailed network model, was undertaken by [37]. This approach assumes that the underlying network uses adaptive routing with virtual channels. Finally, at [38] is shown how a mixed abstraction-level design flow was employed for the design of two different NoC topologies.

## 2.5 Topology Synthesis

One of the most challenging tasks during the NoC design affects the topology synthesis. Specifically, the NoC topology synthesis problem aims to generate at physical level the selected network-based communication infrastructure. As constraints to this procedure are the multiple design objectives posed by system specification, such as performance, power, and area metrics. The tasks that take place at a typical framework for designing NoC architectures are depicted in Fig. 2.17.

In the outer iterations, a number of operational (e.g., frequency) and architectural (e.g., link width) parameters are varied within a set of suitable values. The selected parameters are typically employed during NoC exploration, since the product of the NoC frequency and the link width corresponds to the available bandwidth. In case we perform an exhaustive exploration, then topologies with different numbers of switches should be explored, starting from a topology where all the cores are

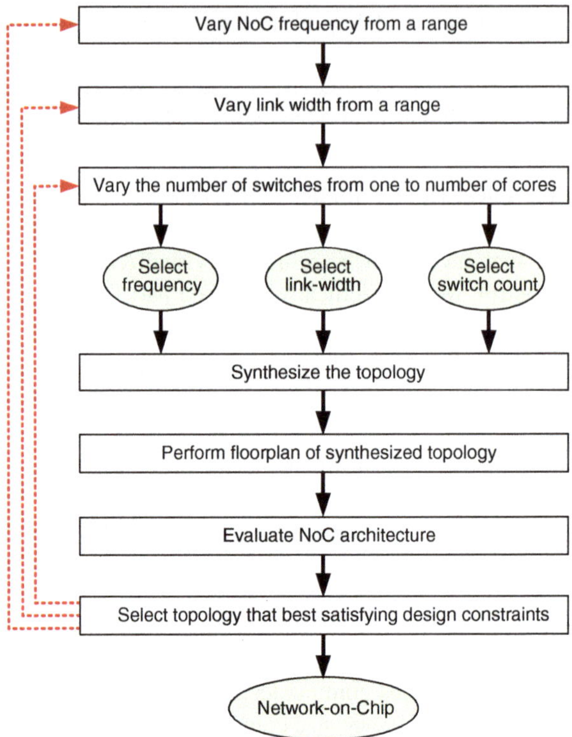


**Fig. 2.17** Tasks for designing NoC architectures

connected to one switch, to one where each core is connected to a separate switch. For each of the architectural parameters of the design space, we perform the topology generation in order to ensure that the traffic on each link is less than or equal to its available bandwidth. Then, an analysis step takes place in order to evaluate the efficiency of derived architecture. For accurate analysis, a floorplaning task to determine the 2D position of the cores and network components has also to be undertaken. Based on the frequency point and the obtained wire lengths, any timing violation on the wires is detected and the power consumption on the links can be retrieved. Eventually, from the set of all synthesized topologies and architectural parameter design points, the topology and the architectural configuration that optimize the user's objectives, while satisfying all the design constraints, are chosen.

The outcome from topology synthesis is either a regular or an irregular NoC topology. Regarding the regular topology, it is suitable for standard architectures (e.g., mesh and torus). However, in case such a topology is applied to a heterogeneous MPSoCs, the derived communication scheme usually exhibits poor performance and large power/area overheads due to the nonoptimal utilization of interconnection resources. In order to overcome this limitation, customized NoC topologies can also be generated. Even though these architectures impose significantly increased design overhead, because the target system included cores of different sizes and shapes, as compared to previously mentioned regular topologies, they result to superior performance in terms of various design parameters under identical performance requirements [39].

Due to the importance of developing application specific NoCs, a number of design methodologies have been proposed, each of which aims to tackle a different optimization problem. A constraint-driven communication synthesis approach based on specifications posed by the point-to-point communication specifications is discussed in [40]. The derived architecture consists of optimized channels obtained by merging, or separating the original point-to-point links. Similarly, a heuristic for the constraint-driven communication synthesis of on-chip communication networks is presented in [41], while [42] describes a design methodology targeting to synthesize NoC architectures, where long-range links are inserted on top of a mesh network.

The advantages of employing a library-based approach to NoC design, where predesigned soft macros are appropriately instantiated and interconnected in order to form arbitrary topologies, are summarized in [43, 44]. However, the full exploitation of customized network topologies requires an ad hoc design methodology spanning different levels of abstraction (ranging from application specification to physical implementation), in order to derive the most efficient NoC configuration for a given application domain. Another network design methodology that supports several parametrization options including topology is discussed in [45]. This approach is based on building a library of components that can be appropriately combined in order to realize the different communication networks.

Apart from the 2D architectures, academia, research centers, and industry focus also on the design of NoCs targeting to 3D SoCs. The synthesis of these interconnection architectures introduces additional challenges, both in hardware, as well as at the algorithmic level, that have to be sufficiently addressed. New switch architectures for 3D NoCs have been presented in [46], and are further discussed in Chap. 3. Efficient application-specific 3D NoC topology synthesis algorithms are studied in [47]. Another 3D topology synthesis algorithm based on a direct extension of the 2D NoC synthesis procedure is discussed in [48]. This approach assumes that each of the NoC's layers is designed separately and then we determine the connectivity of the switches across the layers. Apart from the flexibility of this approach, it does not tackle issues related to the assignment of switches to layers, the placement of TSV macros and network components with minimum perturbation of the input floorplan, etc. A more flexible toolflow for performing application-specific synthesis of 3D NoCs is the SunFloor_3D [49]. Among others, this flow can determines the most candidate NoC topology for a given application, finds the paths for data transfers, assigns the network components onto the 3-D layers and places them.

## 2.6 Application Mapping

After defining the network's topology, the last step involves the application mapping onto the target NoC architecture. As the distance traversed by packets in the network is tightly firmed to the performance of target architecture, the mapping of the SoCs modules into the NoC structure is an important step in the design process. Intuitively, the mapping problem deals with determining how to topologically arrange the selected application's kernels onto the target architecture, such that certain metrics of interest (e.g., area, bandwidth, latency) are optimized. Hence, communicating modules should be placed in close proximity if they exchange large amounts of data, or if they require a short network latency. Constraints to selections performed during this task are posed by the system's specifications.

Even though application mapping is not a NoC-related problem, we will discuss a number of representative approaches found in relevant references. Typically, the inputs to a NoC mapping algorithm are: (*i*) the dataflow graphic (signals that have to be transferred between nodes) and (*ii*) the maximum affordable delay between each pair of nodes. At the conventional graph representation of flow bandwidth, each data link is bounded to a specific sender and receiver without allowing any flexibility. Since the NoC infrastructure enables a sender to communicate with any other node, binding a flow to a particular pair of module instances is an unnecessary and expensive constraint. Alternatively, it is possible to allow mapping algorithm to benefit from the existence of identical modules, by appropriately assigning flows to the best replica within a module class. Consequently, as flows are no longer associated with particular modules within a class, this provides additional flexibility to the mapping algorithm resulting in a more efficient design.

Moreover, the maximum allowed delay between each pair of nodes limits the potential cost reduction in the mapping phase. By specifying only pair-wise node-to-node timing requirements, the application level data flow is ignored and the latency requirements are over constrained. The evaluation of an extended modeling of the SoC which captures the application's latency needs in a more appropriate way is discussed in [50]. In resemblance to re-timing of a logic path in traditional circuit design, authors at [50] allow trading-off delays between flows that compose a stream of information processing. More specifically, throughout that work, authors propose to replace (wherever it is applicable) a "chain" of end-to-end delay constraints with a single, unified constraint, and describing the latency requirement of an application stream. This time is measured from the time the first module in the chain generates the data until the last module receives the data. By using this cross-layer knowledge of the application needs, the mapping algorithm exploits new degrees of freedom in the design space in order to reduce the cost of derived solutions.

The majority of existing mapping algorithms mainly target mesh-based NoC topologies [51–54], because the architectures are most applicable to regular designs. A mapping technique targeting regular NoCs, which is based on a branch-and-bound algorithm for performing design space exploration, is presented in [55]. This solution allows topological mapping of IP cores to network tiles and generation of routing path for each communicating pair aiming to minimize the energy consumption for communication architecture under different routing algorithms. A similar work is discussed in [56], where a branch and bound algorithm aiming to perform mapping of architecture's cores onto a tile-based NoC, having as goal to satisfy the bandwidth constraints and minimize the total energy consumption. On the other hand, there are also research works dealing with core mapping onto heterogeneous architectures. For instance, an algorithm that performs mapping of heterogeneous cores onto a target NoC architecture is discussed in [53]. This algorithm incorporates also an embedded floor planner for performing physical prototyping of the derived architectures.

A two-step heuristic mapping approach is discussed in [57]. Given the graphs of clustered tasks and the distributed architecture, initially this algorithm maps on adjacent nodes the highly communicative tasks, while the remaining tasks (starting from those close to the highly communicative tasks) are mapped to the remaining (unoccupied) nodes. Another two-phase mapping algorithm targeting to map parts of applications onto processing cores is discussed in [57]. As authors claim, such an approach leads to lower communication cost. A heuristic approach for application mapping onto NoC architectures are discussed in [58, 59]. The initial mapping at this approach is retrieved by focusing to the minimization of communication delay, power dissipation, as well as silicon area, while the packet delivery is based on a predefined routing function. Then, by selecting and swapping iterative random nodes, the algorithm aims to achieve a more efficient application mapping.

The previously mentioned solutions rely on multi-step approaches, where the mapping is carried out in advance of the routing phase [52, 59, 60]. However, such a modular (sequential) approach usually leads to nonefficient solutions, because the objectives from mapping and routing algorithms do not necessarily coincide. Hence, the routing phase must adhere to decisions already taken in the mapping phase that invariably limits considerable the routing solution space. In Guerin et al. [61], a unified algorithm, called Unified MApping, Routing and Slot allocation (UMARS), that couples mapping, path selection and time-slot allocation simultaneously, using a single consistent objective, is discussed.

An algorithm for application mapping under bandwidth constraints onto mesh-based architectures is discussed in [58]. This algorithm, named NMAP, results to minimum average communication delay both for single minimum-path routing, as well as split traffic routing. The bandwidth requirements can be further improved by splitting the application's traffic across multiple paths. For demonstration purposes, the efficiency of this algorithm is evaluated under various DSP applications with a cycle accurate simulator in SystemC, using macros from the ×Pipes library [62]. Apart from single goal optimization constraints, there are also mapping algorithms guided by multi-objective cost functions [63]. A mapping algorithm that integrates physical planning and QoS guarantees is discussed in [59], where the design space is explored with a robust tabu search. A mapping algorithm that guarantees to derive a deadlock-free deterministic routing targeting to regular NoC architecture, is presented in [64]. There are also mapping algorithms that provide QoS guarantee assuming static communication flows and traffic that does not vary with input data [65]. An extension of this work is discussed in [52], where the algorithm is extended in order to balance the network load. Reliability related issues are also addressed during NoC mapping. For instance, a algorithm aiming to reduce the potential temperature hotspots and obtain a thermally balanced design in conjunction to the minimization of communication cost, with the usage of a genetic model is discussed in [66].

## 2.7 Benchmarking

The problem of designing efficient NoC architectures is extremely complex, because it requires an almost exhaustive exploration. Since architecture-level exploration usually is a timing-consuming procedure, there are different approaches aiming initially to identify and then to parameterize the architectural components that define a NoC, their properties, and their interactions. To overcome this limitation, architects usually quantify the efficiency of NoC-based architectures with the usage of various test cases, each of which stresses target architecture under different traffic scenario and/or operating condition. These test cases, also known as benchmark suites, intend to cover a wide spectrum of NoC design aspects and provide detailed information to device architects about the performance of different architecture's components. Note that these benchmarks are completely different from those employed for quantifying the performance of computer architectures. More specifically, the benchmarks at NoC domain have to study the impact of numerous (almost) orthogonal design parameters, such as:

- Different network sizes (small, medium, and large).
- Both regular and irregular topologies.
- Traffic models with different spatial and temporal characteristics that correspond to various application domains.
- Alternative requirements for Quality-of-Service (QoS), such as best-effort, guaranteed bandwidth, guaranteed latency, etc.
- Benchmarks with different composition of IP cores (e.g., number and type of processing and memory cores).

Similar to our previous discussion about different traffic models (please refer to Sect. 2.3), the benchmarks can also be categorized as follows:

**Application benchmarks**: The traffic at these benchmarks is generated either from programs, or models, which resemble real applications. These benchmarks evaluate the resources (both computational and communication) of entire platform. Consequently, given the target application(s), or application domain, these benchmarks can be employed from device architects in order to determine the most suitable architectural parameters for the target NoC.

**Synthetic benchmarks**: Instead of using a real application, it is possible to incorporate a synthetic one. These benchmarks are retrieved from a task graph with known computation times and communication loads among the computing cores. The goal of these benchmarks is to test and evaluate in detail both the NoC architecture, as well as the employed design methodology and/or software tools. By using synthetic traffic, it is easier to gain insight into particular design features of communication infrastructure, and consequently it is feasible to optimize these parameters under various architectural and algorithmic (e.g., mapping and routing) constraints.

The usage of both application and synthetic benchmarks exhibits advantages and disadvantages that should be carefully tackled. More specifically, since the NoC design is still in its infancy, generally, companies and institutions are not open to share specifications, models, and other proprietary data regarding NoCs. This limitation can be alleviated with the usage of academic, synthetic benchmarks, which can be shared and used without these limitations. Furthermore, the existence of an open format for benchmarks specifications makes possible for interested research groups to contribute with relevant models and test cases.

On the other hand, application benchmarks offer the best accuracy but they are difficult to be ported to different systems, whereas their simulation is a time intensive task, compared to synthetic benchmarks. Additionally, application benchmarks scale poorly with system size, for example the number of tasks, which means that the maximum number of processing and storage elements, is fixed. In contrast, synthetic benchmarks are more suitable for benchmarking purposes, because they can exploit the properties of particular fixed size application benchmarks, but they can also scale with system size.

Of major importance is the benchmark measurement methodology, which defines parameters of interest and their points of measurement (in time and space) relative to the structure and representation of the NoCs. Typical benchmark suites that contain various representative media applications are the MiBench [67] and Mediabench [23], whereas Dhrystone [68] is another synthetic benchmark suite that represents not a specific but an average of applications. Apart from these NoC platform-based solutions, it is also desirable to have benchmarks able to quantify the efficiency of supporting algorithms, like mapping, routing, etc.

Even though the previously mentioned benchmark suites are able to stress NoCs under various parameters, all of them focus on conventional (2D) NoC architectures. However, the continuously increased number of cores, as well as the requirement for integrating onto single chip different process technologies imposes the usage of 3D integration. Consequently, new benchmarks that are aware of the inherent features and limitations posed by this integration paradigm are absolutely required.

## 2.8 Exercises

**2.1** Draw examples of the most common NoC topologies. Can you summarize the advantages and disadvantages of alternative NoC topologies?

**2.2** Classify the following topologies as direct or indirect: (a) mesh, (b) torus, (c) octagon, and (d) fat-tree

**2.3** What are the differences between direct and indirect topologies?

**2.4** Calculate the total number of routers and links in the following topologies: (a) $3\times4$ mesh, (b) $3\times4$ torus, (c) a 12-node Spidergon, and (d) a 16-node fat-tree.

**2.5** Calculate the average distance between two nodes in the following topologies: (a) $N\times\mathrm{N}$ mesh, (b) $N\times\mathrm{N}$ torus, (c) $N$-node Spidergon, and (d) $N$-node fat-tree.

**2.6** Do you believe that the ordering of nodes at *Bit Rotation* and *N Complement* traffics affects the actual distribution of traffic?

**2.7** Do you think that in case ordering of nodes at *Bit Rotation* and *N Complement* traffics is performed regularly numbed according to their topology, implies that these two traffic patterns are most beneficial to regular topologies?

## Problems

**2.8** Assume that, the maximum distance between a source node and all other nodes for a polygon-based topology is 2.

(a) What is the probability for a node of sending traffic to another node with distance 1?

(b) What is the probability for a node of sending traffic to another node with distance 2?

**2.9** A flit has as source address the "0111". Can you compute the destination address for this flit, if the bit rotation approach is employed?

**2.10** At the bit rotation approach, can you prove that if $N$ is a power of 2, then the destination node uniquely exists? Otherwise, the destination node address is determined by destination address = (bit $\text{rotation}_{\text{source address}}$) modulo $N$.

**2.11** A network has 7 nodes numbered from 0, 1, ..., 6. Can you determine the destination nodes for $\text{node}_0$, $\text{node}_1$, ..., $\text{node}_6$ assuming the $N$ complement approach?

## References

1. R. Thid, M. Millberg, A. Jantsch, Evaluating NoC communication backbones with simulation, in *IEEE NorChip Conference*, pp. 27–30, 2003
2. D. Wiklund, L. Dake Liu, SoCBUS: switched network on chip for hard real time embedded systems, in *Parallel and Distributed Processing Symposium*, p. 8, April 2003
3. I. Saastamoinen, M. Alho, J. Nurmi, Buffer implementation for Proteo network-on-chip. Int. Proc. Circuits Syst. **2**, 113–116 (May 2003)
4. F. Karim A. Nguyen, S. Dey, An interconnect architecture for networking systems on chips. IEEE J. Micro High Perform. Interconnect **22**(5), 36–45 (2002)
5. S. Kumar, A. Jantsch, J.-P. Soininen, M. Forsell, M. Millberg, J. Oberg, K. Tiensyrja, A. Hemani, A network on chip architecture and design methodology, in *IEEE Computer*, pp. 117–124, 2002
6. W.J. Dally, B. Towles, *Principles and practices of interconnection networks*. (Morgan Kaufmann, San Francisco, 2004)
7. Y. Xu, Y. Du, B. Zhao, X. Zhou, Y. Zhang, J. Yang, A low-radix and low-diameter 3D interconnection network design, in *International Symposium on High Performance Computer Architecture (HPCA)*, pp. 30–42, Feb. 2009
8. A. Weldezion, M. Grange, D. Pamunuwa, L. Zhonghai, A. Jantsch, R. Weerasekera, H. Tenhunen, Scalability of network-on-chip communication architecture for 3-D meshes in *International Symposium on Networks-on-Chip (NoCS)*, pp. 114–123, May 2009

9. The Standard Performance Evaluation Corporation, http://www.spec.org/hpg/
10. R. Dick, Embedded System Synthesis Benchmarks Suites (E3S), http://www.ece.northwestern.edu/dickrp/e3s/
11. ITC'02 SOC Test Benchmarks, http://www.hitech-projects.com/itc02socbenchm/
12. K. Puttaswamy, G.H. Loh, Thermal analysis of a 3D die-stacked high-performance microprocessor, in *ACM Great Lakes Symposium on VLSI (GLSVLSI)*, pp. 19–24, 2006
13. A. Bartzas, L. Papadopoulos, D. Soudris, A system-level design methodology for application-specific networks-on-chip. J. Embed. Comput. **3**(3), 167–177 (2009)
14. H. Hua, C. Mineo, K. Schoeniess, A. Sule, S. Melamed, R. Jenkal, W. Rhett Davis, Exploring compromises among timing, power and temperature in three-dimensional integrated circuits, in *Annual Conference on Design Automation (DAC)*, pp. 997–1002, 2006
15. V. Soteriou, H. Wang, L.S. Peh, A statistical traffic model for on-chip interconnection networks, in *Modeling, Analysis, and Simulation of Computer and Telecommunication Systems (MASCOTS)*, pp. 104–116, 2006
16. R. Widyono, The design and evaluation of routing algorithms for real-time channels. TR-94-024, University of California at Berkeley and International Computer Science Institute (1994)
17. G. Brebner, D. Levi, Networking on chip with platform fpgas, Field-Programmable Technology (FPT), pp. 13–20 (2003)
18. Z. Lu, A. Jantsch, Flit ejection in on-chip wormhole-switched networks with virtual channels, in *IEEE Norchip Conference*, pp. 273–276, Nov 2004
19. N. Genko, D. Atienza, G. De Micheli, L. Benini, J.M. Mendias, R. Hermida, F. Catthoor, A novel approach for network on chip emulation', in *International Symposium on Circuits and Systems (ISCAS)*, pp. 2365–2368, 2005
20. R. Thid, I. Sander, A. Jantsch, Flexible bus and NoC performance analysis with configurable synthetic workloads, Digital System Design: Architectures, Methods and Tools (DSD), pp. 681–688(2006)
21. W. Dong, B. Al-Hashimi, M. Schmitz, Improving routing efficiency for network-on-chip through contention-aware input selection, in*Asia and South Pacific Conference on Design Automation (ASP-DAC)*, pp. 36–41, 2006
22. S. Woo, M. Ohara, E. Torrie, J.P. Singh, A. Gupta, The Splash-2 Programs: characterization and methodological considerations, *International Symposium on Computer, Architecture*, pp. 24–36 (1995)
23. L. Chunho, M. Potkonjak, W. Mangione-Smith, Mediabench: a tool for evaluating and synthesizing multimedia and communications systems, *International Symposium on Microarchitecture*, pp. 330–335 (1997)
24. The Standard Performance Evaluation Corporation (2013). http://www.spec.org/
25. T. Bjerregaard, S. Mahadevan, A survey of research and practices of network-on-chip. ACM Comput. Surv. **38**(1), 1–51 (2006) (Article 1)
26. J. Madsen, S. Mahadevan, K. Virk, M. Gonzalez, Network-on-chip modeling for system-level multiprocessor simulation, in *International Real-Time Systems Symposium (RTSS)*, pp. 82–92 (2003)
27. S. Mahadevan, M. Storgaard, J. Madsen, K. Virk, ARTS: a system-level framework for modeling MPSoC components and analysis of their causality, in *International Symposium on Modeling, Analysis, and Simulation of Computer and Telecommunication Systems (MASCOTS)*, pp. 480–483 (2005)
28. J. Xu, W. Wolf, J. Henkel, S. Chakradhar, A methodology for design, modeling, and analysis of networks-on-chip, in *International Symposium on Circuits and Systems (ISCAS)*, pp. 1778–1781 (2005)
29. E. Bolotin, I. Cidon, R. Ginosaur, A. Kolondy, QNoC: QoS architecture and design process for network-on-chip. J. Syst. Archit. **50**(2–3), 105–128 (Feb. 2004)
30. P. Pande, C. Grecu, M. Jones, A. Ivanov, R. Saleh, Effect of traffic localization on energy dissipation in NoC-based interconnect, in *International Symposium on Circuits and Systems*, pp. 1774–1777 (2005)

31. N. Banerjee, P. Vellanki, K. Chatha, A power and performance model for network-on-chip architectures, in *Proceeding of Design, Automation and Testing in Europe Conference (DATE)*, pp. 1250–1255 (2004)
32. H.-S. Wang, X. Zhu, L.-S. Peh, S. Malik, Orion: a power-performance simulator for interconnection networks, in *International Symposium on Microarchitecture*, pp. 294–305 (2002)
33. SystemC, The SystemC Version 2.0.1 (2002). http://www.systemc.org
34. T. Fitzpatrick, System verilog for VHDL users, in *Proceeding of Design, Automation and Testing in Europe Conference (DATE)*, pp. 1530–1591 ( 2004)
35. M. Coppola, S. Curaba, M. Grammatikakis, R. Locatelli, G. Maruccia, F. Papariello, OCCN: a NoC modeling framework for design exploration. J. Syst. Archit. **50**(23), 129–163 (Feb. 2004)
36. B. Juurlink, H. Andwijshoff, A quantitative comparison of parallel computation models. ACM Trans. Comput. Syst. (TOCS) **16**(3), 271–318 (Aug. 1998)
37. R. Vaidya, A. Sivasubramaniam, C. Das, Impact of virtual channels and adaptive routing on application performance. IEEE Trans. Parallel Distrib. Syst. **12**(2), 223–237 (Feb. 2001)
38. T. Bjerregaard, S. Mahadevan, J. Spars, A channel library for asynchronous circuit design supporting mixed-mode modeling, in *International Workshop on Power and Timing Modeling, Optimization and Simulation (PATMOS)*, pp. 301–310 (2004)
39. G. Leary, K. Srinivasan, K. Mehta, K. Chatha, Design of network-on-chip architectures with a genetic algorithm-based technique. IEEE Trans. Very Large Scale Integr. (VLSI) **17**(5), 674–687 (2009)
40. D. Siguenza-Tortosa, J. Nurmi, Vhdl-based simulation environment for proteo noc, in *High-Level Design Validation and Test Workshop*, pp. 1–6 ( 2002)
41. A. Pinto, L.P. Carloni, A.L. Sangiovanni Vincentelli, A methodology for constraint-driven synthesis of on-chip communications. IEEE Trans. Comput Aided Des. Integr. Circ. Syst. **28**(3), 364–377 (March 2009)
42. V. Puente, J.A. Gregorio, R. Beivide, SICOSYS: an integrated framework for studying interconnection network performance in multiprocessor systems, in <error l="74" c="Undefined command " />Proceeding of of Euromicro Workshop on Parallel, Distributed and Network-based Processing, pp. 15–22 (2002)
43. L. Se-Joong, S. Seong-Jun, L. Kangmin, W. Jeong-Ho, K. Sung-Eun, N. Byeong-Gyu, Y. Nam, An 800MHz star-connected on-chip network for application to systems on a chip, in *Solid-State Circuits Conference*, pp. 468–469 (2003)
44. A. Jalabert, S. Murali, L. Benini, G. De Micheli, xpipesCompiler: a tool for instantiating application specific Networks on Chips, in *Proceeding of the Design and Test Europe Conference (DATE)*, pp. 884–889 (2004)
45. L. Kangmin, L. Se-Joong, K. Sung-Eun, C. Hye-Mi, K. Donghyun, K. Sunyoung, L. Min-Wuk, Y. Hoi-Jun, A 51mW 1.6GHz on-chip network for low-power heterogeneous SoC platform, in *Solid-State Circuits Conference*, pp. 152–518 (2004)
46. I. Loi, F. Angiolini, L. Benini, Supporting vertical links for 3D networks on chip: toward an automated design and analysis flow, in *Proceeding of the International Conference on Nano-Networks (Nano-Net)*, Article 15 (2007)
47. S. Yan, B. Lin, Design of application-specific 3D networks-on-chip architectures, in *Proceeding of the International Conference on Computer Design (ICCD)*, pp. 142–149 (2008)
48. Ge-Ming Chiu, The odd-even turn model for adaptive routing. IEEE Trans. Parallel Distrib. Syst. **11**(7), 729–738 (July 2000)
49. S. Murali, G. De Micheli, Bandwidth constrained mapping of cores on to NoC architectures, in *Proceeding of the Design and Test Europe Conference (DATE)*, pp. 884–889 (2004)
50. I. Walter, I. Cidon, A. Kolodny, D. Sigalov, The era of many-modules SoC: revisiting the NoC mapping problem, in *Proceeding of the International Workshop on Network on Chip Architectures (NoCArc)*, pp. 43–48 (2009)
51. J. Kim, C. Nicopoulos, D. Park, R. Das, Y. Xie, V. Narayanan, M. Yousif, C. Das, A novel dimensionally-decomposed router for on-chip communication in 3D architectures, in *International Symposium on Computer Architecture (ISCA)*, pp. 138–149 (2007)

52. T.M. Pinkston, R. Pang, J. Duato, Deadlock-free dynamic reconfiguration schemes for increased network dependability. IEEE Trans. Parallel Distrib. Syst. **14**(8), 780–794 (Aug. 2003)
53. S. Murali, G. De Micheli, SUNMAP: A tool for automatic topology selection and generation for NoCs, in *Proceeding of the Annual Design Automation Conference (DAC)*, pp. 914–919 (2004)
54. T. Ahonen et al, Topology optimization for application specific networks on chip, in *Proceeding of the International Workshop on System Level Interconnect Prediction (SLIP)*, pp. 53–60 (2004)
55. U. Orgas, R. Marculescu, Energy- and performance-driven NoC communication architecture synthesis using a decomposition approach, in *Proceeding of the Design Automation and Test in Europe (DATE)*, pp. 352–357 (2005)
56. J. Hu, R. Marculescu, Energy-aware mapping for tile-based NOC architectures under performance constraints, in *Asia and Sourth Pacific Design Automation Conference (ASP-DAC)*, pp. 233–239 (2003)
57. N. Koziris, M. Romesis, P. Tsanakas, G.Papakonstantinou, An efficient algorithm for the physical mapping of clustered task graphs onto multiprocessor architectures, in *Proceeding of Euromicro Workshop on Parallel and Distributed Processing*
58. S. Murali, G. De Micheli, Bandwidth-constrained mapping of cores onto NoC architectures, in *Proceeding of the Design, Automation and Test in Europe Conference and Exhibition (DATE)* 2004), pp. 896–901 (2004)
59. E. Rijpkema, K. Goossens, A. Radulescu, J. Dielissen, J. van Meerbergen, P. Wielage, E. Waterlander, Trade-offs in the design of a router with both guaranteed and best-effort services for networks on chip. IEE Proc. Comput. Digital Tech. **150**(5), 294–302 (2003)
60. F. Feliciian, S. Furber, An asynchronous on-chip network router with quality-of-service (QoS) support, in *Proceeding of SOC Conference*, pp. 274–277 (2004)
61. R. Guerin, A. Orda, D. Williams, QoS routing mechanisms and OSPF extensions. Global Telecommun Conf. (GLOBECOM) **3**, 1903–1908 (1997)
62. S. Li, L. Peh, A. Kumar, N.K. Jha, Thermal modeling, characterization and management of on-chip Networks, in *International Symposium on Microarchitecture (MICRO-37)*, pp. 67–78 (2004)
63. G. Ascia, V. Catania, M. Palesi, Multi-objective mapping for mesh-based NoC architectures, in *International Conference on Hardware/Software Codesign and System, Synthesis (CODES+ISSS)*, pp. 182–187 (2004)
64. J. Hu, R. Marculescu, Energy- and performance-aware mapping for regular NoC architectures. IEEE Trans. Comput. Aided Des. Integr. Circ. Syst. **24**(4), 551–562 (April 2005)
65. M. Dall'Osso, G. Biccari, L. Giovannini, D. Bertozzi, L. Benini, Xpipes: a latency insensitive parameterized network-on-chip architecture for multi-processor SoCs, in *International Conference on Computer Design (ICCD)*, pp. 45–48 (2012)
66. R. Tamhankar, S. Murali, S. Stergiou, A. Pullini, F. Angiolini, L. Benini, G. De Micheli, Timing-error-tolerant network-on-chip design methodology. IEEE Trans. Comput. Aided Des. Integr. Circ. Syst. **26**(7), 1297–1310 (July 2007)
67. M.R. Guthaus, J.S. Ringenberg, D. Ernst, T.M. Austin, T. Mudge, R.B. Brown, MiBench: a free, commercially representative embedded benchmark suite, in *Annual Workshop on Workload Characterization*, pp. 3–14 (2001)
68. R. Weicker, Dhrystone: a synthetic systems programming benchmark. Commun. ACM **27**(10), 1013–1030 (Oct. 1984)
69. M. Jabbar, D. Houzet, 3D architecture implementation: a survey, in *IP Embedded System Conference (IP-SOC)*, pp. 1–5 (2011)
70. R. Kourdy, M.R. Nouri, Compare performance of 2D and 3D mesh architectures in network-on-chip. J. Comput. **4**(1), 83–87 (2012)
71. A. Gerstlauer, Communication abstractions for system-level design and synthesis. Technical Report TR-03-30, Center for Embedded Computer Systems, University of California, Irvine, CA, 2003

# Chapter 3
# Communication Architecture

**Abstract** The communication infrastructure is the backbone of the NoC system. After determining the NoC topology for the given application, designing the communication infrastructure is the next step. The routing algorithm is selected based on both the selected topology and design constraints. After the selection of routing algorithm and flow control scheme, the router and link design can begin. In this chapter switching techniques, routing algorithms and flow control schemes are discussed and compared, while the design of a generic 2D and 3D router is illustrated and improvements proposed in the literature are discussed.

After topology selection, the next step in the NoC design flow is to select the appropriate switching technique and routing algorithm based on the design constraints. Based on these decisions, the design of the router(s) and channels can begin.

## 3.1 Switching Technique

The switching technique defines how and when the *input channel* of the switch is connected to the *output channel* selected by the routing algorithm. Data are transmitted as *messages* that are splitted into *packets*, which in turn are splitted into *flow control units (flits)* and finally into *physical units (phits)*. Switching technique selection involves selecting the optimal granularity for the above data.

- *physical units (phits)*: the unit of data transferred through the physical link. Essentially, the bit-width or word-length of the channel and, therefore, the number of bits transmitted between routers in a single clock cycle.
- *flow control units (flits)*: The unit of synchronization between routers. At least as large as a phit and often equal.
- *packets*: A set of consecutive flits with the same destination. Routers may store an entire packet before forwarding it, or transmit flits separately.
- *messages*: A set of packets that typically corresponds to a complete data transfer (transaction) between nodes. A message could be for example an entire bus

K. Tatas et al., *Designing 2D and 3D Network-on-Chip Architectures*,

DOI: 10.1007/978-1-4614-4274-5_3, 

transaction from processor to memory. Considering that state-of-the-art bus protocols allow long data bursts, a message may require splitting to many packets.

The two commonly used switching techniques are: (i) packet switching and (ii) circuit switching.

## 3.1.1 Circuit Switching

In circuit switching, the network path between two nodes that have to exchange data is established in advanced (before the data are sent) by allocating the proper hardware resources (links). Circuit switching is performed in three stages: circuit establishment (setup), data transmission, and circuit release (tear-down). The setup procedure requires the head flit (probe) to make its way from source to destination reserving links in its path. When the head flit reaches its destination, an acknowledgement is returned to the sender, unless a link is reserved by another circuit, in which case a negative acknowledgment is sent. After a successful acknowledgment, data transfer begins and lasts until circuit release. Contention can only occur during the setup phase and buffering is only required for the head flit, since the data are wired directly from source to destination. Circuit switching between three links is illustrated in the time-space diagram of Fig. 3.1 $t_s$ is the router switching time and $t_r$ the routing decision time.

An advantage of circuit switching is that since the connection over which all subsequent data are transported is set up first, contention resolution takes place at setup at the granularity of connections, and time-related guarantees during data

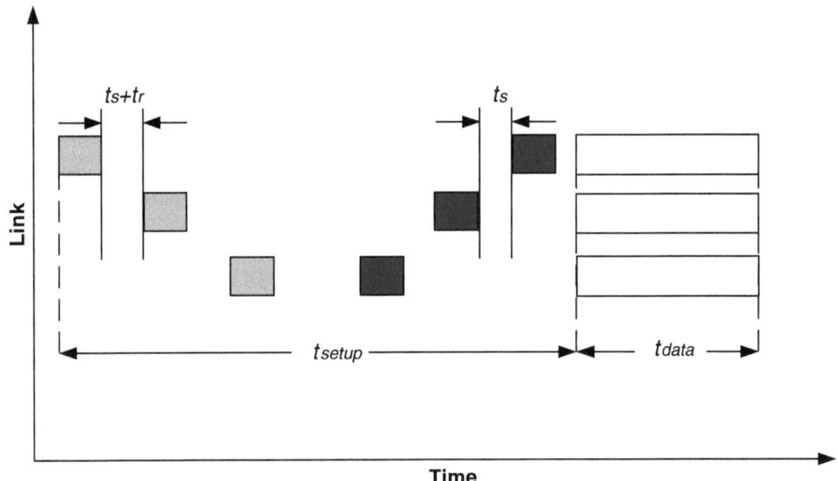


**Fig. 3.1** Circuit switching

transport can be given. On the other hand, when a circuit between two nodes has been established, any other communication on the allocated wires is rejected, until the data transfer is complete, and therefore, other messages may be blocked. Therefore, circuit switching performs well when large amounts of data are transferred, which help to amortize the setup time.

The circuit switching approach delay can be calculated as the sum of the setup time, the data transfer time and the data release time:

$$t_{\mathrm{CS}} = t_{\mathrm{setup}} + t_{\mathrm{data}} + t_{\mathrm{release}} \tag{3.1}$$

Release of the circuit can be done with the last flit, which can be bundled with the data, simplifying Eq. 3.1 by considering $t_{\mathrm{release}} = 0$. It is assumed that the links are not pipelined or multicycle paths.

$$t_{\mathrm{setup}} = D\left(t_r + 2\left(t_s + \frac{\mathrm{ph}}{B}\right)\right) \tag{3.2}$$

$$t_{\mathrm{data}} = (\mathrm{fl} \times \mathrm{ph} - 1) \times \frac{1}{B} \tag{3.3}$$

where $D$ is the distance in hops between source and destination, $B$ is the link bandwidth in bits/s, ph and fl are the phits per flit and flits per packet respectively, assuming one flit in the packet is used for setup.

A number of circuit-switched networks have been implemented such as [1, 2] and [3]. From the above equations, it can be derived that a fast setup process is desirable in order to have an efficient circuit-switched NoC. In [4], the design and implementation of a pipeline circuit-switched switch to support guaranteed throughput was presented. The proposed switch is based on a backtracking probing path setup called backtracking wave-pipelining. In [5], a parallel probe searching setup method, which can search the entire network in constant time, dependent on the network size but independent of the network load was proposed. The proposed method can reduce the search time by up to 20 %.

### 3.1.2 Packet Switching

*Packet switching* [6] allows the packets in a message to be transmitted through different paths. In order to provide efficient routing, the packet switching typically implies some restrictions. More specifically, as soon as a flit (head flit) of a packet is sent over an output port, that output port is reserved for flits of that packet only. Since this approach uses a finer granularity buffer and channel control at the flit level instead of the packet level, it exhibits increased performance, especially when messages are short and frequent. However, in case the head flit of a packet is blocked, the trailing

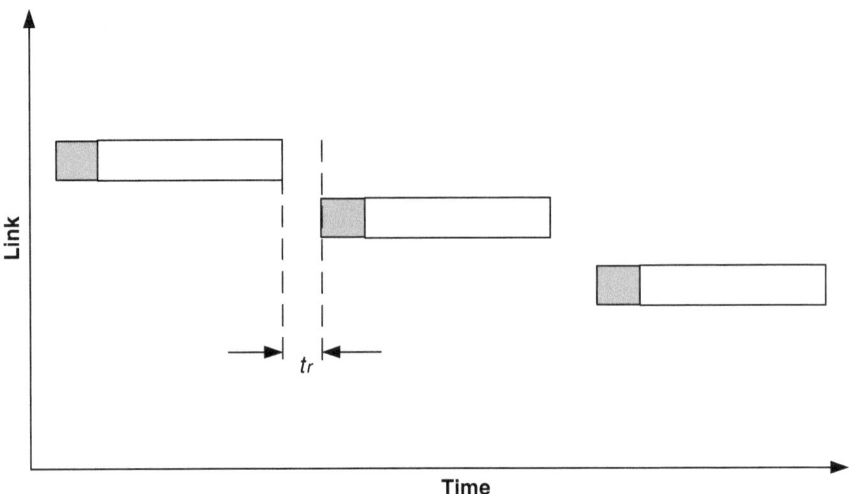


**Fig. 3.2** Store-and-forward switching

flits can therefore be spread over multiple routers, blocking the intermediate links, causing a condition known as deadlock (see Sect. 3.2.1).

Packet switching includes three subcategories, namely: Store-And-Forward (SAF), Virtual Cut-Through (VCT), and Wormhole Switching (WS). More specifically, the store-and-forward approach is based on receiving and storing the whole incoming packet before it is forwarded to the next router as shown in Fig. 3.2. This approach requires more buffer space since the complete packet must be stored, which leads, among others, to a per-router latency of at least the time required for the router to receive the packet.

The delay of a nonblocked packet in SAF switching is computed as follows:

$$t_{\mathrm{SAF}} = D\left(t_r + \left(t_s + \mathrm{ph} \times \frac{l}{B}\right)\right) \tag{3.4}$$

On the other hand, VCT switching is based on forwarding a packet as soon as the next router guarantees that the complete packet will be accepted. In case there are no guarantees, the router must be able to store the whole packet. Consequently, even though such a routing approach requires buffer space for a complete packet, like store-and-forward routing, it allows lower latency communication as shown in Fig. 3.3. Essentially, flits can cut through to the next router input before the packet is completely received in the current switch, and therefore the packet is pipelined through the switch. The delay of a nonblocked packet in VCT switching is computed as follows:

$$t_{\mathrm{VCT}} = \left(t_s + t_r + \frac{\mathrm{ph}}{B} + \mathrm{MAX}\left\{t_s, \frac{\mathrm{ph}}{B}\right\} \times \frac{\mathrm{ph} \times \mathrm{fl} - 1}{B}\right) \tag{3.5}$$

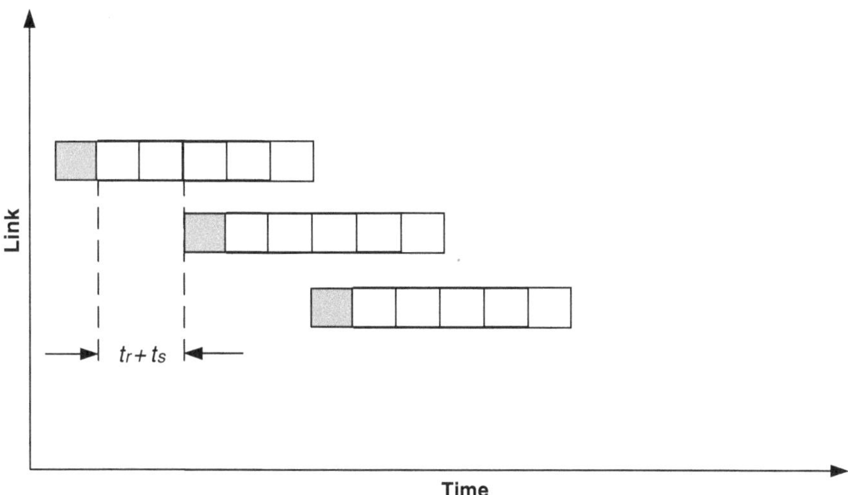


**Fig. 3.3** Virtual Cut-Through (VCT) switching

The worst-case delay of VCT switching is equal to that SAF switching.

Wormhole switching is based on pipelining at the flit level, so the routers are only required to store a few flits instead of a complete packet. The disadvantage of this approach is that if a message gets blocked, its flits will occupy buffer slots in various routers, making it vulnerable to deadlock. A number of NoC architectures are based on wormhole packet routing [7–10], since it allows low-latency communication schemes, while it also requires the least buffering (flits instead of packets) among packet switching methods. This allows relatively small buffers to be used in each router, even for large packet sizes. The wormhole switching space-time diagram is shown in Fig. 3.4. As long as packets are not blocked, wormhole switching achieves the same latency as VCT switching. Since buffers are expensive in on-chip networks (see 3.4.1), wormhole switching is the most common packet switching technique used.

In order to prevent a blocked packet to impede the progress of other packets waiting in line as shown in Fig. 3.5a, the *Virtual Channel* (VC) flow control for wormhole switching was introduced [11]. A VC between two resources A and B is established by allocating time slots (by Time Division Multiplexing - TDM) in each switch on the path between two resources A and B. The [11] approach assigns multiple virtual paths, each of which has its own associated buffer queue, to the same physical channel, in order to allow blocked packets to be passed by other packets as shown in Fig. 3.5b. Virtual channels arbitrate for physical channel bandwidth on a flit-by-flit basis (see 3.4.1).

A hybrid switching technique that dynamically combines virtual cut-through and wormhole switching in order to provide higher achievable throughput values compared to wormhole switching alone, while reducing the buffer space required at the

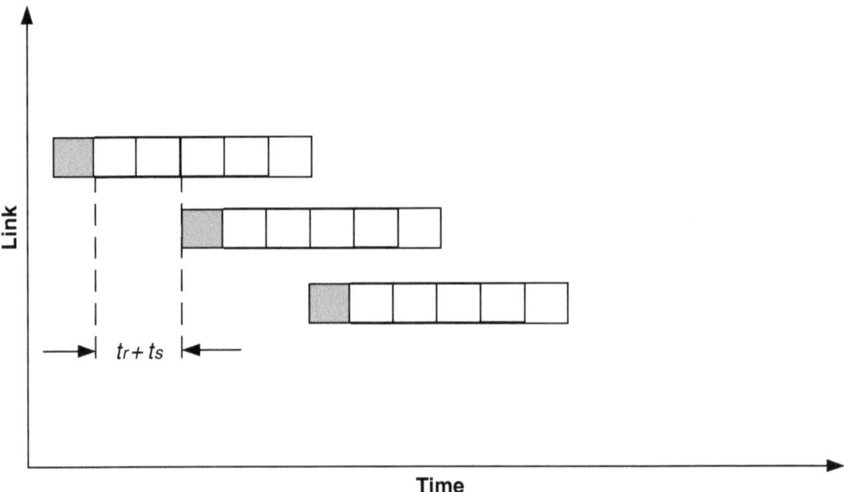


Fig. 3.4 Wormhole switching

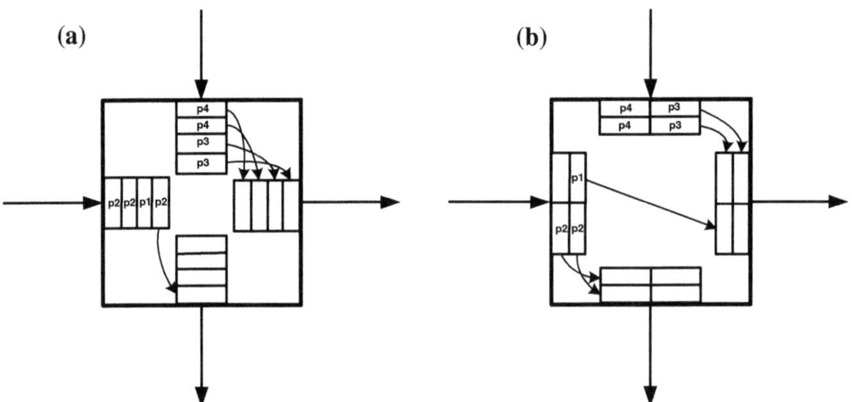


Fig. 3.5 **a** Blocked packet $p_1$ impedes progress of packet $p_2$ **b** packet can progress by using two virtual channels

intermediate nodes when compared to virtual cut-through is discussed in [12]. On the other hand, such a solution results to less efficient link utilization than virtual cut-through routing, since even though wormhole routing allocates storage and bandwidth in flit-sized units, such a flow control holds a physical channel for the duration of a packet.

**Table 3.1** Switching technique comparison

| Switching | Performance | Buffering | Design complexity | Cost | Flexibility |
|---|---|---|---|---|---|
| Circuit switching | High for long messages | 1 flit | Low | Low | |
| SAF | High for short and frequent messages | Packet | Low | High | Low |
| VCT | High | Packet | High | High | Low |
| Wormhole | High | A few flits | Low | Low | Low |
| [15] | High | Low | Medium | High | |
| [4] | High | Low | High | High | |
| [12] | High | High | Low | Low | Medium |
| [4] | High | Very low | Low | Low | High |
| [5] | Very high | Very low | Low | Low | High |
| [13] | High | Low | High | Low | High |
| [14] | High | Low | High | Low | High |

### *3.1.3 Hybrid Switching Techniques*

A hybrid packet-circuit switching technique based on space division multiplexing (SDM) was proposed in [13] in order to efficiently handle both streaming and best-effort traffic generated in real-time applications, accordingly by using a circuit-switched subrouter with both SDM and TDM, and a packet-switched subrouter.

A similar approach was proposed in [14], demonstrating a 22 % improvement in power and 45 % improvement in latency over a conventional packet-switched network.

Table 3.1 compares proposed switching techniques in terms of performance, hardware complexity, buffering, and flexibility.

## 3.2 Packet Routing

As mentioned briefly in Chap. 1, the routing techniques for on-chip communication determine the path selected by a packet to reach its destination. It is obvious that routing is closely linked to the target topology. The first task is to appropriately and uniquely identify the hardware resources that communicate through the network (nodes) using an unambiguous naming and addressing scheme. Typical approaches are:

- By name (e.g. object $X$)
- By address (e.g. object at destination $X$). For example, for up to 8 nodes 3 bits are required to encode addresses, for up to 16, 4 bits, etc.
- With a group identifier (e.g. all objects related to $X$) used to identify a NoC multicast group

- In mesh and torus topologies, it is easy to address a node by using Cartesian coordinates ("$x$", "$y$", and "$z$" in case of a 3-D NoC)

Routing algorithms can be classified into three main groups, namely:

- *Deterministic*: the route between source-destination pairs is fixed. This approach cannot provide solution to dynamic congestion or fault tolerance.
- *Adaptive*: the path choice between source-destination pairs is dynamic.
- *Stochastic*: In this approach the cost among different routing paths varies probabilistically.

Similarly, depending on the degree of adaptiveness provided by the routing algorithms, they can be generally classified into three categories:

- A nonadaptive routing algorithm is deterministic and routes a packet from the source to the destination along a unique, predetermined path.
- A minimal fully adaptive routing algorithm routes all packets through any shortest paths to the destinations.
- A partially adaptive routing algorithm allows multiple choices for routing packets via shortest paths, but it does not allow all packets to use any shortest paths.

In terms of implementation, routing algorithms can be implemented by appropriate logic or commonly a *routing table* is used which is essentially a routing table containing the routing information for all possible destinations. Another technique associated with routing calculation is the distinction between *source routing* and *distributed routing*. In source routing, the entire routing path is computed at the source (typically the NI connected to the source node) and is appended to the packet. The routers do not make any routing decisions, but simply implement the routing path based on the information encoded in the packet. On the other hand, in distributed routing, the routing path is decided in a hop-by-hop basis at each router, even for deterministic routing algorithms. The only information required to be found in the packet is the destination address. The advantage of source routing is that it requires simple routers and can easily support irregular architectures. Its disadvantage is that it does not provide adaptiveness and requires more complex NIs and packets.

Finally, routing algorithms can be classified according to the target topology (regular, irregular, hierarchical, etc.).

A desirable feature of a routing algorithm is its freedom from deadlock and livelock. Freedom from deadlock is especially critical for NoCs, since the implementation of a mechanism which automatically detects and recovers from deadlock may not be affordable in terms of silicon resources; it also may lead to unpredictable delays.

### 3.2.1 Deadlock

Deadlock, in general, is the condition that occurs when two processes are each waiting for the other to finish first, and therefore neither can start. In the network environment, deadlock can occur when circular dependencies exist, for example, a packet $p_1$ holds channel $c_1$ and requests channel $c_2$, which is reserved for packet $p_2$, which in turn requests channel $c_3$, which is reserved for $p_3$, which requests channel $c_4$ etc. and finally packet $p_n$ requests channel $c_1$, and therefore, all packets stay blocked for indefinite time, waiting for an event that will never happen. Wormhole switching is particularly vulnerable to deadlock.

There are two possible solutions to the deadlock problem, namely recovery and avoidance [16]. Deadlock recovery requires fewer resources and may outperform avoidance if deadlocks are infrequent but leads to unpredictable delays and is challenging to implement [17]. Deadlock avoidance generally requires more resources and may restrict routing flexibility and therefore performance. However, it is usually the preferred solution due to its ease of implementation.

#### 3.2.1.1 Deadlock Recovery

More specifically, deadlock recovery requires a run-time deadlock detection mechanism, supported by a deadlock resolution mechanism. Deadlock detection is challenging due to the distributed nature of deadlocks. A common approach is to employ heuristics such as time-out mechanisms [18, 19]. While this approach is certain to detect all deadlocks, it can also lead to blocked packets falsely flagged as deadlocked. The reason is that it is difficult to determine the optimal time-out threshold value, since it is heavily dependent on the network parameters (packet length, network size, topology etc.) as well as the application characteristics (specific traffic).

Concerning deadlock recovery schemes, they can be classified as regressive or progressive. A regressive recovery scheme kills the packet flagged as deadlocked and retransmits it after a time-out [20]. However, retransmission of a packet reduces performance, and usually a progressive recovery scheme which utilizes additional hardware to bypass the suspected packets to their destination [18] is preferred.

In [21], deadlocks are detected using a runtime transitive closure computation scheme to discover the existence of deadlock-equivalence sets, which imply loops of requests. This detection scheme significantly reduces the number of false positives.

#### 3.2.1.2 Deadlock Avoidance

Deadlock avoidance requires proof that the NoC routing scheme is deadlock-free. A necessary and sufficient condition for deadlock-free routing algorithms was introduced in [11], and it is essentially the absence of cyclic dependencies in the resources required by the switching technique and routing algorithm. Therefore, deadlock avoidance is achieved by ensuring that the routing algorithm selected satisfies the above condition. In order to analyze the possible dependencies that can lead to cyclic dependencies and to deadlock, a resource dependency graph can be used. It is a directed graph where nodes are *agents* and *resources* and the arcs are either *hold* or *wait-for* relations that lead from an agent to a resource. In other words, an agent

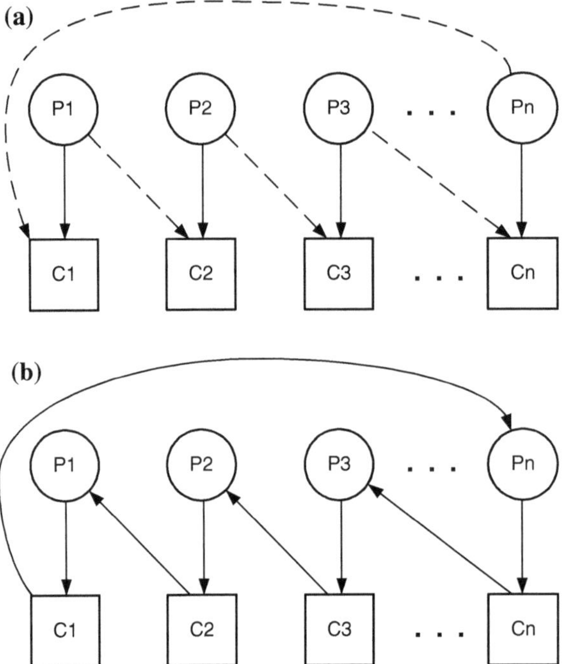


**Fig. 3.6** Resource dependency graph with (**a**) hold and wait-for relations and (**b**) wait-for relations only

either occupies (holds) or waits for a resource in the network to become available. Agents correspond to connections in circuit switching, packets in SAF switching and in flits in wormhole switching. Therefore the deadlock condition described earlier would correspond to the resource dependency graph of Fig. 3.6a. Hold relations are denoted with the dashed line while wait-for relations are denoted with a continuous line. Hold relations can be replaced by equivalent wait-for relations of the opposite direction, since an agent holding a resource is equivalent to a resource waiting for an agent to release it. This leads to the wait-for resource dependency graph of Fig. 3.6b. It can be seen that the graph is cyclic, which is a necessary condition for deadlock.

A cyclic resource dependency graph can be transformed to an acyclic one by simply adding more resources. This is a common approach for ensuring freedom from deadlock in case an algorithms is not proven deadlock-free. Another approach is to prohibit certain turns [22, 23] which comes at some performance penalty due to the routing restrictions, and is also not applicable to irregular architectures.

### 3.2.2 Livelock

Livelock is the situation where a flit or packet is perpetually deflected and, even though not blocked, never reaches its destination either. It can happen in routing algorithms that allow deflection of a packet (such as hot potato routing), and these algorithms must therefore ensure by some mechanism that all packets eventually reach their destination. There are two basic categories of livelock-free algorithms, namely deterministically livelock-free and probabilistically livelock-free. The former are based on adding to each packet additional information used to prioritize packets and not deflect the higher priority ones. Typical metrics used are age, number of deflections, etc. Probabilistically, livelock-free algorithms are based on ensuring that as time approaches infinity the probability of a packet reaching its destination approaches one, meaning that the packet will eventually reach its destination. However, the designer must also ensure that the selected solution satisfies performance constraints, as "eventually" may be unacceptably long, and is also unpredictable.

### 3.2.3 Routing Algorithms for Regular Architectures

Regular architectures are the most commonly used ones, and therefore a number of routing algorithms have been proposed, particularly for the highly popular mesh topology.

Regarding mesh topologies, it is very easy to accomplish a shortest path deterministic routing with reasonable area penalty, by employing a simple variation of dimension order routing [24] such as $XY$ (or $XYZ$ for 3-D NoCs). $XY$ routing ensures deadlock and livelock freedom, but it provides no adaptiveness. This algorithm is a *tableless* routing technique whereby each packet is routed first in the $X$ direction and after it reaches the same $X$ as the destination address, similarly moves along the perpendicular dimension(s). Such a routing approach provides a number of advantages over alternative implementations. Among others, (i) it is shortest path, and hence minimizes the energy spent per information unit transfer, and (ii) it requires a reduced number of gates as compared to routing table-based or source-route schemes.

A number of shortest path, deadlock-free, partially adaptive routing algorithms are based on restricting certain turns as mentioned in Sect. 3.2.1. These are West-first, North-last, and Negative-first. As their names imply, West-first requires that a packet is routed to the west direction first, if it is a productive direction, otherwise any shortest path can be taken. Similarly, North-last requires that if north is a productive direction, to be taken last, while negative first requires that. All these schemes imply that certain turns are prohibited as shown in Fig. 3.7 together with an example of possible routing paths using the above schemes.

As can be seen in Fig. 3.7a, $XY$ routing, by restricting four turns becomes completely deterministic. The partially adaptive routing algorithms restrict only two turns. West first, by requiring that the west direction is taken first if required, allows

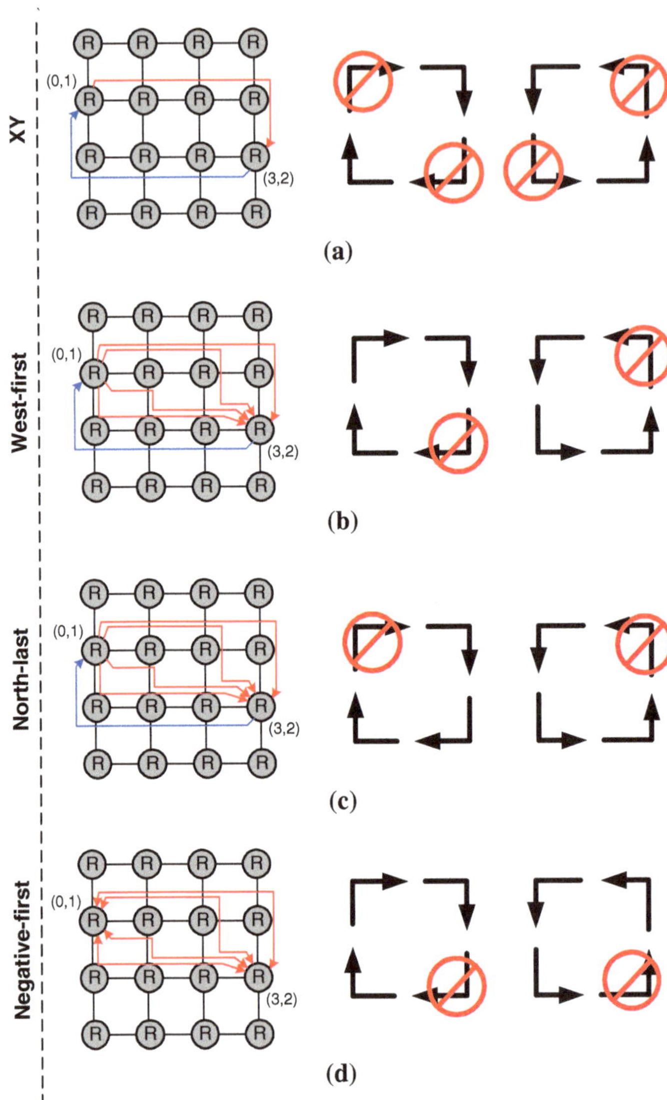


**Fig. 3.7** *XY* and partially adaptive mesh routing algorithm turn model

all possible shortest path when a packet has to travel eastward, but only one possible path when traveling westward (Fig. 3.7b), clearly favoring traffic toward the east direction. Likewise, North-first (Fig. 3.7c) by demanding that traffic going toward the north takes the north direction last, allows all possible shortest paths for traffic going toward the south but again, only a single path for traffic going north. Clearly, this algorithm favors traffic going from north to south. Finally, negative-first requires that in case the packet destination is toward any negative axis, horizontal or vertical, along with any other direction, then the packet should be routed first toward that negative axis direction and afterward toward the other direction. All these schemes imply that certain turns are prohibited as shown in Fig. 3.7 together with an example of possible routing paths using the above schemes.

As mentioned, the above partially adaptive schemes, only allow multiple routes in a subset of source-destination pairs. A different approach for a partially adaptive, deadlock-free routing algorithm targeting to meshes is presented in [23]. Unlike the implementations which rely on prohibiting certain turns in order to achieve deadlock freedom, this approach restricts the locations where some types of turns can be taken. Also, a deterministic routing methodology for high performance without the use of virtual channels, targeting (but not limited) to torus and mesh topologies, is described in [25]. This algorithm, named Segment-based Routing, is based on partitioning a topology into subnets, and then subnets into segments. This allows placing bidirectional turn restrictions locally within a segment, which leads to a larger degree of freedom as compared to similar routing strategies.

A routing algorithm named DyAD, which combines the advantages of both deterministic and adaptive routing schemes, is presented in [15]. This approach is based on the current network congestion since each router in the network continuously monitors its local network load and makes decisions based on this information. When the network is not congested, the DyAD router works in a deterministic mode, thus enjoying the low routing latency enabled by deterministic routing. On the contrary, when the network becomes congested, the DyAD router switches back to the adaptive routing mode, and thus avoids the congested links by exploiting other routing paths (this result to higher network throughput which is highly desirable for applications implemented using the NoC approach).

Adding virtual channels allows the design of highly adaptive routing algorithms. In fact, it is impossible to produce a deadlock-free fully adaptive routing algorithm for a mesh without addition of virtual channels [26]. However, adding virtual channels to meshes is not free. It involves adding buffer space and complex control logic to routers, thus communication performance of the network and reliability of the routers may be affected. In [27], the authors propose a routing algorithm that achieves deadlock-free, minimal, and adaptive layered routing using virtual channels for regular ($k$-ary $n$-cubes) networks. However, we have to take into consideration a limitation of this approach, since the need for virtual channels grows exponentially with $n$.

Regarding torus topologies, the straightforward routing approach is a modification of $XY$ or $XYZ$ routing that takes into account the wraparound links. It is fairly simple for a router or NI to calculate whether the shortest path is through a wraparound link or not, based on the packet destination address. However, this approach requires more calculations than mesh topology $XY$ routing, or using a routing table. In [28], the authors propose a different encoding for the nodes in a torus topology based on Johnson coding [29] which implies the relation among neighboring nodes and the global information of routing, allowing route calculation with three to six logic operations.

### 3.2.4 Routing Algorithms for Irregular Architectures

The previously discussed routing algorithms exhibit two major drawbacks when applied in real life NoC architectures. More specifically, some of them (i.e., the $XY$ routing) favor specific paths of NoC architecture, which might be over utilized, whereas some other parts of the device might be under-utilized. Additionally, the existing VLSI layouts are irregular (due to modules' shape and size variability in VLSI layouts), and hence a mesh topology cannot be applied without significant modifications.

In order to alleviate the limitation of shortest turn-based routing algorithms, a number of alternative implementations were studied. In [30], multiple instantiations of shortest path $XY$ routing are examined for a SoC-based NoC with an irregular mesh topology. Since the considered algorithms are static shortest path schemes, which all have equal link power dissipation, the comparative metric is the total implementation gate-count.

A deadlock-free routing algorithm targeting to irregular topologies, called Layered Routing, is presented in [31]. This approach groups virtual channels into network layers and to each layer it assigns a limited set of source/destination address pairs. Such a separation of traffic yields a significant increase in routing efficiency, as it achieves load balancing and guaranteeing shortest path routing, without requiring any features in the switches other than the existence of a modest number of virtual channels.

The issue of efficient routing in irregular networks was initially addressed in [32, 33], where a source routing method guarantees that messages always use shortest paths. This algorithm attempts to alleviate the restrictions imposed by up*/down* routing by letting intermediate nodes eject messages completely and later re-inject them, breaking the dependencies of illegal turns. In [34], the same idea was used in an InfiniBand setting, but here letting the packets that use these turns ascend to a higher layer of virtual channels broke illegal turns. Another approach is based on providing a routing with layers. More specifically, based on this approach packets escape from possible deadlocks in a higher layer by making a transition down to a lower level. If the lowest layer is deadlock-free, the entire system will also be [35–39].

In order to support NoC architectures with irregular (i.e., non mesh) topologies, both source routing and distributed routing have been employed. As mentioned, regarding the source routing, each flit carries a sequence of routing instructions, whereas in the distributed routing the destination is looked up at each intermediate router. In general, both SR and DR make extensive use of routing tables. Regarding the SR, the tables are located at each source (they are also indexed by packet destination addresses, but they contain sequences of routing commands, one for each hop along the routing path), while at the DR approach the routing tables are located at each router (they are indexed by the packet destination address and contain output port values). Usually, the routing table implementations have as many entries as there are nodes in the network. However, such a communication scenario, where any node can communicate to any other node is very unlikely to happen, since the actual set of destinations used at each source is typically a small fraction of the number of nodes. However, even for these architectures with limited routing info, the area overhead of routing tables usually is not negligible. Consequently, an important factor to optimize in such routing schemes is the overall size of routing tables. Up to now, a number of different approaches have been discussed in order to achieve area savings. Among others are reduced-size ROM and Boolean logic implementations. Two techniques for further optimization of the cost of both SR and DR routing algorithms, named $XY$-Deviation Table (XYDT) Routing and Source Routing for Deviation-Points (SRDP), are presented in [30]. The performance of irregular networks can further improved by employing methods, such as virtual channel multiplexing, adaptivity, and shortest path routing combined with escape paths [40–42].

Further research in this area should address not only missing nodes in the mesh topology, but also the insertion of long serial router-to-router wires, which bypass intermediate routers [43]. Such long wires can reduce the power dissipation of long distance traffic, since the communication cost over a wire does not increase linearly with its length (see Sect. 3.5).

The previously discussed algorithms can perform without problems when the target applications, as well as their behavior are known at priori. However, NoC architectures often have to support dynamic applications with real-time constraints. A typical approach takes traffic information into account in order to reduce power consumption [44, 45].

### *3.2.5 Topology-Agnostic Routing Algorithms*

A limitation of the routing algorithms mentioned in the previous sections is that they are sensitive to topology changes (i.e., a faulty switch or link will degrade the topology into an irregular one and then the algorithms will fail). To overcome this limitation, a topology-agnostic routing algorithm can be used. Increasingly unreliable technology nodes have led to an increase in interest for topology-agnostic algorithms. Several high performance topology- agnostic routing algorithms for interconnection networks exist, such as Layered Shortest Path (LASH) [46], Transition Oriented Routing (TOR) [34], their combination called LASH-TOR [47], the Descending Layers (DL) algorithm [48], and multiple virtual networks [49]. Common for all these algorithms are that they require virtual channels, a feature that not all technologies support. In fact, some topology-agnostic routing algorithms require increasing number of VCs as the size of the network grows [50], making them prohibitive for the NoC environment. Furthermore, if virtual channels happen to be supported the number of available channels is often limited, and often dedicated to certain purposes such as quality of service. There are, however, also topology-agnostic routing algorithms that do not rely on virtual channels (up/down [51], DFS [52], FX [49]).

The most prominent algorithm example is the up*/down* [51] routing algorithm, which is based on turn restrictions. This algorithm relies on generating a breadth-first spanning tree of the network, and then a direction to each link (either up or down) is assigned. It was first presented in connection with Autonet [53] and later used in Myrinet [54]. This routing algorithm restricts turns by applying the up*/down* rule [51] according to which a packet may never traverse a link in the up direction after having traversed one in the down direction. Authors in [52] propose an improved up*/down* routing scheme, where instead of using the usual breadth-first traversal in the generation of the spanning tree, a depth-first search that optimizes the spanning tree is applied. This modification gives greater flexibility during the generation of the routing tables, and therefore leads to improved performance.

Since the up*/down* routing approach uses bidirectional turn restrictions, a methodology which avoids deadlock by prohibiting a subset of all turns in the network, it may cause an uneven distribution of traffic by having many paths crossing the same link, which results in lower performance. Also, this approach is unable to guarantee shortest path routing and unable to exploit any regularity in the underlying topology. A solution to this problem is tried by the FX routing scheme, which improves also the performance by introducing unidirectional routing restrictions [49].

In [50], topology-agnostic routing algorithms are evaluated for a variety of scenarios including regular mesh and torus topologies and regular topologies with 1, 3, and 5 % failures.

### 3.2.6 Bufferless Routing Algorithms

Apart from routing approaches that based on storing temporally packets inside switches, there are also approaches where buffers are removed completely. There are two approaches in bufferless routing algorithms. Packet dropping and retransmission [55, 56] and "hot potato" or deflection routing, first introduced in [57]. The fundamental idea behind hot potato routing is to route all incoming packets (or flits) to an output port of the router, regardless of whether or not that output port results in a lower distance to the destination of the packet. Since it is not always possible to route all packets to an output port that reduces the distance to the destination node (called a *productive* direction) as it may be already allocated to another packet, it is possible for packets to be deflected or "misrouted" [58]. The output port allocation in deflection-based bufferless NoC architecture are based on flit priority ratings. These can be: furthest to go, nearest to go, time of creations (age), time of departure (age), as well as user defined priorities.

In [58], a set of simple and practical bufferless routing algorithms, applicable to almost any NoC topology, is evaluated against baseline buffered algorithms. Based on the results shown in this chapter, bufferless routing can yield substantial reductions in energy consumption, while incurring little extra latency (versus buffered algorithms) if the average injected traffic into the network is low, i.e., below the network saturation point.

An important issue with bufferless routing algorithms is that flits in a packet may arrive out of order due to deflections leading to different paths for each flit with

**Table 3.2** Routing algorithm comparison

| Algorithm | Flexibility | Minimal | Buffering | Topology | Design Complexity |
|---|---|---|---|---|---|
| XY/XYZ | Deterministic | Yes | Yes | Mesh | Low |
| Turn model ([22]) | Partially adaptive | Yes | No | Regular | Low |
| DyAD ([15]) | Deterministic/adaptive hybrid | Yes | Yes | Regular | Medium |
| up/down ([51]) | Adaptive | No | Yes | Irregular | Low |
| [27] | Adaptive | No | Yes | Regular | Medium |
| [31] | Deterministic/adaptive | Yes | Yes | Irregular | |
| [59] | Deterministic | Yes | Yes | Regular (torus) | Low |
| [32] | Deterministic after initialization | Yes | Yes | Irregular | Low |
| [44] | Adaptive | Yes | Yes | Agnostic | Medium |
| [45] | Partially Adaptive | Yes | Yes | Irregular | Low |
| LASH ([47]) | Partially Adaptive | Yes | Yes | Irregular | Low |
| [49] | Partially Adaptive | No | Yes | Irregular | medium |
| [58] | Adaptive | No | No | Regular | High |

no intermediate storage. Solving this issue requires flit reordering at the destination (reassembly buffers).

### *3.2.7 Routing Algorithm Comparison*

It is obvious that no routing algorithm is suitable for every application and topology. In this section, there is an attempt at classification and qualitative comparison of the routing algorithms mentioned in the previous sections (Table 3.2).

## 3.3 QoS, Congestion Control and Flow Control

It is common in real-life applications to be hard or soft time constraints (deadlines), which need to be met for proper application execution. The guarantee of these timing constraints is usually addressed as *Quality of Service* (QoS), which is another important design issue in NoC architectures [60]. Quality-of-Service (QoS) guarantees independent design and validation of every part of the SoC by ensuring that real-time application requirements are met under any circumstances [61]. Typical examples of time-related guarantees are the throughput guarantees, as well as the latency bounds.

Since the NoC architecture is essentially a shared medium, *contention protocols* are used to decide how the bandwidth will be allocated to demand. If it is handled inefficiently, *congestion* may occur, a condition in which buffers overrun, packets are dropped and QoS degrades. *Flow control* generally refers to the policies that allocate resources to packets, and can be seen as a resource allocation or contention resolution problem [16]. It can be either *centralized* meaning a central controller makes the decisions or *distributed* meaning that each router makes decisions. The latter is the most common approach in NoCs.

## 3.3.1 Flow Control

Flow control can be intra-switch, switch to switch and end to end, though the term most commonly refers to switch-to-switch flow control. In this section, we examine switch-to-switch flow control, while end-to-end flow control is discussed in Chap. 5. Switch-to-switch flow control schemes can be classified into bufferless and buffered flow control. In bufferless flow control, the packets are deflected if required as described in the section on bufferless routing algorithms.

A simple buffered flow control protocol used in NoCs is acknowledged/not acknowledged (ACK/NACK) [62]. The ACK/NACK approach is basically a handshaking protocol [29]. When a sender puts data on the link, it activates a VALID signal. When the receiver is ready to consume the valid data, it activates the corresponding ACK signal. If the data are corrupt or there is no buffer space to store them, a NACK signal is activated instead. Upon receipt of a NACK, the sender starts resending flits starting from the not acknowledged one (Fig. 3.8). This flow control scheme has the advantage that it inherently supports fault tolerance, while its main disadvantage is the additional buffer space required to keep sent flits in case retransmission is required.

Other similar handshake-based protocols have been used for flow control such as the flow control mechanism NoCGEN which uses a request, grant, and ready handshake [63]. Similarly, in the flow control in the SoCIN NoC architecture [64],

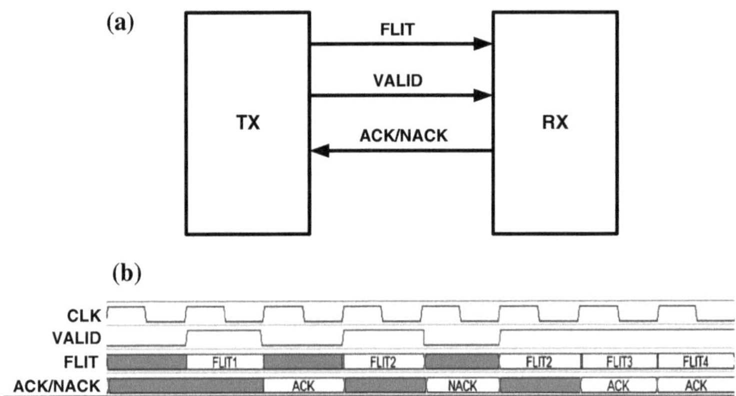


**Fig. 3.8** ACK/NACK flow control (**a**) signals and (**b**) timing diagram. (Note retransmission of Flit2)

when a sender puts data on the link, it activates the related VALID signal. When the receiver is ready to consume the validated data, it activates the corresponding ACK signal.

ON/OFF [16] is another simple protocol and STALL/GO its most common implementation in the NoC environment. It requires just two control wires: one going forward, signifying data availability, and one going backward and signaling either a condition of buffers filled ("STALL") or of buffers free ("GO") (Fig. 3.9).

T-Error [65] is a much more complex protocol aiming at improving performance. In credit-based flow control [16], the transmitter has a "credit" counter that is initialized to the value of empty buffer slots of the receiver and decrements it every time a flit is sent. The credit counter must be updated in case the receiver consumes or forwards a flit and therefore increases its buffer space. This is accomplished by a credit value that is sent back to the transmitter to be added to the current value of the credit counter. The transmitter stalls when the credit value is zero and resumes when its value increases again. The signals and timing diagram of a credit-based flow control scheme are shown in Fig. 3.10. Both handshake- and credit-based flow control are supported in the revised SoCIN architecture called ParIs [66].

Table 3.3 compares the above flow-control schemes in terms of performance, buffering requirements, logic requirements, and fault tolerance support.

A more detailed evaluation of the above flow-control schemes is presented in [67].

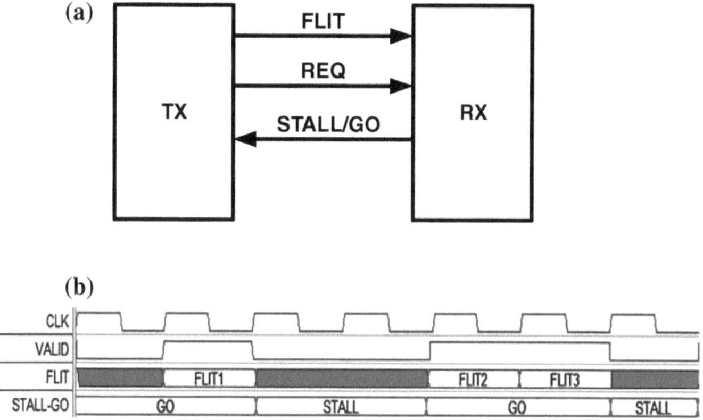


**Fig. 3.9** STALL/GO flow control signals

**Table 3.3** Flow control comparison

| Flow control | Performance | Buffering | Logic | Fault tolerance |
|---|---|---|---|---|
| ACK/NACK | High for long messages | Very low | Low | Low |
| STALL/GO | Medium | High | Low | High |
| Credit-based | High | High | Low | High |
| T-error | High | High | Low | Low |

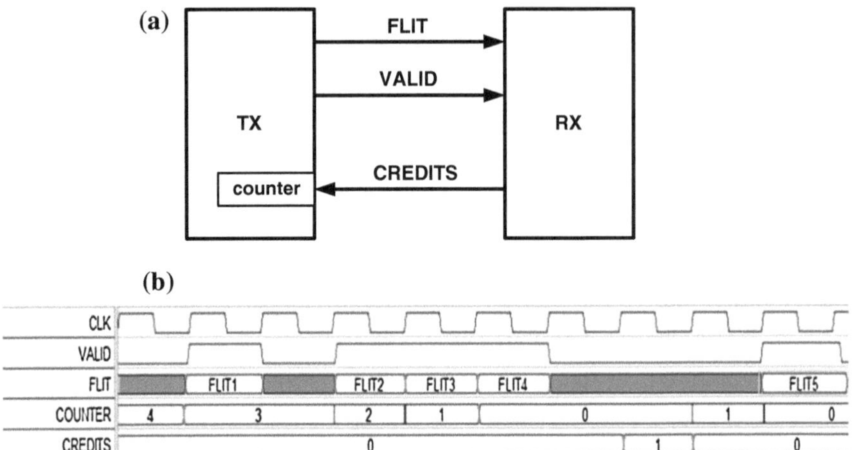


**Fig. 3.10** Credit-based flow control (**a**) signals and (**b**) timing diagram

## *3.3.2 Quality of Service (QoS)*

Unlike computer networks that are built for ongoing expansion, future growth and standards compatibility, on-chip networks are designed and customized for a priori known set of computing resources and precharacterized traffic patterns among them. Consequently, various design parameters of the network architecture (i.e., buffer size, link bandwidth allocation, etc) can be tuned for specific applications in order to provide a required QoS (for known traffic patterns). Typical approaches to QoS in NoC architectures are *best effort* QoS and *guaranteed services* QoS.

### 3.3.2.1 Best Effort QoS

Best Effort (BE) services do not reserve any resources, and hence provide no guarantees while it also leads to smaller delay. Even though this approach uses fewer resources, it results in quite efficient network communication since this kind of service is typically designed for average-case scenarios instead of worst-case scenarios. However, its limitation lies in its unpredictability.

### 3.3.2.2 Guaranteed Services QoS

NoCs with guaranteed services take provisions in their architectures to offer connections with guaranteed performance, such as the absence of data loss, minimum throughput, and maximum latency. Examples are [24, 68, 69], using packet switching with time-division-multiple-access (TDMA) schemes and [1, 4] using pure circuit switching.

## 3.4 Router Design

In this section, the design of each module of a generic (often termed baseline or conventional) NoC router is detailed first and modifications and improvements proposed in the literature are discussed later. Further examples on router design are given in the NoC case studies examined in the second part of the book.

It is obvious that router design is closely related to all NoC parameters discussed so far, namely architecture, routing algorithm, switching technique, flow control, and QoS. This implies that there is a variety of possible router implementations. Router architectures have dominated early NoC research, and the first NoC designs [8, 9] proposed the use of simplistic routers, with deterministic routing algorithms. Gradually, researchers have explored the impact of various design parameters such as:

- *number of ports (radix)*: This is typically equal to three for octagon topologies, five for 2D mesh and torus topologies, seven for 3D mesh topologies, and any number for irregular topologies. Generally, routers do not scale well with the number of ports as will be discussed extensively in a following section.
- *link word-length (phit size)*: The number of parallel lines in each router link. Typically, this is common for all router ports. Typically, routers do scale well with phit size.
- *buffer placement and organization* : Common approaches are input queueing, output queueing, both, no queuing (bufferless routing), and using VCs.
- *buffer size* : The number of flits stored in the router. Buffers are costly in terms of area and power consumption, therefore only a few flits are stored typically. However, NoCs with little buffering may feature degraded performance under heavy traffic conditions.

The input/output of a typical router includes the following types of signals:

- *global control signals*: Clock and reset inputs for typical synchronous routers.
- *data signals*: Inputs and outputs used to transmit data. Equal to phit size.
- *flow control signals*: DATA VALID, REQ, ACK/NACK, STALL/GO, or credits signals depending on the flow-control protocol. In bufferless routing only a data valid signal is required.

In terms of RTL design, since the router is a component that is very likely to be used in future versions of the system, and its architecture options may be either revised or even different instances may coexist in the same architecture (heterogeneous NoCs), it should be designed as a reusable IP block [70]. The parameters mentioned above should be configurable through a package or include file, and the router should be verified for a number of values of these parameters.

A detailed block diagram of a typical 5-port switch/router is shown in Fig. 3.11. The ports of a 5-port mesh/torus router are commonly designated as North, South, East, West, and Local or PE for the local node. In other topologies ports are simply numbered. The basic router components are:

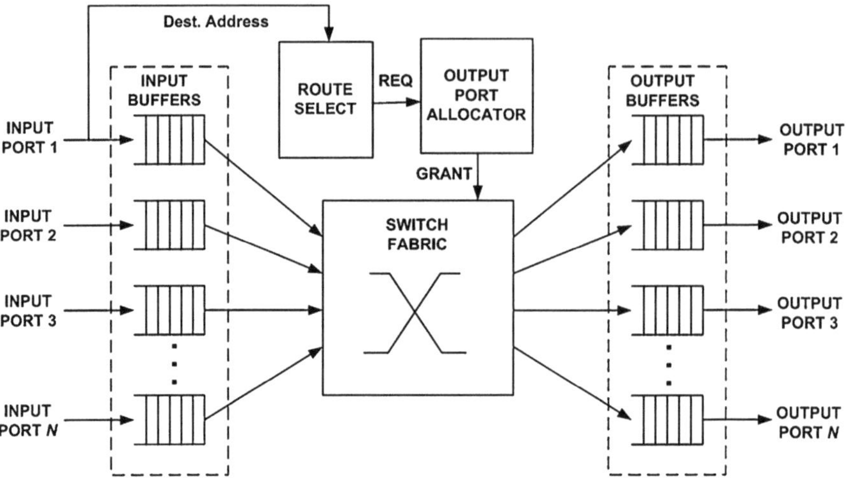


**Fig. 3.11** Generic NoC router block diagram

1. the input/output buffers which temporarily store flits,
2. the output port allocation logic which selects the output port for each flit/packet,
3. the switch fabric which makes the physical connection from input to output port, and
4. the control logic responsible for overall synchronization.

During typical router operation, an incoming flit is first successfully received and possibly stored in the input queue. Second, the output port request for the incoming flit is determined based on the flit destination address according to the routing algorithm. Third, the output port allocator receives the flit requests and allocates the output ports according to priority. Finally, as soon as a flit is granted a port it is wired through the switch fabric to the granted output port or queue finally is transmitted through the link to the neighboring router or node.

Because of the tight performance, area and power constraints of many MPSoC applications, careful analysis, and design of all router components are important. Typical performance metrics for routers are operating clock frequency and throughput. Generally, it is not feasible to increase clock frequency by adding pipeline stages, as it is common in other digital systems, since low latency is important in NoC also. Additional metrics used to compare/evaluate router implementations are power-delay and power-delay-fault products, also taking into account power consumption and fault tolerance, respectively.

### 3.4.1 Buffers

Buffers are the greatest power consumers. Thus, efficient buffer design is critical for achieving a good performance/area/power trade-off. To minimize the implementation cost, the on-chip network has to be implemented with little area overhead. Thus, unlike off-chip networks which feature large memories, ofter DRAMs, NoCs typically use small registers for buffering. Another advantage of using registers over large memories is that the address decoding/encoding latency and the access latency can be significantly reduced. This is critical for those latency sensitive applications that are typical for many SoCs.

Buffer design can be broken down to *sizing* and *organization*. Buffer sizing is important since the area occupied by an on-chip router is dominated by the buffers [8, 71], while these buffers are the largest leakage power consumers in a NoC router, consuming about 64 % of the total router leakage power [72]. In addition to that, buffers consume significant dynamic power while this consumption increases rapidly as packet flow throughput increases [73]. Previous studies have shown that storing a packet in a buffer consumes far more energy than transmitting the packet [73]. However, decreasing buffer size in order to save power consumption and silicon area often is not a viable solution, since it also results in network performance degradation, especially under heavy traffic [58]. This is why good buffer organization is important. The fact that buffers are the major power consumers in NoCs has led the scientific community to focus on buffer design and optimization. A disadvantage of the ACK/NACK flow-control scheme is that it requires additional buffering in order to maintain a copy of the transmitted flit in case retransmission is required.

Buffer design for virtual channels requires multiplexers/demultiplexers in the input ports of the router as shown in Fig. 3.12. The de-multiplexer assigns the appropriate queue based on the Virtual Channel ID in the packet. In this distributed, private buffer implementation, even though increasing the number of VCs improves performance, increasing the number of buffers accordingly quickly leads to unacceptable power consumption. Therefore, buffer sharing schemes between VCs have been proposed [74].

Buffer organization in early NoC implementations was limited to input-output queueing [75], which means that the router stored incoming or outgoing traffic, respectively. Each input-output port had its own private buffers. However, since then various attempts at buffer organization were proposed. In regular topologies, usually all routers and ports within a router have the same buffer size. Still optimal buffer size and number of VCs needs to be determined. In custom irregular architectures though, each router and port buffer size can be tuned according to the application traffic requirements. Due to the impact of buffers in the router power consumption and performance, buffer organization/allocation has been given the greatest attention by the research community. Approaches are broadly classified as software based and hardware based for convenience in the next two sections, they are however, based on one of the following strategies:

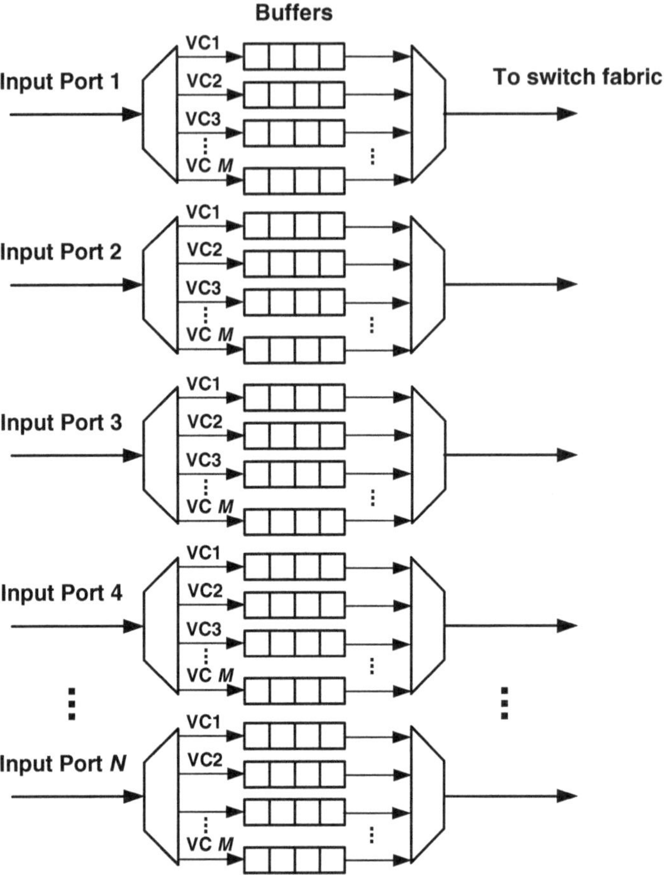


**Fig. 3.12** Virtual Channel distributed buffer organization with $M$ virtual channels

1. *Static approaches* [76, 77]: Optimal buffer organization is determined at design time using traffic information. The hardware is usually based on configurable baseline routers. This approach, though optimal for a specific application, is not general and therefore lacks flexibility.
2. *Run-time buffer allocation or centralized buffer* [74, 78–80]: Centralized/shared buffers that are allocated dynamically to VCs according to real-time traffic requirements.
3. *Buffer bypassing* [81]: Reducing buffer sizing together with flit storing on the links in order to bypass router buffers as much as possible.
4. *Deflection routing* [58, 82]: Reducing buffer size or eliminating buffers completely by miss-routing (deflecting) incoming packets.

#### 3.4.1.1 Software-Based Solutions

Previous studies show that nonoptimal design/usage of buffers may result rather than having an efficient network (a network is assumed efficient when it occupies at least the 80 % of its theoretical capacity), to implement a network policy with about 30 % capacitance utilization [83]. The buffer optimization can also improve the area of NoC. The work in [84] analyzes the properties of on-chip buffers and reports gate-level area estimates and buffer utilization across the network. A similar approach is discussed in [85], where the impact of FIFO sizing on the interconnect throughput for single-source, single-sink interconnect scenarios is discussed.

An iterative methodology for assigning buffer space for structured NoC architectures with grid topology is described in [71, 86]. This methodology initially assigns to all the network buffers a depth of "1" and then by using appropriately selected sophisticated queuing models for routers, it determines the most likely channel to become congested. This likelihood is expressed as a function of the architectural (e.g., network topology, packet service time at routers, etc.) and application-specific (e.g., injection rate at each IP, probability distribution of packet destinations, etc.) parameters. Then, the buffer size increased for buffers connected to the most congested channels of NoC. This procedure is repeated until the total buffering space is used up. A similar work is discussed in [87], where a NoC cost minimization process by exploring the influence of increasing the number of wormhole buffers versus decreasing link bandwidth (by reducing the number of wires), is presented.

Another approach, as mentioned in the section on bufferless routing, is reducing the size of buffers or eliminating them completely by appropriate deflection routing or even packet dropping.

#### 3.4.1.2 Hardware-Based Solutions

Centralized buffer organizations have been studied extensively in the macronetwork realm, but the solutions proposed are not amenable to resource constrained on-chip implementations. During the last years, buffer sizing has been investigated in [71]. However, these works adopt a static approach, where optimal buffer sizes are predetermined at design time based on a detailed analysis of application-specific traffic patterns. Such an approach results to an application-specific solution, which is also optimized for a specific hardware mapping. A more advanced solution involves architectures that incorporate techniques to dynamically alter the buffer organization at run-time, since it is more desirable for general purpose and reconfigurable SoC executing different workloads.

In particular, a unified and dynamically allocated buffer structure was originally presented in [88] in the form of the Dynamically Allocated Multi-Queue (DAMQ) buffer. The DAMQ project spawned a few other designs, which aimed to simplify the hardware implementation and lower overall complexity. Two notable instantiations of these designs were the DAMQ with self-compacting buffers [78] and [74], which aims at area and power constrained, ultra low latency on-chip communication.

Also using repeaters in inter-router channels as buffers along the channel when required has been proposed [81].

Apart from this design approach, where buffers are used in order to store incoming (or outgoing) traffic, there are implementations that perform packet switching with very limited buffering. More specifically, the output switches in the Nostrum NoC [69] under normal operation conditions try to send packets to the direction of their destination. However, if the corresponding output link is blocked by another packet, then the packet is deflected into another direction and effectively moved away from its destination. In other words, switches in Nostrum do not temporarily store packets, and therefore they have no internal buffers (they only have an input and an output buffer for each link). Next we try to point out the main features of achieving traffic or communication scheduling through appropriate buffer sizing in software and hardware levels.

## 3.4.2 Switch Fabric

Essentially, the switch fabric multiplexes all inputs to each of the outputs. The straightforward approach to the switch fabric is implementing a crossbar (also known as a matrix) connecting each input queue output to any of the queue outputs depending on the grant signals provided by the output port allocation logic. There are two common implementations of crossbar switches, using pass transistors and using multiplexers (Fig. 3.13). The pass transistor implementation exhibits reduced area and power consumption, while the multiplexer implementation has the advantage that

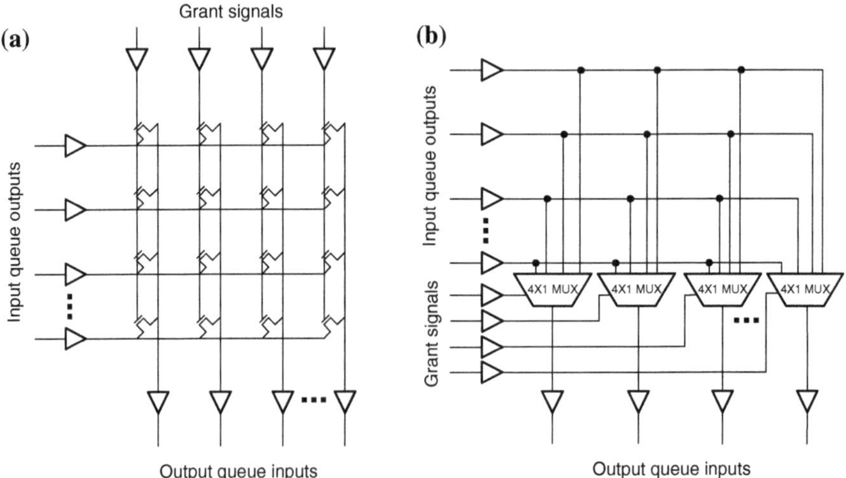


**Fig. 3.13** Crossbar implementation with (**a**) pass transistors, (**b**) multiplexers

can be automatically synthesized from an RTL description and can be used in FPGA devices also.

In theory, crosssbar area grows with the square of the number of ports, and indeed automatic placement and routing using EDA tools leads to prohibitive area above 32 ports [75]. It has been shown, however, that with carefully tuned placement, taking into account the regularity of the design, even 128 tiles can be connected using a single $128\times128$ crossbar occupying a mere 6 % of the total area [89].

In order to reduce the complexity of the $5\times5$ crossbar, a row-column decoupled router architecture was proposed in [90]. The typical $5\times5$ crossbar is replaced by a pair of $2\times2$ crossbars (one for row-mode or East-West routing and one for column-mode or North-South routing) and an ejection multiplexer for the local PE.

### 3.4.3 Port Allocation

The output port allocator is basically a scheduler that receives requests from incoming flits or flits pending in the input buffers and grants the output port according to a scheduling policy. It must efficiently arbitrate among flits for the same output port and among flits from different virtual channels for the same output channel.

More formally, an allocator performs a *matching* between resources (output ports and/or physical channels) and requesters (flits pending for transmission) according to two basic constraints: (i) Resources are only granted to requesters if a corresponding request exists and (ii) at most one resource is assigned to each requester and vice versa.

Possible scheduling policies are:

- *Round-robin*: Most widely used [75, 91, 92] because it is fair and prevents starvation. Typically implemented using a ring counter token and priority encoder-based arbiters as shown in Fig. 3.14.
- *Pseudo-LRU*: A scheduling policy that is biased toward flits that have not been granted output lately. It was implemented in [93, 94], with good results.
- *Fixed priority*: Not commonly used because it is unfair and can cause starvation. It can be useful in custom topologies where the designer desires a peripheral to always have priority over another. Certain router implementations give priority to flits requesting to move in the same direction over flits requesting to turn.

A basic round-robin port allocator is shown in Fig. 3.14. The ring counter is used to produce the round-robin token. It is initialized to a value containing a single logic "1" and it counts each time a flit is transmitted, by using the VALID, GO, or ACK signal as a count enable depending on the flow control scheme used. The arbiter blocks are priority encoders with decoded outputs as shown in the truth table of Fig. 3.14. The token produced by the ring counter is used to enable only one of the arbiters each time, giving priority to a different request each time in a round-robin manner.

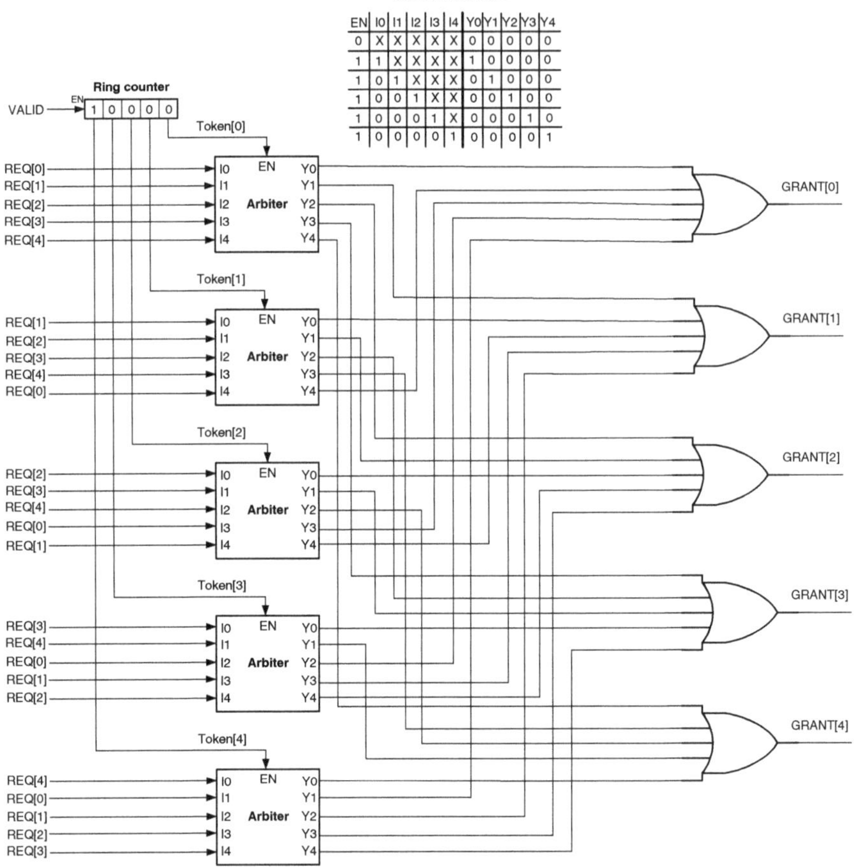


| EN | I0 | I1 | I2 | I3 | I4 | Y0 | Y1 | Y2 | Y3 | Y4 |
|---|---|---|---|---|---|---|---|---|---|---|
| 0 | X | X | X | X | X | 0 | 0 | 0 | 0 | 0 |
| 1 | 1 | X | X | X | X | 1 | 0 | 0 | 0 | 0 |
| 1 | 0 | 1 | X | X | X | 0 | 1 | 0 | 0 | 0 |
| 1 | 0 | 0 | 1 | X | X | 0 | 0 | 1 | 0 | 0 |
| 1 | 0 | 0 | 0 | 1 | X | 0 | 0 | 0 | 1 | 0 |
| 1 | 0 | 0 | 0 | 0 | 1 | 0 | 0 | 0 | 0 | 1 |

**Fig. 3.14** Round-robin output port allocator detailed diagram

High-radix arbiters impose significant latency and area comparable to the switch fabric. Therefore, in the case of multiple virtual channels, two-stage arbitration is preferred [16]. In the first stage, a output virtual channel is assigned to the outgoing packet (VC allocation) and in the second stage the physical channel is granted to one virtual channel (switch allocation).

A different approach is the wavefront allocator [95] implemented in [96] which uses a regular array of tiles leading to linear delay and quadratic area scaling. In [97], the symmetric structure of the round-robin scheduler is exploited to produce an area-efficient design by folding the scheduler onto itself, reducing its area roughly by 50 %. In [90], early ejection of flits whose destination is the local PE (without going through the switch fabric), reduces the size and number of the required arbiters. In [98], a pessimistic mechanism for speculative port allocation was proposed that reduces switch allocator delay by up to 23 % compared to a conventional implementation without increasing the router's zero-load latency.

### *3.4.4 Bufferless Router Design*

Eliminating buffers altogether is not straightforward and imposes a number of modifications to the baseline router described in the previous sections. The block diagram of a bufferless router is shown in Fig. 3.15. First of all, due to the lack of buffers, no flow control signals exist. Second, eliminating livelock requires sorting incoming flits according to some priority metric, such as the number of previous hops.

More specifically, as shown in Fig. 3.15 incoming flits are prioritized (sorted) through the sorting network based on the selected metric. After the flits are sorted by priority, the routing information (typically comparison of local address and flit destination address) is used to generate the prioritized output port requests that are fed into the output port allocator.

As shown, bufferless router implementation entails two inherent complexities not present in buffered routers. First, the sorting network which is composed of magnitude comparators and multiplexers. Second, the output port allocator must allocate a resource (output port) to all requests, based on their priority. Therefore the highest priority flit is granted the requested output port, which is then eliminated

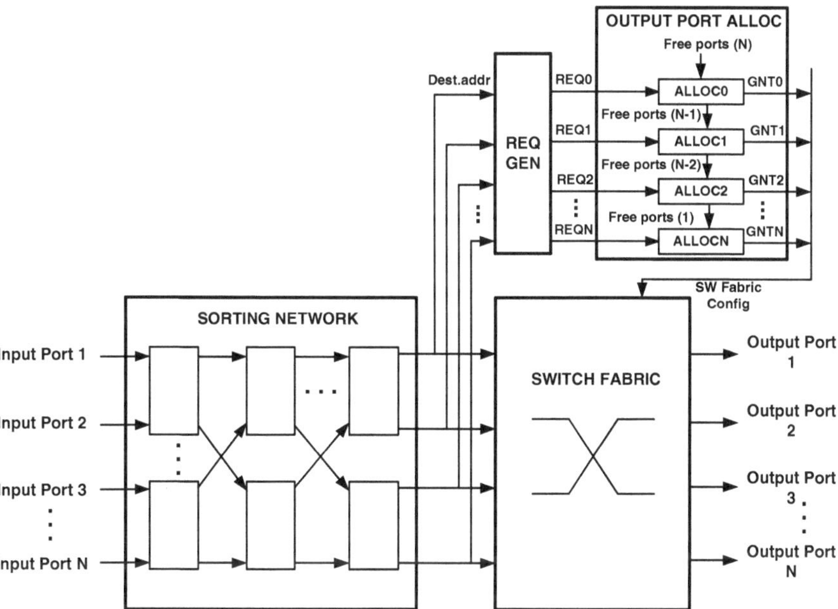


**Fig. 3.15** Bufferless router block diagram

from the list of available resources. The second highest priority flit gets the most desirable (productive) output port from the remaining ones and so on. In the case, the router must forward flits from all directions, all output ports must be assigned, and therefore the lowest priority flit will be left with the last remaining output port. This inherently serial nature of the sorting network and output port allocator combines to a long critical path responsible for the lower clock frequency or higher latency of bufferless routers in comparison to buffered ones.

The router proposed in [99] efficiently resolves both complexities imposed by the baseline bufferless router by replacing the sorting, allocation, and crossbar by a permutation network and only ensuring that the highest priority packet is granted its requested port. This has some impact on routing efficiency but still ensures livelock freedom.

In [82], a minimally buffered deflection router was proposed. It uses a small "side buffer" to briefly store some network traffic that would have otherwise been deflected to reduce deflections. Results show performance that approaches the conventional input-buffered router with area and power close to that of a bufferless router.

### *3.4.5 3-D NoC Router Design*

The straightforward extension of a NoC router for a 2D mesh topology to the third dimension is to add two more ports, one for the layer above and one for the layer below, increasing its radix to seven, as shown in Fig. 3.16. Typically, the two additional ports of this type of router are designated as UP and DOWN. This implies extending the switch fabric to 7×7 as well as the buffers, and port allocators.

The advantage of such a baseline 3D router is its simplicity. However, since routers do not generally scale well with the number of ports in terms of area due to the quadratic growth of the crossbar (see also next section), a baseline 3D router can be significantly larger than a 2D router [100]. Furthermore, the inherent symmetry of

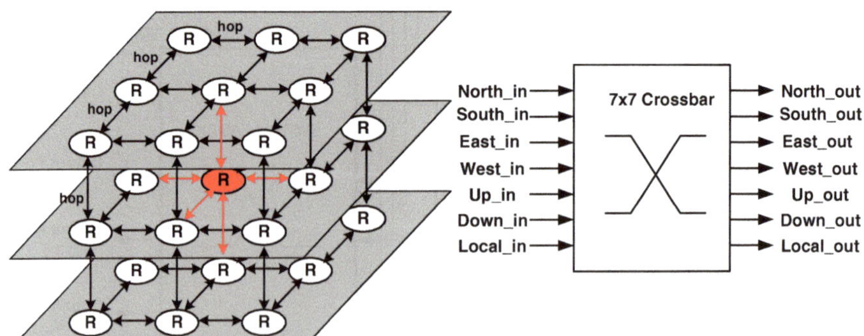


**Fig. 3.16** 3D Mesh 7-port Router

this design treats horizontal and vertical links as similar. However, the short vertical links make traversing all layers in the 3D chip feasible in a single hop [100], while using baseline 3D routers imposes one hop from one layer to the next. Moreover, it ignores possible different design constraints between layers. Furthermore, process variations across layers in 3D ICs can cause much more pronounced differences in delay among identical routers than in conventional 2D ICs [100]. Lastly, TSV pads required for the vertical links occupy significant chip area and in large numbers they can have a negative effect on the yield of 3D IC [101].

For the above reasons, 3D NoC router architectures that deviate from the symmetrical NoC router were explored in order to take into advantage the inherent asymmetry of the vertical links. In [102], a NoC-bus hybrid architecture was proposed. The bus link was used in the vertical dimension in order to take advantage of the fast vertical links. This allows single hop data transfer in the third dimension and the additional benefit of only a $6 \times 6$ crossbar, since the bus requires one additional port to the generic 2D $5 \times 5$ crossbar instead of two (Fig. 3.17). However, flits from different layers wishing to move up/down now have to arbitrate for access to the shared medium. This limits bandwidth between layers which may be unacceptable in many applications. An improved 3D NoC-Bus Hybrid [103] based on bypassing the router when the flit travels in the vertical dimension (according to a rule called the *LastZ* rule), enables better optimization of the inter-layer communication architecture and leads to $5 \times 6$ size routers.

Another approach is decoupling inter-layer and intra-layer communication. Such an architecture is the dimensionally decomposed router described in [104]. While it extends the idea of the row-column decoupled router of [90] in the third dimension, the true novelty is fusing the crossbars of all the routers in the same vertical "column" to a true physical 3D crossbar. Also, in [105], a router composed of two totally decoupled modules, one for inter-layer communication and one for intra-layer communication was presented.

Another option is the implementation of a true 3D crossbar as presented in [106]. The authors designed a 3D crossbar to route any permutation between a set of $N \times N$

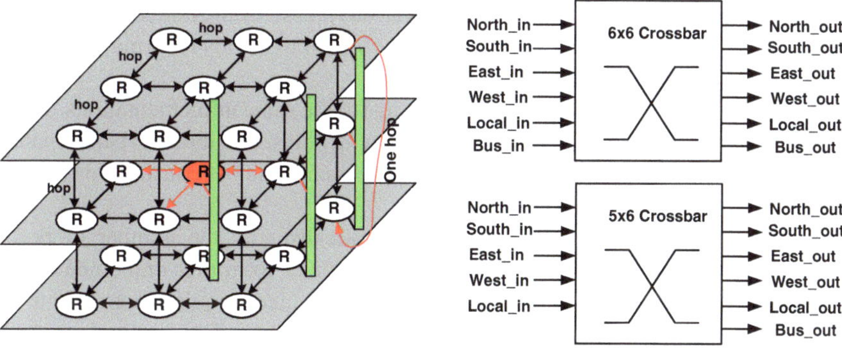


**Fig. 3.17** 3-D router-bus hybrid

I/O terminals on one layer of a 3D chip and a second set of $N \times N$ I/O terminals on another layer using intermediate layers called Crossbar-Switch Layer Sets (CLSs) sandwiched by the two I/O layers. This approach provides high bandwidth, but adding too many CLSs could lead to defects.

A multi-layered 3D router was proposed in [107]. It is in fact a router occupying all layers in the 3D chip. An important advantage besides reduced power consumption and increased performance, is that it is compatible with similarly multilayered cores [108].

In order to take advantage of the fast vertical links and at the same time alleviate the problems related to TSVs, in [109] the authors present a Quasi Delay Insensitive (QDI) asynchronous logic 7-port router with serial vertical links. This implementation requires serializer/deserializer circuits to forward a flit from a "planar" port to a vertical one and vice versa. Investigation of the optimal serialization ratio showed that serializing vertical signals can be advantageous for medium density TSVs for any CMOS technology node. However, for high density TSVs the serial vertical links are advantageous only from 32 nm.

Another approach for reducing vertical link footprint was proposed in [110]. The authors proposed a 3-D NoC architecture based on Bidirectional Bisynchronous Vertical Channels (BBVC), which can be dynamically self-configured to transmit flits in either direction, improving bandwidth utilization, routability at each layer and area footprint by 47 % at 65 nm technology node.

Bufferless routing has also been proposed for 3D routers in [111]. Utilizing a three-stage permutation network instead of an allocator and crossbar, a single cycle, 1.25 GHz was achieved in a 65 nm technology.

### 3.4.6 Router Scaling

Since adopting an irregular topology may lead to large-radix routers, this section contains an exploration of how routers scale with phit size (word-length or bit-width) and number of ports (radix) in terms of performance and area. Xilinx FPGA technology was used to prototype routers of various phit sizes and radices. Figure 3.18 shows how a 7-port router scales with phit size from 4 to 64 bits in terms of area. As expected, the growth is linear.

Figure 3.19 shows how performance scaling for the same implementations. It can be seen that the router degrades very gracefully with phit size in terms of performance (clock frequency). In fact, the clock frequency remains almost constant for all phit sizes.

As mentioned, routers scale roughly quadratically with the number of ports. Figure 3.20 shows the slices required in Xilinx devices for router implementations for five to 16 ports. It can be seen that the 16-port router requires approximately nine times the slices required by the 5-port router, which is close to the theoretical $\frac{16^2}{5^2} = \frac{256}{25} = 10.24$.

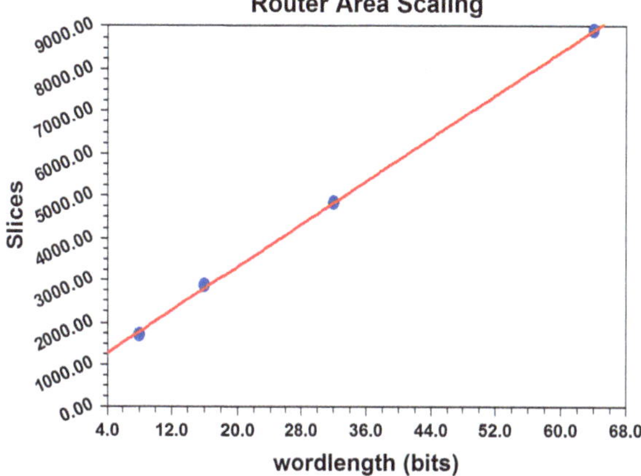


**Fig. 3.18** 7-port router area scaling with word-length (phitsize)

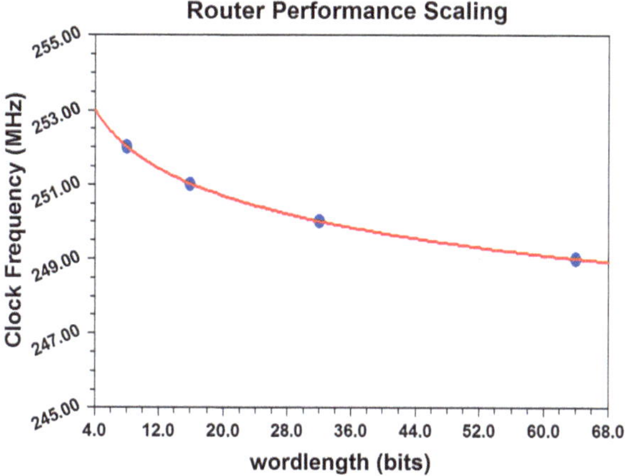


**Fig. 3.19** 7-port router performance scaling with word-length (phitsize)

## 3.5 Network Link Design

The link that physically connects the ports of two adjacent routers or a router and the NI of a node is realized by wires. The number of wires required is the phit size plus the wires needed for the flow control information that must be exchanged (see Sect. 3.1). Multiple physical links [112] can be placed between a transmitter and a receiver. This complicates floor planning but it can be a viable alternative to virtual channels, multiplexing flits in space instead of time.

In order to achieve the potential gains of employing a NoC-based architecture (i.e., link energy savings, lower traffic delay among high-capacity links, etc), careful

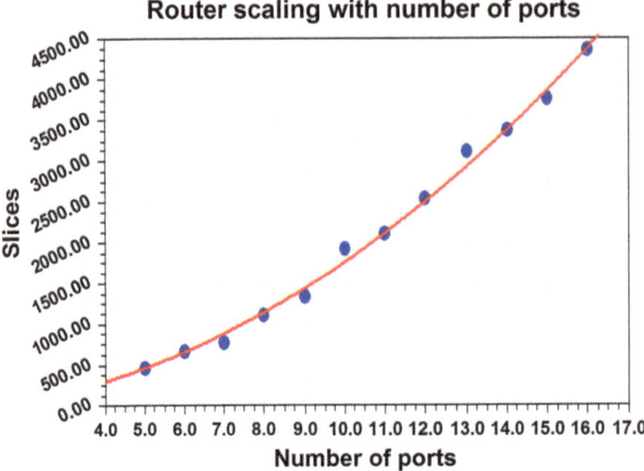


**Fig. 3.20** Router scaling with the number of ports

design of links should be done to overcome penalties due to uneven load distribution across links.

The typical parameters that characterize the links of an NoC architecture can be summarized as follows:

- *direction*: it gives the orientation of message transfer. Possible values for this parameter are simplex, half-duplex, or duplex. For half-duplex, control messages are needed to switch the message transfer direction. Full-duplex is the usual choice for NoCs for applications with demanding bandwidth constraints.
- *burstiness*: it corresponds to the traffic characteristics, which can be either periodic or not. The first approach (i.e., periodic) is suitable for modeling a channel with constant data rate. On the other hand, a periodic traffic is characterized by minimum burst interval and maximum burst length.
- *latency*: it describes the absolute (or relative) time of a single unit of data transmitted from the sender to the receiver. This parameter is constrained by the target application.
- *bandwidth*: it defines the channel ability of transferring data.
- *reliability*: guarantee that data are sent and received correctly (i.e., without corruption, loss, duplication. etc.).

## *3.5.1 Planar Link Design*

Figure 3.21a shows a link between a transmitter (TX) and a receiver (RX). The link is composed of a number of wires, the phit (data) and the flow control signal wires. The objective of NoC link design is the same as general wire design in VLSI technologies, namely to arrive at an implementation with low delay and power consumption with acceptable noise levels. Wire delay is due to the inherent resistance and capacitance of

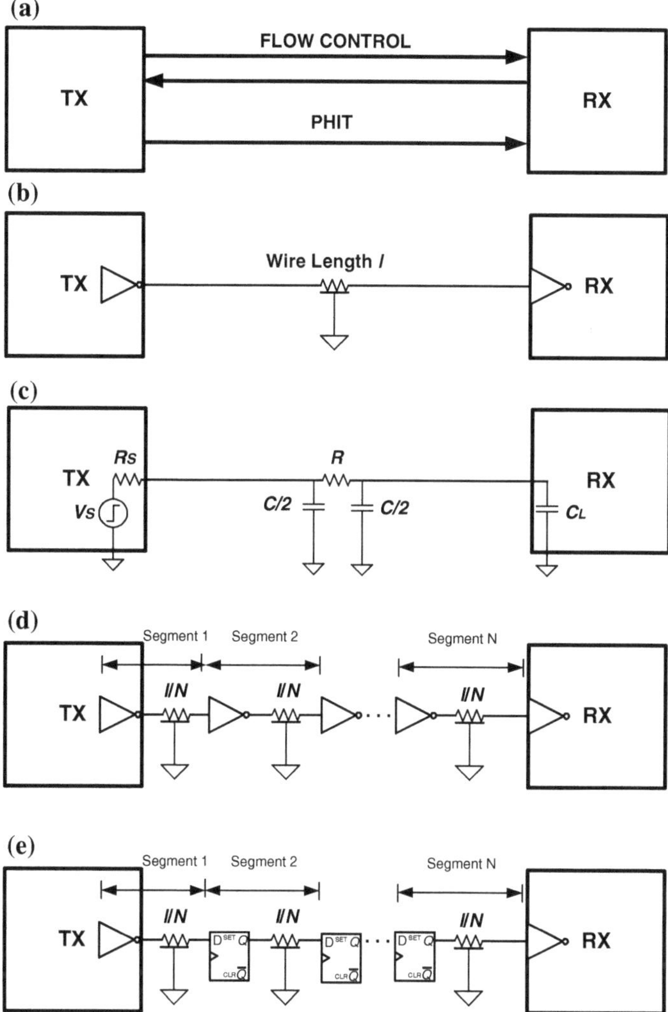


**Fig. 3.21** Link designs options (**a**) link signals, (**b**) wire circuit equivalent with lumped resistance and capacitance, (**c**) Equivalent $\pi$ model circuit (**d**) wire broken into segments using inverters and (**e**) using registers

the wire (known as RC delay) [113]. An equivalent RC circuit for a single wire of the link of Fig. 3.21a with lumped resistance and capacitance is shown in Fig. 3.21b. Since both wire resistance and capacitance are directly proportional to wire length, wire delay is proportional to the square of the wire length, making long wires prohibitively slow. The equivalent circuit known as the $\pi$ model is shown in Fig. 3.21c.

A simple model for the wire delay is the Elmore delay model [114]. It can be further simplified if we consider the $R_s$ and $C_L$ as negligible. In that case the wire *flight time* equals $\frac{RC}{2}$ [113].

A common wire design technique is to break a long wire of length $l$ into $N$ shorter segments, each $\frac{l}{N}$ long, by inserting $N-1$ repeaters, each driving the next segment. The repeaters can be inverters as shown in Fig. 3.21d, noninverting buffers, registers as shown in Fig. 3.21e or even FIFOs. The repeaters introduce delay themselves (as well as additional area and power consumption), and in the case of registers or FIFOS, latency which must remain in acceptable levels for the target application. Using FIFOs requires flow control signals among repeaters as well as shown in Fig. 3.21e. On the other hand, register repeaters act as both distributed buffering and pipeline stages, possibly increasing throughput. It must be noted, however, that the above techniques should be no substitute for good floor planning that leads to reasonable wire lengths, but rather a complementary (orthogonal) optimization. Automatic synthesis from RTL especially with poorly selected design constraints will lead to prohibitive wire lengths, and performance that cannot be sufficiently improved by pipelining links.

Inserting repeaters introduces delay, area and power overheads. If the repeaters are too few (and therefore the segments relatively long) the delay is dominated by the wire delay. On the other hand, inserting too many repeaters will again increase delay because of the repeater delay. From the above, it can be inferred that there is an optimal number of repeaters that should be inserted. In the case of inverter repeaters, the optimum segment length is given by the formula [113]:

$$\frac{l}{N} = 0.77 \times \sqrt{\frac{FO4}{RC}} \tag{3.6}$$

while the propagation delay per unit of length is given approximately by [113]:

$$\frac{t_{pd}}{l} = 1.67 \times \sqrt{FO4 \times RC} \tag{3.7}$$

where FO4 is the delay of a fan-out of four inverter in the target technology.

However, the energy per unit of length for sending a single bit for this minimum propagation delay is 87 % increased compared to the energy per unit of length required for sending a single bit with no repeaters [113].

Furthermore, A wire also has high capacitance to its neighbors [113]. Because of that capacitance, when a neighbor switches either from 1 to 0 or 0 to 1, the wire tends to switch too. This effect is known as capacitive coupling or crosstalk and can have significant impact on noise on nonswitching wires and therefore reliability, as well as increased delay on switching wires [113]. Considering the typically high number of wires in a link, crosstalk effects should be carefully analyzed in NoC link design. Common design approaches for reducing crosstalk effects increase the distance of neighboring wires and shielding data wires from crosstalk on one or both sides using VDD and ground wires. An alternative, if the application constraints allow it, is using a serial link as in [109], where an asynchronous bit-serial interconnect structure, comprising encoder, serializer, deserializer, and decoder circuits was proposed.

Besides the general design goals for performance, reliability and power consumption, links are used as distributed buffers in a number of research efforts. iDEAL (inter-router Dual-function Energy and Area-efficient Links) [115] proposed to reduce buffer size while minimizing performance degradation due to the reduced buffer size by using already existing repeaters along the inter-router channels as buffers along the channel when required. This approach was combined with completely bypassing the router buffers when possible in [81]. Repeaters can also be designed to sample and hold data values thereby storing values on the channels [116].

Apart from conventional link implementations, there are also approaches trying to improve the efficiency of communication infrastructure in NoCs. Among others is Photonic NoC [117], which can provide high bandwidth photonic links for high payload transfers. The limitation of employing such a technique is the complex design for switch architectures that incorporate more than four ports. Also, there are approaches for replacing long interconnects with on-chip RF/wireless connections [118].

In [119], long link insertion between nonneighboring nodes in a mesh topology was proposed to increase performance by traversing otherwise multihop distances in a single hop through the long links, together with an appropriate routing algorithm. Similarly, long links are also required for the wraparound links in torus topologies. In [120], instead of utilizing repeaters for such long links, a current mode signaling scheme was proposed.

### *3.5.2 Vertical Link Design*

Vertical links differ significantly from planar links, creating an inherent asymmetry in 3D routers as already discussed. There are three technologies that allow inter-die interconnect: Two are contactless (capacitive and inductive coupling) while the third is Through Silicon Vias (TSVs) [101]. TSVs are the most promising technology for NoC architectures.

As already discussed, TSV vertical links create an inherent asymmetry in 3D NoC architectures, due to their shorter length compared with horizontal (planar) links. Vertical links are significantly faster (about two orders of magnitude). There are, however, certain challenges associated with 3D link design. TSVs have different resistance and capacitance characteristics than planar links. Therefore the wire connecting transmitter and receiver of Fig. 3.22 has three distinct segments as shown in Fig. 3.22a in the case of a single interplane via. The equivalent circuit that can be used with the Elmore delay model is shown in Fig. 3.22b. According to the Elmore delay model the propagation delay is:

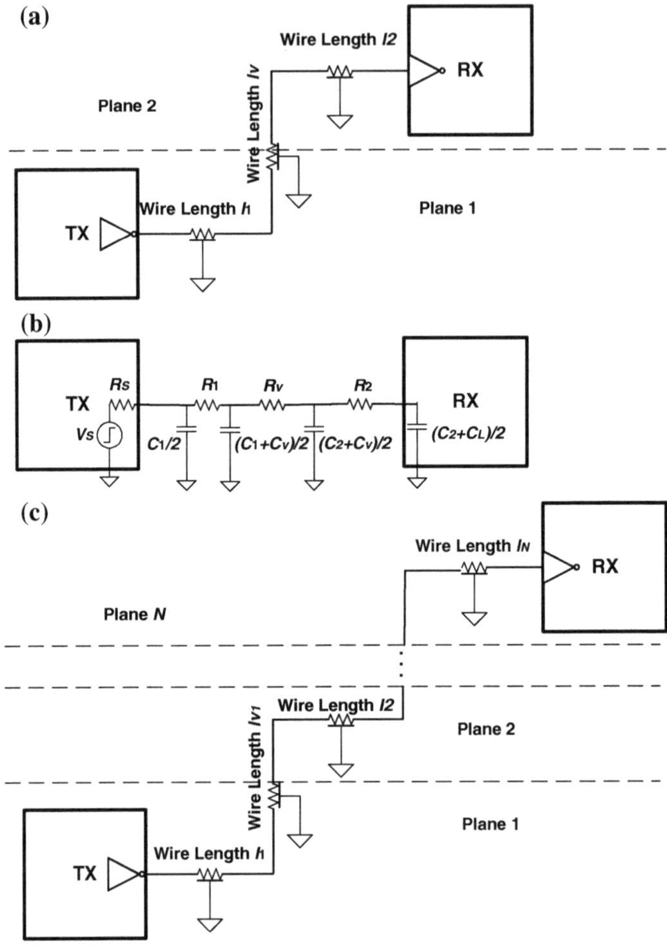


**Fig. 3.22** Link designs options (**a**) single inter-plane TSV link, (**b**) $\pi$-model equivalent circuit, and (**c**) multiple inter-plane TSV link

$$t_{pd} = \frac{R_s C_1}{2} + \frac{(R_s + R_1)(C_1 + C_v)}{2} + \frac{(R_s + R_1 + R_v)(C_2 + C_v)}{2} + \frac{(R_s + R_1 + R_v + R_2) \times (C_2 + C_L)}{2} \quad (3.8)$$

In the case of multiple interplane vias, the equivalent circuit is shown in Fig. 3.22c. Even using the simple Elmore delay model, optimal via placement in order to minimize total interconnect delay is not straightforward, particularly for multiple interplane vias [101]. For that reason, heuristic algorithms have been proposed.

In [121], the authors present a design methodology and performance evaluation for a hierarchical small-world NoC with on-chip millimeter (mm)-wave wireless channels as long-range communication links.

## Questions

**3.1** What is the difference between circuit switching and packet switching?

**3.2** What is the difference between virtual cut-through and wormhole switching?

**3.3** Name the basic approaches for solving the deadlock problem.

**3.4** What are deadlock and livelock? Why is it important for NoC routing algorithms be deadlock and livelock-free?

**3.5** Which are the basic classes of NoC routing algorithms? Are they prone to deadlock, livelock or both?

**3.6** Describe the basic blocks of a baseline NoC router and their function.

**3.7** What are the advantages and disadvantages of bufferless routing?

**3.8** How can the baseline 2D mesh router be extended to 3D mesh? Why is that design naive?

**3.9** What are the main design challenges in link design?

## Problems

**3.10** Assuming a minimum required data rate between a source and destination of 2 Mb/s with a distance $D$ of four hops, a phit size of 32 bits and routing and switching times of 5 ns evaluate the following switching schemes:

1. circuit switching
2. SAF packet switching
3. wormhole switching
4. pipelined packet switching

**3.11** Assuming a 3×3 torus NoC, calculate the minimum number of hops (shortest distance) required to reach destination node (2, 2) from source node (1, 0). Give the exact trace of hops for $XY$ routing.

**3.12** Likewise for a 3D NoC with a 3×3×3 mesh topology and $XYZ$ routing from source node (3, 1, 2)to destination node (2, 1, 1).

**3.13** Write pseudocode for the $XY$ (and $XYZ$) routing algorithm.

**3.14** Prove that $XY$ (or $XYZ$) routing is livelock-free.

**3.15** Assume a bufferless deflection algorithm in a mesh topology that prioritizes flits based on distance from their destination in other words favors flits that are away from their destination. Is this algorithm deterministically livelock-free?

**3.16** Create the routing table for router (1, 1) for the topology of Fig. 2.1.

**3.17** Assuming a 5-port hot potato router in a mesh topology that prioritizes flits depending on an age (number of hops) counter. Allocate the output ports for the following flits that arrive simultaneously:

| Flit | Requested port | Hops |
|---|---|---|
| 1 | W | 3 |
| 2 | N | 2 |
| 3 | W | 2 |
| 4 | S | 1 |
| 5 | W | 4 |

**3.18** Derive an expression for the complexity in gates of the multiplexer implementation of an $n$ input crossbar.

**3.19** Similarly for an $n$-input-output port allocator.

**3.20** Assuming a 5-port router has an area of $S$ and the target MPSoC in the same technology has 16 cores, estimate which of the following NoCs is more area efficient without using any CAD tools.

(a) A $4\times4$ mesh.
(b) A single $16\times16$ crossbar.
(c) A custom topology using one $10\times10$ crossbar and two $2\times2$ meshes.

**3.21** Consider a NoC with a link bandwidth requirement of 200 Gb/s, and a 3-cycle maximum latency requirement between routers. The link features a phit size of 64 bits and a link wire delay of 1 ns without repeaters. The FO4 delay is 10 ps.

(a) Does a link without repeaters meet the bandwidth requirement?
(b) Design a link with appropriate stages and repeaters to meet the design constraints.

## Projects and Lab Exercises

**3.22** Design and simulate a $4\times4$ pass transistor crossbar with 4 bits per port.

**3.23** Design a configurable $5\times5$ MUX-based crossbar with 8, 16, 32, and 64 bits per port. Implement all configurations in a technology of your choice. Notice the scaling.

**3.24** Similarly, a configurable $5\times5$ port allocator with 8, 16, 32, and 64 bits per port.

**3.25** Design and simulate a 5-port NoC wormhole router with two VCs in VHDL or Verilog. Use the designs of labs 9.11 and 9.12.

**3.26** Design and simulate a 7-port NoC wormhole router with four VCs in VHDL or Verilog.

## References

1. D. Wiklund, L. Dake, SoCBUS: switched network on chip for hard real time embedded systems. in *Parallel and Distributed Processing, Symposium* (2003), p. 8
2. P. Wolkotte, G. Smit, G. Rauwerda, L. Smit, An energy-efficient reconfigurable circuit-switched network-on-chip. in *Parallel and Distributed Processing, Symposium* (2005), p. 155a
3. D. Truong, W. Cheng, T. Mohsenin, Y. Zhiyi, A. Jacobson, G. Landge, M. Meeuwsen, C. Watnik, A. Tran, X. Zhibin Xiao, E. Work, J. Webb, P. Mejia, B. Baas, A 167-processor computational platform in 65 nm CMOS. IEEE J. Solid-State Circ. **44**(4), 1130–1144 (2009)
4. P. Phi-Hung, P. Jongsun, M. Phuong, K. Chulwoo, Design and implementation of backtracking wave-pipeline switch to support guaranteed throughput in network-on-chip. IEEE Trans. Very Large Scale Integration (VLSI) Syst. **20**(2), 270–283 (2012)
5. S. Liu, A. Jantsch, Z. Lu, Parallel probing: dynamic and constant time setup procedure in circuit switching NoC. in *Proceedings of Design, Automation and Test in Europe (DATE)* (2012), pp. 1289–1294
6. W.J. Dally, C.L. Seitz, The torus routing chip. J. Distrib. Comput. **1**(3), 187–196 (1986)
7. E. Bolotin, I. Cidon, R. Ginosar, A. Kolodny, QNoC: QoS architecture and design process for network on chip. J. Syst. Archit. **50**(2–3), 105–128 (Feb. 2004)
8. W. Dally, B. Towles, Route packets, not wires: on-chip interconnection networks. in *Design Automation Conference (DAC)* (2001), pp. 684–689
9. P. Guerrier, A. Greiner, A generic architecture for on-chip packet-switched interconnections. in *Design, Automation and Test in Europe Conference and Exhibition (DATE)* (2000), pp. 250–256
10. E. Rijpkema, K. Goosens, P. Wielage, A router architecture for networks on silicon. in *Workshop on Embedded Systems (PROGRESS)* (2001), pp. 1–8
11. W. Dally, Virtual-channel flow control. IEEE Trans. Parallel Distrib. Syst. **3**(2), 194–205 (1992)
12. K. Shin, S. Daniel, Analysis and implementation of hybrid switching. IEEE Trans. Comput. **45**(6), 684–692 (1996)
13. M. Modarressi, H. Sarbazi-Azad, M. Arjomand, A hybrid packet-circuit switched on-chip network based on SDM. in *Design, Automation and Test in Europe Conference and Exhibition (DATE '09)* (2009), pp. 566–569
14. A. Lusala, J. Legat, A hybrid router combining SDM-based circuit swictching with packet switching for on-chip networks. in *International Conference on Reconfigurable Computing and FPGAs (ReConFig)* (2010), pp. 340–345
15. J. Hu, R. Marculescu, DyAD - smart routing for networks-on-chip. in *Design Automation Conference (DAC)* (2004), pp. 260–263
16. W.J. Dally, B. Towles, *Principles and Practices of Interconnection Networks* (Morgan Kaufmann, San Francisco, 2004)
17. A. Lankes, T. Wild, A. Herkersdorf, S. Sonntag, H. Reinig, Comparison of deadlock recovery and avoidance mechanisms to approach message dependent deadlocks in on-chip networks. in *International Symposium on Networks-on-Chip (NOCS)* (2010), pp. 17–24
18. K. Anjan, T. Pinkston, DISHA: a deadlock recovery scheme for fully adaptive routing. in *Parallel Processing, Symposium* (1995), pp. 537–543
19. S. Lee, A deadlock detection mechanism for true fully adaptive routing in regular wormhole networks. Comput. Commun. **30**(8), 1826–1840 (2007)
20. J. Kim, L. Ziqiang, A. Chien, Compressionless routing: a framework for adaptive and fault-tolerant routing. IEEE Trans. Parallel Distrib. Syst. **8**(3), 229–244 (1997)
21. R. Al-Dujaily, T. Mak, X. Fei, A. Yakovlev, M. Palesi, Run-time deadlock detection in networks-on-chip using coupled transitive closure networks. in *Design,Automation and Test in Europe Conference and Exhibition (DATE)* (2011) pp. 1–6
22. C. Glass, L. Ni, The turn model for adaptive routing. in *International Symposium on Computer, Architecture* (1992), pp. 278–287

23. C. Ge-Ming, The edd-even turn model for adaptive routing. IEEE Trans. Parallel Distrib. Syst. **11**(7), 729–738 (2000)
24. K. Goossens, J. Dielissen, A. Radulescu, AEthereal network on chip: concepts, architectures, and implementations. Des. Test Comput. **22**(5), 414–421 (2005)
25. A. Mejia, J. Flich, J. Duato, S.-A. Reinemo, T. Skeie, Segment-based routing: an efficient fault-tolerant routing algorithm for meshes and tori. in *Parallel and Distributed Processing, Symposium*(2006), pp. 25–29
26. L. Ni, P. McKinley, A survey of wormhole routing techniques in direct networks. IEEE Comput. **26**(2), 62–75 (1993)
27. D. Linder, J. Harden, An adaptive and fault tolerant wormhole routing strategy for k-Ary n-Cubes. IEEE Trans. Comput. **40**(1), 2–12 (1991)
28. Y. Xiaoqiang, D. Huimin, H. Jungang, Research on node coding and routing algorithm for network on chip. in *International Colloquium on Computing, Communication, Control, and Management (CCCM)* (2008), pp. 198–203
29. M. Mano, R. Kime, *Logic and Computer Design Fundamentals*, 4 edn. (Prentice Hall, 2007)
30. E. Bolotin, I. Cidon, R. Ginosar, A. Kolodny, Routing in irregular meshes. TR CCIT 554 Department of Electrical Engineering, Technion, Sept. 2005
31. O. Lysne, T. Skeie, S. Reinemo, I. Theiss, Layered routing in irregular networks. IEEE Trans. Parallel Distrib. Syst. **17**(1), 51–65 (Jan. 2006)
32. J. Flich, M.P. Malumbres, P. Lopez, J. Duato, Performance evaluation of a new routing strategy for irregular networks. in *International Conference on Supercomputing (ICS)* (2000), pp. 34–43
33. J. Flich, P. Lopez, M. Malumbres, J. Duato, Boosting the performance of myrinet networks. IEEE Trans. Parallel Distrib. Syst. **13**(7), 1166–1182 (2002)
34. J. Sancho, A. Robles, J. Flich, P. Lopez, J. Duato, Effective methodology for deadlock-free minimal routing in infiniband networks. in *International Conference on Parallel Processing*, (2002), pp. 409–418
35. K. Anjan, T. Pinkston, An efficient, fully adaptive deadlock recovery scheme: DISHA. in *International Symposium on Computer, Architecture* (1995), pp. 201–210
36. J. Duato, A necessary and sufficient condition for deadlock-free adaptive routing in Wormhole networks. IEEE Trans. Parallel Distrib. Syst. **6**(10), 1055–1067 (1995)
37. J. Duato, A necessary and sufficient condition for deadlock-free routing in cut-through and store-and-forward networks. IEEE Trans. Parallel Distrib. Syst. **7**(8), 841–854 (1996)
38. J. Duato, T. Pinkston, A general theory for deadlock-free adaptive routing using a mixed set of resources. IEEE Trans. Parallel Distrib. Syst. **12**(12), 1219–1235 (2001)
39. L. Ziqiang, A. Chien, Hierarchical adaptive routing: a framework for fully adaptive and deadlock-free wormhole routing. Symp. Parallel Distrib. Process. 688–695 (1994)
40. J.C. Sancho, A. Robles, Improving minimal adaptive routing in networks with irregular topology. in *International Confenerence on Parallel and Distributed Computing Systems (PDCS)* (2000), pp. 1–9
41. F. Silla, J. Duato, On the use of virtual channels in networks of workstations with irregular topology. IEEE Trans. Parallel Distrib. Syst. **11**(8), 813–828 (2000)
42. F. Silla, J. Duato, High-performance routing in networks of workstations with irregular topology. IEEE Trans. Parallel Distrib. Syst. **11**(7), 699–719 (2000)
43. A. Jantch, H. Tenhunen, *Networks on Chip* (Springer, Berlin, 2003)
44. T. Pinkston, R. Pang, J. Duato, Deadlock-free dynamic reconfiguration schemes for increased network dependability. IEEE Trans. Parallel Distrib. Syst. **14**(8), 780–794 (2003)
45. S. Murali, G. De Micheli, Bandwidth-constrained mapping of cores onto NoC architectures. in *Design, Automation and Test in Europe Conference and Exhibition (DATE)* (2004), pp. 896–901
46. O.T. Skeie, I. Theiss, Layered shortest path (LASH) routing in irregular system area networks. Sympos. Parallel Distrib. Process. 162–169 (2002)
47. T. Skeie, O. Lysne, J. Flich, P. Lopez, A. Robles, J. Duato, LASH-TOR: a generic transition-oriented routing algorithm. in *International Conference on Parallel and Distributed Systems (ICPADS)* (2004), pp. 595–604

48. M. Koibuchi, A. Jouraku, K. Watanabe, H. Amano, Descending layers routing: a deadlock-free deterministic routing using virtual channels in system area networks with irregular topologies. in *International Conference on Parallel Processing* (2003), pp. 527–536
49. J. Flich, P. Lopez, J.C. Sancho, A. Robles, J. Duato, Improving infiniband routing through multiple virtual networks. in *International Symposium on High Performance Computing (ISHPC)* (2002), pp. 49–63
50. J. Flich, T. Skeie, A. Mejia, O. Lysne, P. Lopez, A. Robles, J. Duato, M. Koibuchi, T. Rokicki, J. Sancho, A survey and evaluation of topology-agnostic deterministic routing algorithms. IEEE Trans. Parallel Distrib. Syst. **23**(3), 405–425 (2012)
51. M. Koibuchi, A. Funahashi, A. Jouraku, H. Amano, Lturn routing: an adaptive routing in irregular networks. Int. Conf. Parallel Process. (ICPP) **3**, 374–383 (2001)
52. J. Sancho, A. Robles, J. Duato, An effective methodology to improve the performance of the Up*/Down* routing algorithm. IEEE Trans. Parallel Distrib. Syst. **15**(8), 740–754 (2004)
53. M. Schroeder, A. Birrell, M. Burrows, H. Murray, R. Needham, T. Rodeheffer, E. Satterthwaite, C. Thacker, Autonet: a high-speed, self-configuring local area network using point-to-point links. IEEE J. Sel. Areas Commun. **9**(8), 1318–1335 (1991)
54. N. Boden, D. Cohen, R. Felderman, A. Kulawik, C. Seitz, J. Seizovic, S. Wen-King, Myrinet: a gigabit-per-second local area network. IEEE Micro **15**(1), 29–36 (1995)
55. C. Gomez, M. Gomez, P. Lopez, J. Duato, Reducing packet dropping in a bufferless noc. in *International Euro-Par Conference* (2008), pp. 899–909
56. M. Hayenga, N. Jerger, M. Lipasti, SCARAB: a single cycle adaptive routing and bufferless network. in *International Symposium on Microarchitecture (MICRO-42)* (2009), pp. 244–254
57. P. Baran, On distributed communications networks. IEEE Trans. Commun. Syst. **12**(1), 1–9 (1964)
58. T. Moscibroda, O. Mutlu, A case for bufferless routing in on-chip networks. in *International Symposium on Computer, Architecture* (2009), pp. 196–207
59. W. Dally, *Scalable Switching Fabrics for Internet Routers* (White paper for Avici Systems Inc., 2004)
60. R. Widyono, *The Design and Evaluation of Routing Algorithms for Real-Time Channels*, TR-94-024, (University of California at Berkeley and Int'l Computer Science Institute, Berkeley, 1994)
61. F. Feliciian, S. Furber, An asynchronous on-chip network router with quality-of-service (QoS) support. in *International SOC Conference* (2004), pp. 274–277
62. D. Bertozzi, L. Benini, Xpipes: a network on chip architecture for gigascale systems-on-chip. IEEE Circ. Syst. Mag. **4**(2), 18–31 (2004)
63. J. Chan, S. Parameswaran, NoCGEN:a template based reuse methodology for networks on chip architecture. *International Conference on VLSI Design* (2004), pp. 717–720
64. C. Zeferino, A. Susin, SoCIN: a parametric and scalable network-on-chip. in *Symposium on Integrated Circuits and Systems Design* (2003), pp. 169–174
65. R. Tamhankar, S. Murali, G. De Micheli, Performance driven reliable link design for networks on chips. Asia South Pacific Des. Autom. Conf. (ASP-DAC) **2**, 749–754 (2005)
66. C. Zeferino, F. Santo, A. Susin, ParIS: a parameterizable interconnect switch for networks-on-chip. in *Symposium on Integrated Circuits and Systems Design* (2004), pp. 204–209
67. A. Pullini, F. Angiolini, D. Bertozzi, L. Benini, Fault tolerance overhead in network-on-chip flow control schemes. in *18th Symposium on Integrated Circuits and Systems Design* (2005), pp. 224–229
68. M. Winter, G. Fettweis, Guaranteed service virtual channel allocation in NoCs for run-time task scheduling. in *Design, Automation and Test in Europe Conference and Exhibition (DATE)* (2011), pp. 1–6
69. M. Millberg, E. Nilsson, R. Thid, A. Jantsch, Guaranteed bandwidth using looped containers in temporally disjoint networks within the nostrum network on chip. in *Design,Automation and Test in Europe Conference and Exhibition (DATE)* (2004), pp. 890–895
70. M. Keating, P. Bricaud, *Reuse Methodology Manual for System-On-A-Chip Designs*, 3rd edn. (Springer, Berlin, 2007)

71. J. Hu, R. Marculescu, Application-specific buffer space allocation for networks-on-chip router design. in *International Conference on Computer Aided Design (ICCAD)* (2004), pp. 354–361
72. C. Xuning Chen, P. Li-Shiuan, Leakage power modeling and optimization in interconnection networks. in *International Symposium on Low Power Electronics and Design (ISLPED)* (2003), pp. 90–95
73. T. Ye, L. Benini, G. De Micheli, Analysis of power consumption on switch fabrics in network routers. in *Design Automation Conference (DATE)* (2002), pp. 524–529
74. C. Nicopoulos, P. Dongkook, K. Jongman, N. Vijaykrishnan, M. Yousif, C. Das, ViChaR: a dynamic virtual channel regulator for network-on-chip routers. in *International Symposium on Microarchitecture (MICRO-39)* (2006), pp. 333–346
75. L. Benini, G. De Micheli, *Networks on Chips: A Circuit and Systems Perspective* (Morgan Kaufmann, San Francisco, 2006)
76. F. Jafari, L. Zhonghai, A. Jantsch, M. Yaghmaee, Optimal regulation of traffic flows in networks-on-chip. *Design, Automation and Test in Europe Conference and Exhibition (DATE)*(2010), pp. 1621–1624
77. F. Jafari, L. Zhonghai, A. Jantsch, M. Yaghmaee, Buffer optimization in network-on-chip through flow regulation. IEEE Trans. Comput. Aided Des. Integr. Circ. Syst. **29**(12), 1973–1986 (2010)
78. J. Park, B. O'Krafka, S. Vassiliadis, J. Delgado-Frias, Design and evaluation of a DAMQ multiprocessor network with self-compacting buffers. in *Proceedings of Supercomputing* (1994), pp. 713–722
79. R. Ramanujam, V. Soteriou, B. Lin, P. Li-Shiuan, Design of a high-throughput distributed shared-buffer NoC router. in *International Symposium on Networks-on-Chip (NOCS)* (2010), pp. 69–78
80. L. Wang, J. Zhang, X. Yang, D. Wen, Router with centralized buffer for network-on-chip. in *Great Lakes Symposium on VLSI (GLSVLSI)* (2009), pp. 469–474
81. A. Kodi, A. Louri, J. Wang, Design of energy-efficient channel buffers with router bypassing for network-on-chips (NoCs). in *Quality of Electronic Design (ISQED)* (2009), pp. 826–832
82. C. Fallin, G. Nazario, Y. Xiangyao, K. Chang, R. Ausavarungnirun, O. Mutlu, MinBD: minimally-buffered deflection routing for energy-efficient interconnect. in *International Symposium on Networks on Chip (NoCS)* (2012), pp. 1–10
83. P. Li-Shiuan, W. Dally, A delay model for router microarchitectures. IEEE Micro **21**(1), 26–34 (2001)
84. I. Saastamoinen, M. Alho, J. Nurmi, Buffer implementation for Proteo network-on-chip. Int. Symp. Circ. Syst. (ISCAS) **2**, 113–116 (2003)
85. V. Chandra, A. Xu, H. Schmit, L. Pileggi, An interconnect channel design methodology for high performance integrated circuits. in *Design, Automation and Test in Europe Conference and Exhibition (DATE)* (2004), pp. 1138–1143
86. J. Hu, U. Orgas, R. Marculescu, System-level buffer allocation for application-specific networks-on-chip router design. IEEE Trans. Comput. Aided Des. Integr. Circ. Syst. Vl. **25**(12), 2919–2933 (2006)
87. E. Bolotin, I. Cidon, R. Ginosar, A. Kolodny, Cost considerations in network on chip. Integr. VLSI J. **38**(1), 19–42 (2004)
88. Y. Tamir, G. Frazier, High-performance multiqueue buffers for VLSI communication switches. in *International Symposium on Computer Architecture* (1988), pp. 343–354
89. G. Passas, M. Katevenis, D. Pnevmatikatos, A 128x128x24Gb/s crossbar interconnecting 128 tiles in a single hop and occupying 6% of their area. in *International Symposium on Networks-on-Chip (NOCS)* (2010), pp. 87–95
90. K. Jongman, C. Nicopoulos, P. Dongkook, V. Narayanan, M. Yousif, C. Das, A gracefully degrading and energy-efficient modular router architecture for on-chip networks. in *International Symposium on Computer Architecture (ISCA)* (2006), pp. 4–15
91. H. Shih-Hsun, L. Yu-Xuan Lin, J. Jer-Min, Design of a Dual-Mode NoC router integrated with network interface for AMBA-based IPs. in *Asian Solid-State Circuits Conference (ASSCC)* (2006), pp. 211–214

92. L. Kangmin, L. Se-Joong Lee, K. Sung-Eun, C. Hye-Mi, K. Donghyun, K. Sunyoung, L. Min-Wuk, Y. Hoi-Jun, A 51 mW 1.6 GHz on-chip network for low-power heterogeneous SoC platform. in *International Solid-State Circuits Conference (ISSCC)* (2004), pp. 512–518
93. K. Lee, S.-J. Lee, H.-J. Yoo, A high-speed and lightweight on-chip crossbar switch scheduler for on-chip interconnection networks. in *European Solid-State Circuits Conference (ESSCIRC)* (2003), pp. 453–456
94. L. Kangmin, L. Se-Joong Lee, Y. Hoi-Jun Yoo, A distributed crossbar switch scheduler for on-chip networks. in *Custom Integrated Circuits Conference* (2003), pp. 671–674
95. Y. Tamir, H.C. Chi, Symmetric crossbar arbiters for VLSI communication switches. IEEE Trans. Parallel Distrib. Syst. **4**(1), 13–27 (1993)
96. J. Delgado-Frias, G. Ratanpal, A VLSI crossbar switch with wrapped wave front arbitration. IEEE Trans. Circ. Syst. I Fundam. Theor. Appl. **50**(1), 135–141 (2003)
97. R. Maroofi, V. Nitnaware, S. Limaye, Area-efficient design of scheduler for routing node of network-on-chip. Int. J. VLSI Des. Commun. Syst. (VLSICS) **2**(3), 111–118 (2011)
98. D. Becker, W. Dally, Allocator implementations for network-on-chip routers. in *Conference on High Performance Computing, Networking, Storage and Analysis*, Article No. 52 (2009)
99. C. Fallin, C. Craik, O. Mutlu, CHIPPER: a low-complexity bufferless deflection router. in *High Performance Computer Architecture (HPCA)* (2011), pp. 144–155
100. L. Carloni, P. Pande, Y. Xie, Networks-on-chip in emerging interconnect paradigms: advantages and challenges. in *International Symposium on Networks-on-Chip (NoCS)* (2009), pp. 93–102
101. V. Pavlidis, E. Friedman, *Three-Dimensional Integrated Circuit Design* (Morgan Kaufmann, Elsevier, 2008)
102. L. Feihui, C. Nicopoulos, T. Richardson, X. Yuan, V. Narayanan, M. Kandemir, Design and management of 3D chip multiprocessors using network-in-memory. in *International Symposium on Computer Architecture (ISCA)* (2006), pp. 130–141
103. A. Rahmani, P. Liljeberg, J. Plosila, H. Tenhunen, An efficient hybridization scheme for stacked mesh 3D NoC architecture. in *International Conference on Parallel, Distributed and Network-Based Processing (PDP)* (2012), pp. 507–514
104. J. Kim, C. Nicopoulos, D. Park, R. Das, Y. Xie, V. Narayanan, M. Yousif, C. Das, A novel dimensionally-decomposed router for on-chip communication in 3D architectures. in *International Symposium on Computer Architecture (ISCA)* (2007), pp. 138–149
105. W. Lafi, D. Lattard, A. Jerraya, An efficient hierarchical router for large 3D NoCs. in *International Symposium on Rapid System Prototyping (RSP)* (2010), pp. 1–5
106. K. Nomura, K. Abe, S. Fujita, A. DeHon, Novel design of three-dimensional crossbar for future network on chip based on post-silicon devices. in *International Conference on Nano-Networks and Workshops (NanoNet)* (2006), pp. 1–5
107. P. Dongkook, S. Eachempati, R. Das, A. Mishra, X. Yuan, N. Vijaykrishnan, C. Das, MIRA: a multi-layered on-chip interconnect router architecture. in *Internatioanl Symposium on Computer Architecture (ISCA)* (2008), pp. 251–261
108. K. Puttaswamy, G. Loh, Thermal herding: microarchitecture techniques for controlling hotspots in high-performance 3D-integrated processors. in *International Symposium on High Performance Computer Architecture (HPCA)* (2007), pp. 193–204
109. R. Dobkin, I. Cidon, R. Ginosar, A. Kolodny, A. Morgenshtein, Fast asynchronous bit-serial interconnects for network-on-chip. CCIT TR529 EE Department, Technion (2005)
110. A. Rahmani, P. Liljeberg, J. Plosila, H. Tenhunen, BBVC-3D-NoC: an efficient 3D NoC architecture using bidirectional bisynchronous vertical channels. in *IEEE Computer Annual Symposium on VLSI (ISVLSI)* (2010), pp. 452–453
111. C. Feng, Z. Lu, A. Jantch, M. Zhang, A 1-cycle 1.25 GHz bufferless router for 3D network-on-chip. IEICE Trans. Inform. Syst. **E95D**(5), 1519–1522 (2012)
112. Y. Yoon, N. Concer, M. Petracca, L. Carloni, Virtual channels Versus multiple physical networks: a comparative analysis. in *Design Automation Conference (DAC)* (2010), pp. 162–165
113. N. Weste, D. Harris, *Principles of CMOS VLSI Design*, 4th edn. (Addison Wesley, 2010)

114. W. Elmore, The transient response of damped linear networks with particular regard to wide-band amplifiers. J. Appl. Phys. **19**(1), 55–63 (1948)
115. A. Kodi, A. Sarathy, A. Louri, W. Janet, Adaptive inter-router links for low-power, area-efficient and reliable Network-on-Chip (NoC) architectures. in *Asia and South Pacific Design Automation Conference (ASP-DAC)* (2009), pp. 1–6
116. M. Mizuno, W. Dally, H. Onishi, Elastic interconnects: repeater-inserted long wiring capable of compressing and decompressing data. in *International Conference on Solid-State Circuits (ISSCC)* (2001), pp. 346–347
117. A. Shacham, K. Bergman, L. Carloni, Photonic networks-on-chip for future generations of chip multiprocessors. IEEE Trans. Comput. **57**(9), 1246–1260 (2008)
118. M. Chang, J. Cong, A. Kaplan, M. Naik, G. Reinman, E. Socher, S. Tam, CMP network-on-chip overlaid with multi-band RF-interconnect. in *International Symposium on High Performance Computer Architecture (HPCA)* (2008), pp. 191–202
119. U. Ogras, R. Marculescu, It's a small world after all: NoC performance optimization via long-range link insertion. IEEE Trans. Very Large Scale Integr. (VLSI) Syst. **14**(7), 693–706 (2006)
120. E. Nigussie, T. Lehtonen, S. Tuuna, J. Plosila, J. Isoaho, High-Performance long NoC link using delay-insensitive current-mode signaling. VLSI Des. **2007**, Article ID 46514 (2007)
121. S. Deb, K. Chang, A. Ganguly, Y. Xinmin, C. Teuscher, P. Pande, H. Deukhyoun, B. Belzer, Design of an efficient NoC architecture using millimeter-wave wireless links. in *International Symposium on Quality Electronic Design (ISQED)* (2012), pp. 165–172

# Chapter 4
# Power and Thermal Effects and Management

**Abstract** As semiconductor processes scale to smaller and smaller feature sizes, manufacturing reliable digital designs is challenging how systems are traditionally designed. Specifically, the shrinking of transistor and wire size imposes that these components simultaneously are becoming more prone to complete, or parametric, failure at manufacturing time. Additionally, the derived systems are increasingly expensive to produce and less likely to function correctly for as long as intended. In order to address these challenges, the NoC-based systems have to be designed with reliability and fault tolerance features in mind. Toward this goal, a number of design techniques and methodologies are available that promise to provide sufficient fault coverage with controllable overhead in terms of hardware redundancy and performance (e.g., delay/power) degradation. This chapter studies the origin of faults in modern technologies and explains the classification to transient, intermittent, and permanent faults. A survey of fault tolerance methods is presented to demonstrate the diversity of available methods. Fault tolerance methods for NoCs are studied at different layers of the OSI reference model.

## 4.1 Introduction

In this chapter, we discuss why power consumption has become one of the main (if not the main) design concerns in today's complex digital integrated circuits. The importance of this problem is also highlighted from the continuously increased interest both from industry and academia for proposing solutions at different level of abstractions aiming to address the consequences of the excessive amount of power consumed by MPSoC platforms. Toward this direction, various roadmaps (e.g., ITRS [28]) indicate that in the near future these concerns are most likely not to go away; in contrast, CMOS scaling only seems to make the problem even worse.

There are many reasons why designers worry about power dissipation. One concern that has come consistently to the foreground in recent years is the need for

K. Tatas et al., *Designing 2D and 3D Network-on-Chip Architectures*,
DOI: 10.1007/978-1-4614-4274-5_4, © Springer Science+Business Media New York 2014

"green" electronics. While the power dissipation of electronic components until recently was only a small fraction of the overall electrical power budget, this picture has changed substantially in the last few decades. More specifically, if Moore's law would continue unabated in the future and the computational needs would keep on doubling every year, the total energy of our galaxy would be exhausted in the relatively low time span of 180 years (even if we assume that every digital operation is performed at its lowest possible level).

There exist very compelling reasons why a further increase in power density at MPSoCs should be avoided at all costs. As shown in Fig. 4.1, power densities for on-chips can become excessive and lead to degradation or failure, unless extremely expensive packaging techniques are used [48]. To drive the point home, power density levels of some well-known processors are compared to general world examples, such as hot plates, nuclear reactors, rocket nozzles, or even the sun's surface. Surprisingly, high-performance ICs are not that far off from some of these extreme heat sources.

In advance of discussing in more detail the impact of increased power consumption to reliability degradation, some words about useful metrics are necessary. So far, we have used the terms power and energy quite interchangeably. Yet, each has its specific role depending upon the phenomena that are being addressed, or the constraints of the application at hand. Average power dissipation is the prominent parameter when studying heat removal and packaging concerns of high-performance processors. On the other hand, peak power dissipation is the parameter to watch when designing the complex power supply delivery networks for integrated circuits and systems. In contrast, for some types of devices (e.g., mobile platforms), the type of energy source determines which property is the most essential. Specifically, in a battery-powered system, the energy supply is finite, and hence energy minimization is crucial. Finally, by dividing power dissipation into dynamic (proportional to

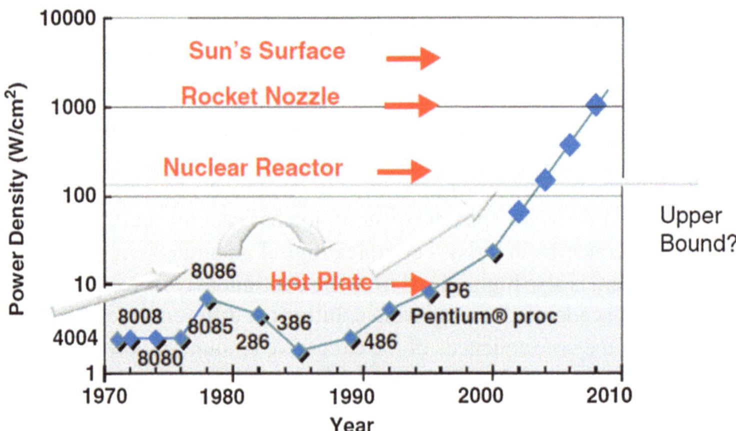


Fig. 4.1 Power density for different processors [48]

activity) and static (independent of activity) is crucial in the design of power management scenarios exploiting the operational modes of the system.

Even though concerns about power density may seem quite recent to most designers, this issue has surfaced numerous times in the design of electronic components before. This occurs mainly because power density is tightly firmed to the on-chip temperature values, which might affect the performance of target MPSoCs. In the past, die sizes were small enough, and hence the activity distribution over the die was quite uniform. This translated into an almost flat temperature profile at the surface of the die. However, with the advent of SoCs, more and more diverse functionality is integrated in close proximity, which very often leads to considerable variations in terms of workloads and activity profiles. For instance, a typical NoC integrates multiple levels of cache memories on the die, just next to the processing computing cores. As the data path of this NoC is clocked at the highest speed and is kept busy almost 100 % of the time, its power dissipation is substantially higher than that of the cache memories.

This results in the creation of thermal hotspots over the die, which in turn impacts the long-term reliability of the part and complicates the verification of the NoC-based design. Moreover, execution speed and propagation delay are indeed strongly dependent on temperature values. With temperature gradients over the die, which may change dynamically depending upon the operation modes of the processors, simulation can now not be performed for a single temperature, as was the common practice.

The impact of temperature on long interconnects in a NoC framework is discussed in [26]. Based on circuit simulations in different CMOS process technologies (65 nm, 45 nm and 32 nm), there is an increasing impact of both spatial and temporal thermal variations on interconnect delay and energy. Temperature effects are particularly challenging due to their dynamic nature, their dependency on the external environment, and the circular relationship with device leakage. Additionally, temporal and spatial temperature variations on the physical layers can significantly impact the performance and reliability of the overall NoC. Thus, it is necessary to conduct a thorough study both for spatial and temporal variations of thermal impacts on NoC interconnect in order to meet design budgets.

Performance degradation and reliability issues due to thermal variation have been an ongoing problem in chip design. Furthermore, the communication bandwidth in future network on chip architectures will probably be limited by prohibitive levels of power consumption [45]. In order to address the previously mentioned temperature imbalance, a complex package has to be constructed, which allows for the heat to spread over a wider area, thus improving the heat removal process. However, packaging cost can then become an important (if not dominating) fraction of the total cost. This imposes that design techniques and methodologies that help to mitigate the packaging problems, either by alleviating the gradients, or by reducing the power density of selected subsystems, are absolutely essential. Moreover, with power density and hence cooling costs rising exponentially, processor packaging can no longer be designed for the worst case, and there is an urgent need for runtime processor-level techniques that can regulate operating temperature when the package's capacity is exceeded.

To make matters worse, regarding the 3D NoCs, heat transfer is further restricted by the low thermal conductivity bonding interfaces and thermal obstacles across multiple IC layers. A viable solution to dramatically reduce the operating temperature of 3D architectures is based on using liquid cooling on microfluidic channels. In order to deliver currents to 3D architectures, while suppressing the power supply noise to an acceptable level, designers use a highly complex hierarchical power distribution network in conjunction with decoupling capacitors. However, the usage of such solutions poses major challenges to routing completion and congestion.

This chapter discusses in more detail the impact of increased power and temperature values on reliability degradation for NoC-based MPSoCs. Toward this direction, we present an overview of different methodologies applied at different layers of abstraction. Since reliability modeling and reliability degradation analysis are two computational intensive tasks, algorithms and tools that automate this procedure are also discussed throughout this chapter.

### 4.1.1 Models for Power, Energy, and Temperature

There are numerous works dealing with the problem of power/temperature modeling and analysis. The most known energy models for NoC architectures can be summarized, as follows:

- *Bit Energy of Packet*: When a packet travels on the interconnect network, both the wires and logic gates on the data path will toggle as the bit-stream flips its polarity. A typical approach toward this direction is described in [68], where the authors aim to estimate the energy consumption for the packets traveling on the network, by following a similar approach to [43]. More specifically, the energy consumed for each bit (in wires and logic gates inside the switch) is computed every time a bit flips its polarity (from previous bit) in the bit stream.
- *Packets and Hops*: When the source and destination nodes are not placed adjacent to each other on the network, a packet needs to travel several intermediate nodes (i.e., hops) until reaching the destination. Additionally, depending on the traffic scenario, packets with the same source and destination nodes may not travel through the same number of hops, and they may not necessarily travel on the data path with the minimum number of hops. Thus, the number of hops a packet travels greatly affects the total energy consumption needed to transport the packet from source to destination. Based on this model, as the packet travels between nodes, the corresponding interconnect wires will be charged and discharged, whereas the logic gates inside the node switches will toggle. An example of applying this model in order to estimate the energy consumption can be found in [68]. More specifically, authors in this work assume a tiled floor-plan implementation for

MPSoC and calculate the total energy consumed per packet by multiplying the number of hops a packet travels by the energy consumed by one packet per hop.

Apart from these, in the literature there are a lot of models that deal with power/temperature issues both for 2D and 3D architectures. For instance, TEMPEST [11] is a thermal model based on an equivalent RC circuit. However, this approach contains only a single RC pair for the entire chip, giving no localized information. The importance of using a detailed thermal model that includes localized heating, thermal diffusion, and coupling with the thermal package is discussed in [8]. The derived model is applied to evaluate a variety of techniques for supporting Dynamic Thermal Management (DTM). A fast and accurate high-level power model based on an empirical power model of links and switches for Nostrum NoC is presented in [65]. This model was validated with the Synopsys Power Compiler, while the experimental results shown that it allows a fast power analysis with accuracy within 5 %. As compared to Power Compiler for similar-sized NoC, there are works (e.g., the one proposed by [65]) that claim considerable speedup ranging up to 500 × faster execution. Hotspot [62] is another accurate yet fast model based on an equivalent circuit of thermal resistances and capacitances that correspond to microarchitecture blocks and essential aspects of the thermal package.

One challenge of utmost importance about 3D architectures is the heat dissipation and the thermal management [59]. In order to tackle this issue, several analysis techniques have been proposed the last years [14, 29, 44, 49].

### 4.1.2 Wear-Out Mechanisms

Defects at VLSI designs are tightly coupled to the operating conditions. As manufacturing processes scale, permanent faults occur more frequently due to wear out because of increased strain on ever smaller transistors and wires. Wear out is a time-dependent process, whereby over the course of normal operation the integrity of a portion of a device degrades and eventually fails to behave as originally intended, resulting in a permanent wear out induced fault. More specifically, as the power density increased over the last decades due to the technology scaling, it had a consequence that also total power/energy consumption was raising, which in turn leads to higher on-chip temperature values. This has a direct impact on the system's reliability, since the failure rates rise exponentially with the temperature increase.

There can be distinguished three periods of device reliability, as they are depicted in Fig. 4.2. In the very beginning during the burn-in period and in the end during the wear out period, the failure rates are high enough. On the other hand, during normal working period, the failure rate remains constant on a rather low level.

Wear out phenomenon occurs for a variety of reasons and in a variety of ways. Important wear out mechanisms include [31]:

- *Negative Bias Temperature Instability* (NBTI), which is the degradation of transistor performance as charge is implanted in the gate, resulting in timing failures;

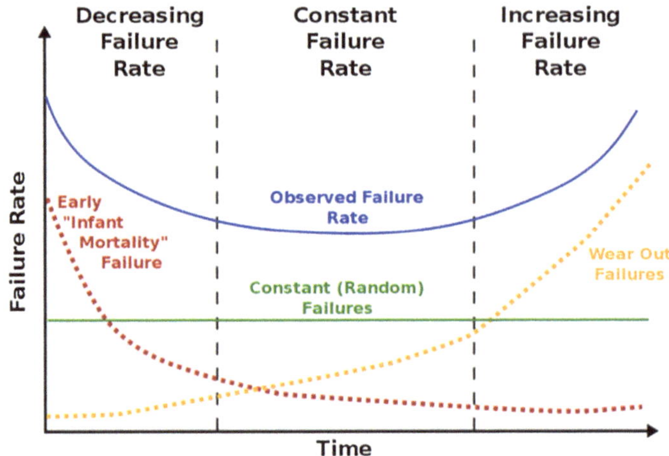


**Fig. 4.2** Failure rates during product's lifetime

- *Time-Dependent Dielectric Breakdown* (TDDB) represents the destruction of gate oxide that occurs when sufficient charge is implanted in the gate to result in a junction-gate short, due to the high electric field strength;
- *Electro-Migration* (EM) models the movement of metal atoms away from where they had been deposited, induced by the repeated collision of high-energy electrons, eventually resulting in short- or open circuits;
- *Thermal Cycling* (TC) affects the damage accumulated as a result of uneven expansion and contraction of different parts of the system due to uneven heating and cooling of the system;
- *Stress Migration* (SM) is the movement of metal atoms (much like EM), induced by the uneven expansion and contraction of different materials in the system.

The previously mentioned failure mechanisms have a strong dependence on operating conditions, and more specifically on the temperature values. Since the on-chip temperature increases with technology scaling, the wear out problem is expected to become far more important in the close future [28]. Additionally, due to leakage power, both the supply, as well as the threshold voltage, no longer scale ideally. As a consequence, power density is increasing, and therefore so is system temperature. Among others, the increase at temperature values also imposes higher mean-time-to-failure (MTTF) rates because the EM, TDDB, SM, and NBTI decreases exponentially, while the MTTF due to TC decreases proportionally to $(1/T)^{2.35}$ [31].

## 4.2 Classification of Faults

System failure occurs when, due to one or more faults, the system is no longer able to implement the function for which it was designed. For instance, it is possible after a fault, a system not to have the necessary amount of hardware resources in order to complete a task (i.e., the amount of memory is not sufficient to perform a computation). Alternatively, a system may simply lack the resources to complete a task before a given deadline.

The three fundamental terms in the present context are fault, error, and failure. While these terms are often used synonymously in the literature, they are not identical but rather related by the following cause-and-effect relationship: faults are the cause of errors, and errors are the cause of failures. A different view of the cause-and-effect relationship mentioned above is the classification into different abstraction levels at which faults, errors and failures occur. More specifically, failures occur in the external level, errors occur in the informational level, while faults occur in the physical level.

The classical fault model for digital circuits and busses can be summarized as follows:

- *Single Stuck-at Fault Models*: It is one of the first introduced fault models which is common up to now. In this model, faults are represented as a node having a fixed logic value (stuck-at-0 or stuck-at-1). Such a kind of faults is permanent, while the basic functionality of the circuit is not altered. The main advantage of employing this model is its conceptual simplicity. Current research efforts still examine the single stuck-at fault model. However, this model does not provide an accurate representation of the physical defect's behavior [2].
- *Bridging Fault Models*: It models two signals shorted together. Such kind of fault may change the circuit's sequential behavior, while the voltage level at the end of two shorted wires can massively depend on the location of the bridge. A problem with this kind of fault is that they cannot be predicted in advanced. For instance, a strong driver will for sure overpower a smaller transistor if their outputs are shorted but two equally strong drivers will probably generate an unpredictable value as their common output.
- *Open Fault Models*: Similar to bridging faults, this kind of defect is common in CMOS processes, since latest architectures incorporate more metal layers (and hence much more vias). Additionally, the copper interconnects make reliability issues even more critical [1]. This kind of fault fixes the gate of a transistor at the open value, and hence the transistor cannot be switched on. Predicting the behavior of a circuit with a (resistive) open is a difficult task, which has not yet fully understood [2].

In addition to that, it is possible to categorize the system faults depending on their duration. More specifically, based on the duration of the faults, they can be classified into three categories: (i) transient, (ii) intermittent, and (iii) permanent.

Transient faults are single errors in logic caused by an event external to that logic (e.g., cosmic radiation, contaminated package) [5]. The impact of transient fault to the proper system's functionality is temporary, whereas it is almost very limited the probability for a hardware resource to be affected by two consecutive transient faults. However, since the consequences of transient faults usually cannot be estimated, recalculation is almost assured to result in a correct output. Specifically, whether, or not, a fault results in failure depends on whether and how the error propagates to the system (e.g., shared memory, or output).

The occurrence of transient faults is tightly firmed to the scaling of process technology, since as transistors shrink, less charge (deposited due to the external event) is necessary to trigger a fault [7]. This problem becomes far more important with the continued increase of number of transistors that are integrated to the MPSoC architectures. Research work was already devoted to the study, detection, and correction of transient faults.

The available design approaches incorporate techniques in order to reduce the likelihood of transient errors to result to enormous output. For instance, transient errors in processing cores are mitigated through the careful instantiation of functional units with variable resilience so that reliability, performance, and cost constraints are met [60]. Similarly, there are also software tools that automatically provide sufficient protection to digital systems against single event upsets. The most widely accepted of these tools is based on appropriately triplicating the system's functionality mapped both on logic and memory resources, which is candidate to be affected by these faults [24]. Regarding the interconnect structures, there are available design approaches that include various recovery mechanisms for handling transient errors in on-chip networks [42].

These faults can be represented on multiple levels of abstraction. For instance, on the device level, transient faults are represented as voltage, or current, sources. Device level models considering cosmic particles have already been studied with SPICE simulations [40], that were used to represent alpha particle collected charges. Similarly, at the logic level, transient faults are modeled as bit-flip of the current propagating signal, whereas transient faults in memory/storage elements are represented as changes in logic values.

The second class of upsets affects the intermittent faults. There are two differentiations between these faults and those occurred due to transient upsets. Specifically, the intermittent faults are reproducible, while the conditions causing the fault remain in force [15]. For instance, in case the on-chip temperature is higher than a particular threshold, then it is possible to violate the setup time for a flip-flop along the system's critical path. As long as the system is too hot, the fault exists and the system fails. On the other hand, if the system cools down, it returns to the previous (correct) behavior. There are some works dealing with efficient ways to predict when permanent failure will occur based on the detection of intermittent faults [57]. Device degradation tends to precede total failure, and as a result, online test can be used to monitor gradual changes in performance and use this information to predict when timing failure is imminent.

The last class of upsets affects the permanent faults. In contrast to previous approaches, if such a fault occurs onto a device, it is irreversible because it leads to a physically broken device. The only possible way to overcome these faults and provide a correct operation for the architecture is by replacing the faulty component, or working around it. Permanent faults may be further categorized depending on their occurrence (either at manufacturing time, or later). Since it is not possible to overcome from these faults with conventional design and/or algorithmic approaches, throughout this chapter we will not discuss permanent faults any further.

### *4.2.1 Fault Tolerant Systems*

In order to prevent, or at least alleviate, the consequences of reliability degradation, a number of fault tolerant mechanisms have been proposed. The term fault tolerant corresponds to a design able to continue its operation, possibly at a reduced level, rather than failing completely, when some part of the system fails. Even though fault tolerance could be though as a prerequest for the existing architectures, the excessive mitigation cost makes it affordable only for mission critical systems. However, there are numerous applications that can afford lower fault coverage for significantly reduced mitigation cost.

Up to now, a number of architectures and design methodologies able to provide nondistributed device operation have been proposed at different levels of abstraction. Specifically, in literature, there are two mainstream approaches for designing fault-tolerant systems. The first of them deals with the design of new hardware elements, which are fault tolerant enabled, whereas the desired fault masking at the second approach is provided at software level with the usage of specialized CAD tools.

Both approaches exhibit advantages and disadvantages. The (re-)designed hardware blocks can either replace the existing components at conventional NoCs, or new fault tolerant architectures can be designed to improve robustness. The drawback of applying such a strategy is the increased design complexity, while the derived NoC provides also a static (defined at fabrication time) fault tolerant mechanism. Typical instantiations of this approach involve the usage of spare logic and routing resources.

On the other hand, the software-based fault masking combines the required dependability level with the low cost of commodity NoC platforms. However, the software-based fault tolerant systems assume that the designer is responsible for protecting the design. Since this approach does not impose any hardware modifications, it is widely accepted for research and product development. Among others, algorithms that provide different system mapping and/or routing under fault tolerant and/or reliability constraints have been proposed.

## 4.3 Fault Tolerance Metrics

Fault tolerance metrics allows us to measure the system design in terms of its reliability, availability, safety, maintainability and dependability against its failure rate, mean time between failure (MTBF), and mean time to repair (MTTR). Specifically, failure rate ($\lambda$) is the expected number of failures a system can suffer during a specific time period. Its reciprocal is the mean time to failure (MTTF), where $MTTF = \frac{1}{\lambda}$. The relation between reliability and failure rate for a constant $\lambda$ is $R(t) = e^{(-\lambda t)}$ [33]. Similarly, the MTTF is the estimated time for which the system is expected to perform correctly before the first error occurs, whereas MTBF is the average time to next failure.

$$\text{MTBF} = \frac{\text{Total operating time}}{\text{Number of failure occurs}} \quad (4.1)$$

A similar parameter is the MTTR (Mean Time to Repair) which denotes the average time required to repair the system. MTTR provides the average time required between injecting an error to the system and repairing the system.

$$\text{MTTR} = \frac{\text{Time spend for repair}}{\text{Number of repairs}} \quad (4.2)$$

Having these two metrics, it is possible to calculate the system's availability, which denotes the impact of these failures on the system, as follows:

$$\text{Availability} = \frac{\text{MTBF}}{\text{MTBF} + \text{MTTR}} \times 100\,\% \quad (4.3)$$

While useful at system level, these metrics may overlook important properties of fault-tolerant NoC subsystems. One such property that is misrepresented by use of MTBF is the capability of a NoC medium to rapidly recover from failures. Even in the case when the number of failures is high (which indicates a low, undesired MTBF), if the recovery can be performed quickly (e.g., through flit-level recovery [21, 42]), the impact of failures may be minimal, and therefore it may not affect the application at all. For the same reason, the availability of the NoC subsystem in such condition is misrepresented by the last equation.

Another drawback of these generic metrics is that they represent average values. In the case of NoC fabrics that must meet tight quality of service (QoS) requirements in the presence of failures, the average values are not useful since the performance constraints (in terms of guaranteed latency per message or available throughput) have to be met for all possible instances, not only on an average basis.

While there is little doubt that fault tolerance is a desirable and useful property of NoCs, designers need simple, readily available, and self-sufficient metrics to be able to characterize fault tolerant methods in the context of the specific NoC implementation. We introduce these metrics relative to the NoC's ability to detect the occurrence of faults and recover from failures. The employed metrics in this work aim to address the issue of characterizing the effectiveness of fault tolerance schemes for NoC communication subsystems in the context of the specific QoS requirements that designers face in their implementation. They are not intended to substitute the existing metrics, but to complement them by offering a more detailed view of the properties of different fault-tolerance methods. An exhaustive analysis of the classic fault-tolerance metrics is outside the scope of this chapter. Instead, we choose to illustrate how the metrics defined here can help designers gain more insight on the actual performance of fault-tolerant implementations related to NoC architectures.

### *4.3.1 Countermeasures For a Fault-Tolerant System*

One of the most important challenges in designing efficient MPSoC systems affects the insurance that the communication infrastructure is available and operational despite the possibility of faults. Toward this direction, methodologies and tools have already proposed that try to derive fault tolerant systems.

There are five key elements in a comprehensive approach to fault-tolerant design, as it is depicted in Fig. 4.3: avoidance, detection, containment, isolation, and recovery. Ideally, these are implemented in a modular, hierarchical design, and encompassing an integrated combination of hardware and software techniques. Moreover, the fault-tolerant techniques can be applied at different layers from the set of ISO/OSI layers [27] that the NoC may implement, resulting in numerous possibilities for fine tuning the performance of the fault tolerance implementation by combining the (sub) set of fault tolerant elements with the (sub) set of NoC layers. As an example, error detection may be implemented in the data layer, and recovery may be realized either in the data layer (e.g., if an error correcting code is used), or at the application layer.

In a more generic approach, the partitioning and derivation of requirements, and the partitioning and implementation of fault/failure management techniques must be realized in a hierarchical fashion. For each hierarchical level, the existence of appropriate metrics allows the designers to have full control and understanding of the

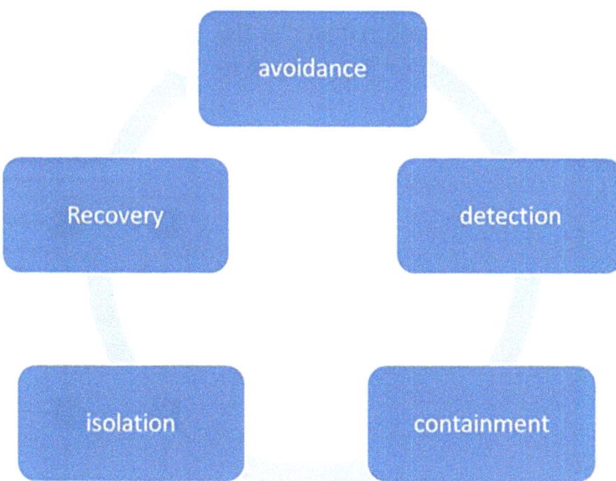


**Fig. 4.3** Countermeasures for a fault-tolerant system

implications that a particular fault tolerant implementation will have on the operation of a NoC subsystem.

As we have already mentioned, these mechanisms are applicable either at hardware, or software level. Even though the hardware-based approaches are much faster compared to the software solutions, the increased fabrication cost, as well as the additional effort required for design modifications, makes the software-based solutions most widely accepted.

## 4.3.2 Fault Tolerance at Different Layer of Abstractions

For illustrating the set of metrics discussed in this chapter and their usability, we consider a simple example of an application running on a NoC-based multiprocessing system. The application employs two processes $P_1$ and $P_2$ on two different processing cores, as shown in Fig. 4.4. Processes $P_1$ and $P_2$ use the NoC subsystem to communicate with each other along the path shaded in gray.

The communication in a NoC-based architecture is performed with a hierarchal fashion. This allows categorizing the available fault-tolerance solutions at different levels of abstraction, ranging from the physical up to the application layer. Using a layered representation of the data communication is a NoC-based system, and considering a subset of the standard OSI layers, Fig. 4.5 shows the propagation of faults from the physical level (faults affecting low-level device functionality) to the application level (faults affecting the software application running on the NoC-based system).

At each level in the hierarchy, faults can be characterized by type, source, frequency of occurrence, and impact. At the lowest level (physical), it is assumed that a fault results in a degraded operation of the respective component. If it is not always cost-effective to detect and recover at the lowest level, the fault manifests itself as a local error/failure which propagates to the next higher level, where the

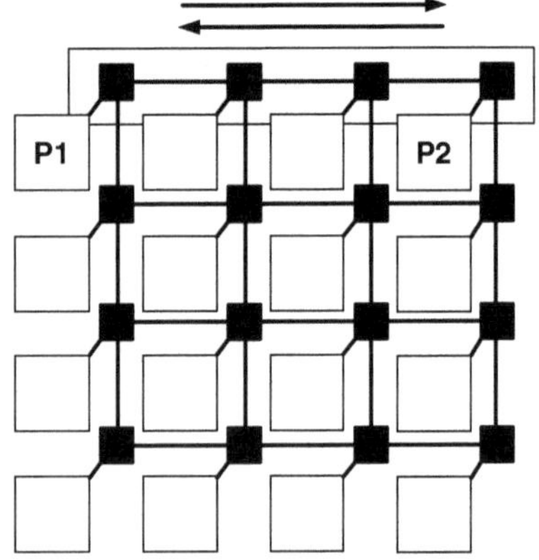


**Fig. 4.4** Processes communicating across a NoC fabric

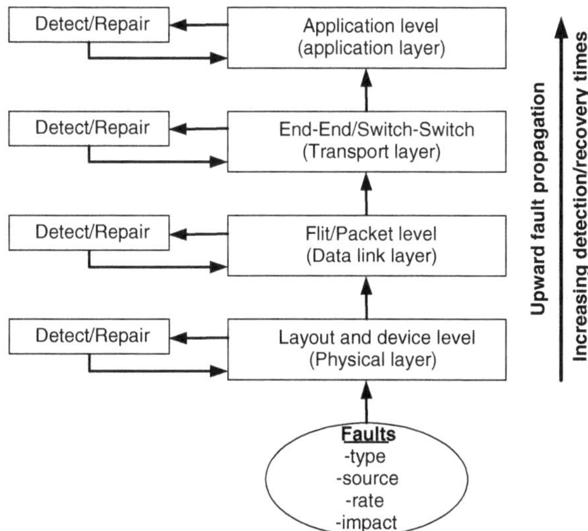


**Fig. 4.5** Fault detection and repair at different hierarchal levels

corresponding fault-tolerant technique is evaluated again relative to performance and cost-effectiveness. The hierarchical approach can provide back up at higher levels for faults which, for any reason, are not handled at lower levels. Generally, the higher the level in the hierarchy, the longer it takes to contain and/or recover from the effect of a failure, but there are certain advantages with respect to area cost and power dissipation.

For real-time NoC systems, time is the critical factor for specifying the performance of the fault-tolerant implementation. Designers must decide on the effectiveness of a particular method by knowing how quickly faults must be detected, how quickly they have to recover from the occurrence of a fault, how long an error can exist in the NoC infrastructure without impairing/compromising system performance.

From a hierarchical point of view, it is possible to apply fault detection and correction mechanisms at different level of abstraction, whereas the selection of most appropriate layer is based on the system's specifications. An optimal fault-tolerance solution should include parts from several abstraction layers. A fault should be tackled at the level where it is most cost-efficient in terms of power consumption increase and performance decline, whereas in order to have an efficient system level approach, the methods used in each layer should fit into the whole, i.e., the methods should provide means for interlayer information passing.

More specifically, in case a fault occurs at the lowest level of the hierarchy, it will affect the proper functionality of a component, as compared to a fault that occurs at the system level, where the overall system's functionality might be endangered. Additionally, since not all faults that may occur in a digital architecture are of the same importance, or there are variations in terms of required fault coverage, the incorporated fault-tolerant techniques have to be applied with different aggressiveness.

For instance, in case a fault is propagated to a higher level, it is possible to be detected with a more power efficient control circuit. However, the higher the level, the longer it takes to detect and correct a failure.

Based on the example application in Fig. 4.4 and considering a hierarchical implementation as in Fig. 4.5, we define metrics for the five elements of comprehensive fault-tolerant methods.

#### 4.3.2.1 Fault Avoidance

The goal of this task is to reduce or eliminate, if possible, the chances for an error to occur. Different mechanisms are possible to be employed toward this goal, while the majority of them are based on some form of redundancy. Unfortunately, hardware redundancy imposes performance degradation in terms of delay, power consumption and area utilization, due to additional infrastructure that imposes. Even though such a limitation might be affordable based on the system's specifications, in advance of applying any hardware-based solution targeting to fault tolerance at NoC systems, designers have to study carefully the impact of their decisions onto the system's QoS requirements. On the other hand, in case there is a demand for realizing fault avoidance at software level, the most appropriate way is by incorporating error detection and error correction codes. These approaches insert to the transmitted data either parity information or the error correcting codes. Then, in case the receiver identifies an error at the transmitted data, this error can be overcome. This is possible by requesting data retransmission, or by incorporating dedicated hardware targeting to also support error correction functionality.

Usually, fault avoidance techniques can be realized through information redundancy (by means of error correcting codes for NoCs), or hardware redundancy ($N$-modular redundancy being a typical example). Depending on the targeted number of errors to correct, coding/decoding hardware blocks may require a certain number of cycles to perform their operation. The associated time overhead adds to the total latency of the data being transported across the NoC fabric. We define $T_{av,ov}$ as the time overhead of an avoidance scheme, and compute it as the difference between data latency with ($Lat_{av}$) and without ($Lat$) fault avoidance:

$$\mathrm{T_{av,ov}} = \mathrm{Lat_{av}} - \mathrm{Lat} \tag{4.4}$$

The difference between various implementations of this concept can be significant relative to this metric. In the example in Fig. 4.4, if coding/decoding functions are implemented at each switch on the path between $P_1$ and $P_2$ (switch-to-switch avoidance), the resulting time overhead will be significantly higher than in the case where only end-to-end avoidance is implemented (i.e., data are encoded at the source and decoded at destination).

#### 4.3.2.2 Fault Detection

The next hierarchical level in a fault-tolerant design is the detection of faults that were not handled by the avoidance mechanism. Detection is built in most error-correcting codes, which generally can provide information regarding the number of un-corrected faults, when the correction mechanism fails (or is not even present in the particular implementation). Fault detection is then used to assess the need for recovery from potentially uncorrected fatal failures. The quicker the detection mechanism signals the presence of un-corrected faults, the quicker the recovery can be initiated.

The error detection at NoC-based architectures can be applied with two alternative approaches: either at switch level, or with an end-to-end approach. Specifically, the first approach assumes that error detection is performed at each switch of the NoC's architecture, in contrast to the second approach where errors are detected at the end of a path (i.e., destination node). Even though the second approach (end-to-end check) provides higher fidelity for data transmission without faults, it imposes an increased latency whenever a fault is detected, and consequently the corrupted data have to be resent. Hence, the error detection between consecutive switches is usually employed for real designs.

We define the detection latency $T_{lat}$ as the amount of time between the moment a fault occurs and the moment it is detected. Going back to our example in Fig. 4.4, fault detection may be performed by the processes $P_1$ and $P_2$ whenever data are received (end-to-end detection), at the input ports of the intermediate switches (switch-to-switch detection), or at each switch input–output port (code-disjoint detection).

#### 4.3.2.3 Containment

Fault containment is concerned with limiting the impact of a fault to a well-defined region within the NoC. Error containment refers to avoiding the propagation of the consequences of a fault, the error, out of this defined region. The definition of these regions should be carefully applied in order to ensure that there is no overlap between them. Otherwise, faults that occur at a given region will affect hardware resources assigned to the rest (overlapped) regions.

Fault containment regions (FCR) may be defined with variable resolutions, directly correlated with the quality and resolution of the fault detection mechanism. For the case of Fig. 4.4, and assuming an end-to-end detection mechanism, the fault containment region can be defined as the entire shaded route between the cores where processes $P_1$ and $P_2$ are being executed. It is essential that FCR are independent, in the sense that a fault occurring in a FCR does not affect a different FCR. In this respect, if two routes between cores/processes $P_1$ and $P_2$ can be found that are on independent FCRs, and a fault is detected on one of the routes, the other route can be used to provide an alternative path between processes $P_1$ and $P_2$.

#### 4.3.2.4 Isolation

The independency of fault containment regions can only be achieved if an effective isolation method can be provided, which can guarantee that the effect of a fault occurring in a FCR does not propagate to another FCR. At the physical layer, in the case of permanent faults, isolation can be accomplished by marking or disconnecting the faulty NoC components (links, switches, and routes) and avoiding their use until, eventually, hardware recovery/repair can be performed through reconfiguration. At higher layers, erroneous data packets can be dropped on the fly or at the destination process, such that they are not allowed to interfere with the rest of the data and propagate at application level.

#### 4.3.2.5 Recovery

The ultimate goal of fault-tolerant schemes is to provide means to recover from occurrence of failures. For fault-tolerant and QoS constrained NoCs, it is important to recover from failures within the time budget allowed by the QoS specifications. Late recoveries, even when successful, are not acceptable, since they lead to out-of-specification behavior of the NoC subsystem. Consequently, we define recovery time ($T_{\text{rec}}$) as the amount of time that passes between the detection of a fault and recovery from the corresponding failure. A simple form of attempting recovery of erroneous data flowing between processes $P_1$ and $P_2$ is to provide a hardware correction at lower layers, or a retransmission mechanism at higher layers where, upon detection of an error, an automated retransmission request (ARQ) is generated and an error-free copy of original data is resent from the source process.

## 4.4 Error Control Coding for On-chip Signaling

Error control coding (ECC) is a prominent fault-tolerance method for on-chip signaling. In this paragraph, we describe the terminology and principles of coding together with promising coding approaches targeting to on-chip signaling.

In error control coding, a data word of length $k$ is encoded to form a code-word of length $n$, where $n \geq k$. The added $n - k$ bits are called check bits and in a separable, also called systematic, code they are appended to the end of the data word to form the code word. Separable codes are beneficial to be used in on-chip signaling, since the data bits are directly accessible in the code word. More specifically, rather than many data networks and wireless transmissions, where the common form of transmission is serial, in on-chip signaling it is typical to provide parallel transmissions. Specifically, in on-chip signaling a set of parallel wires usually forms a link, thus the check bits mean additional wires. However, such an architectural selection enables a code word at a time to be transmitted over the link.

The efficiency of a code can be measured by its error detection and/or correction capability and the relation of data word and code-word widths, the code rate $R = \frac{k}{n}$. The code distance $d$ is the minimum number of distinct bits in two code words. In other words, the code distance corresponds to the minimum number of bits that must be flipped to get another code word. The code's error detection capability can be calculated from the distance by $t_d = d - 1$, whereas the corresponding error correction capability is defined as $t_c = \lfloor \frac{d-1}{2} \rfloor$. If a single code is used to correct some number of errors and detect some more errors, the relation to distance is $d = 2t_c + t_d + 1$. Since the on-chip links are commonly formed by a set of parallel wires, the code rate indicates the number of additional wires required for providing the desired functionality. The lower the code rate, the more additional wires are required, which in turn leads to area and power overheads.

A typical error that occurs in on-chip networks affects the burst errors, which have impact on multiple adjacent bits in a code word. In parallel transmissions, a burst error therefore means an error affecting two or more adjacent wires. This kind of error situation can be a result of many phenomena in a chip (e.g., different kinds of couplings between wires are likely to cause burst errors). An efficient way to cope with burst errors is by interleaving. This approach requires partitioning the data word into parts and encoding each of them separately. Then, the final code word is formed by taking one bit at a time from each encoded part.

For instance, if we name the inputs as $i_{0,\dots,k-1}$, the check bits as $c_{0,\dots,n-k-1}$ and there are three interleaving sections, the coding proceeds as follows: the check bits $c_0, c_3, c_6, \dots$ are calculated from inputs $i_0, i_3, i_6, \dots$, the check bits $c_1, c_4, c_7, \dots$ are calculated from inputs $i_1, i_4, i_7, \dots$, and the check bits $c_2, c_5, c_8, \dots$ are calculated from inputs $i_2, i_5, i_8, \dots$. Since the number of interleaving sections in this example is three, the obtained code has burst error detection capability $t_d^{\text{burst}} = 3 \times t_d$ and correction capability $t_c^{\text{burst}} = 3 \times t_c$, where $t_d$ and $t_c$ are the error detection and correction capabilities of the underlying code, respectively. We have to mention that interleaving in parallel transmission links practically imposes reordering of the wires, which usually is a very cost-efficient approach. Therefore, such a coding scheme suits very well for on-chip realizations.

The interleaving affects mainly the burst error tolerance but it also has an effect on tolerance against multiple single errors affecting the same code word. Multiple single errors can be corrected if they affect separate interleaving sections.

### *4.4.1 Linear Block Codes*

Linear block codes are a well-known and widely used set of codes. The name indicates that a set of bits, a block, or a data word, is encoded at a time. A code word is always a linear combination of the other code words. The encoding and decoding of these codes is in most cases straightforward and fast, which makes them well suited for on-chip realizations.

Linear block codes can be presented by their generator matrix G, which is of size $k \times n$, where $k$ is the data word length and $n$ the code word length. The encoding of a linear block code can be done with matrix multiplication $\vec{c} = \vec{d}\,G$, where $\vec{d}$ is a data word vector of length $k$, $G$ is the $k \times n$ generator matrix and $\vec{c}$ is a code word vector of length $n$. The code is systematic if the generator matrix contains an $k \times k$ identity matrix, i.e., the code word includes the data word unaltered.

Hardware realizations of systematic (separable) code encoders can be simplified, because only $n - k$ code word symbols have to be calculated and the rest are just the symbols of the data word. Binary matrix multiplication is simply calculating parity bits, which in hardware are implemented with trees of XOR gates.

Syndrome decoding is a common technique for decoding linear block codes. A syndrome can be considered to contain information on the errors of a transmitted code word in a compressed form. It is calculated by matrix multiplication $\vec{s} = \vec{u}\,H^T$. At this equation, the $\vec{u}$ is a received code word vector of length $n$ ($\vec{u} = \vec{c} + \vec{e}$, where $\vec{c}$ is a transmitted code word and $\vec{e}$ an error vector, both of length $n$), $H^T$ is the transpose of the $(n - k) \times n$ parity check matrix and $\vec{s}$ is a syndrome vector of length $n - k$. The parity check matrix can be constructed from the generator matrix and vice versa. The syndrome gives the index to the table of minimum weight error vectors, so the error vector $\vec{e}$ can be easily determined. The correction is done by $\vec{c} = \vec{u} + \vec{e}$, eliminating the error from the received data word.

### *4.4.2 Hamming Codes*

The most common linear block codes are the Hamming codes. A Hamming code fulfills the rule $2^{(n-k)} \geq n + 1$, where $n - k$ is the number of check bits and $n$ is the length of the code word. The minimum distance of a Hamming code is 3, so it can correct a single error in each code word ($t_c = 1$), or it can be used to detect double errors ($t_d = 2$). The Hamming code is also described as using the concept of overlapping parity, where there are multiple parity bits and every data bit is involved in calculating several of them.

The overlapping parity concept is a direct result of the matrix multiplication encoding presented above. A modified Hamming code for both correcting single errors and detecting double errors can be achieved by adding one more check bit, which is used as the parity bit of the whole code word [33, 37]. The method of adding one more parity bit to each code word is called extending and it can be applied to other linear block codes as well. The opposite approach to extending is called shortening, where a number of data bits is set always to zero, and therefore can be left out of the code word. Thus, shortening with a factor of $s$ results in a code which has data word length $k - s$ and code word length $n - s$. Shortening decreases the rate since $\frac{k-s}{n-s} < \frac{k}{n}$, for $s > 0$. We have to stress that shortening is extensively used in on-chip error control coding, because a bus width seldom directly matches the data word width of the wanted code.

Hamming codes are the most widely used codes in the research on interconnect link error protection [72]. For instance, in [39], parity and Hamming coding have been used in adaptive error detection system which is built to obtain a more energy efficient design without affecting the error detection capabilities. The system monitors the noise level of the transmission channel and dynamically changes to a code that has better error detection capabilities in the case of an increased noise level, and respectively changes to a weaker code when the noise level is lower. In the design, parity, Hamming double error detection, and extended Hamming triple error detection codes are used. A set of codes similar to Hamming codes are the Hsiao codes. They differ from Hamming codes in the way the generator and parity check matrices are constructed [37].

### 4.4.3 Cyclic Codes

Cyclic codes are a set of codes, where a cyclic shift of a code word generates another code word. Because of this property, efficient realizations of these codes can be achieved using linear feedback shift registers (LFSR). In the standard form this creates nonseparable codes, but the codes can also be made separable by small changes in the generation process [33]. Cyclic codes are normally presented with their generator polynomial $g(x)$ instead of the generator matrix. The generator polynomial is of degree $n - k$ and the encoding is proceed by $c(x) = a(x)g(x)$, where $a(x)$ is the data word and $c(x)$ the code word, both in their polynomial forms.

A class of cyclic codes, the cyclic redundancy checks codes (CRC) are often used for detecting errors. These codes are able to detect single errors and adjacent multiple errors, which make them extremely suitable for detecting burst errors. The number of adjacent errors that can be detected is $n-k-1$, where $k$ is the number of data bits and the coded word contains $n$ bits. Toward this scope, a generator polynomial of degree $n - k$ is used. For instance, CRC-8 (8 for the degree of the generator polynomial) is used in a self-calibrating design to detect errors on the transmission channel, where self-calibration means that the voltage swing for the transmission channel is scaled dynamically to obtain minimum energy consumption [66].

### 4.4.4 BCH Codes

As the probability for multiple errors increases when scaling further into the nanometer regime, error correcting codes capable of correcting several errors are needed. Popular linear block codes for multiple error correction are the Bose-Chaudhuri-Hocquenghem (BCH) codes [63], which are also cyclic codes. They can be easily constructed according to the specifications for correcting as many errors as they are required. The BCH code is similar to Hamming code when used as a single error correcting code. Actually, the Hamming codes are BCH codes with $t_c = 1$

(whenever the $t_c \geq 2$, the codes are called as BCH). In on-chip signaling, where only two logic states are possible, only binary BCH codes are of interest.

The syndrome decoding method explained above is limited by the size of the minimum error vector table. When the number of check bits $n - k$ is high the table is impractically large, and therefore alternative decoding methods should be used. The decoding of BCH codes can be done using the Berlekamp-Massey (B-M) algorithm [6]. This algorithm is iterative requiring $2t_c$ iterations, where $t_c$ is the error correction capability of the code. The calculations are done in Galois Field (GF) $2m$, where $m$ depends on the length of the code ($2m \geq n$). In addition to the actual algorithm, a preprocessing circuit is also needed. Using Fourier transform in GF($2m$) $2t_c$ syndromes is calculated. For binary BCH codes, this imposes quite similar trees of XOR gates in hardware, similar to the syndrome calculation explained above. The error vector $\overrightarrow{e}$ is extracted by using a method called Chien search to find the zeros of the error-locator polynomial $\Lambda(x)$ obtained from the B-M algorithm [6].

## *4.4.5 Reed-Solomon Codes*

The Reed-Solomon codes are other cyclic codes that can be used to correct multiple errors. The codes are nonbinary, which means that instead of bits, groups of $m$ bits (e.g., $m$=8, a byte) are used as symbols for the codes. The Reed-Solomon codes are optimal meaning that they provide the maximum distance at the used number of check symbols. If a word contains $k$ groups of data and its length is $n$ groups, at most $\lfloor \frac{(n-k)}{2} \rfloor$ errors can be corrected. Because on-chip signals are binary, each symbol must be coded by binary bits. These codes are especially effective in correcting burst errors, since the detection and correction is based on symbols which consist of many adjacent bits. On the other hand, their effectiveness in single error tolerance is limited, because a single bit fault in a binary-coded symbol takes the whole correction capability of that symbol. Binary codes, such as binary BCH, provide the same tolerance against single faults with a lower number of check bits [6].

Decoding of Reed-Solomon codes is also based on the Berlekamp-Massey algorithm. The main difference is that in addition to the error vector $\overrightarrow{e}$ also the error values are needed to perform the error correction. The error vector $\overrightarrow{e}$ points the erroneous GF($2m$) symbol and the actual correction is done by adding the error value of that particular symbol and the transmitted symbol itself. The error values can be extracted using the Forney algorithm with inputs obtained from the B-M algorithm. Also the Fourier transform requires slight changes compared to the one used for BCH decoding. In Reed-Solomon decoding, all the calculations are done with GF($2m$) symbols while for binary BCH codes the syndrome calculation is just calculating parities of different sets of bits [6].

A special feature of Reed-Solomon decoding is that it can be constructed, so that it provides information on correctness of the decoding. This kind of information could be very important in safety-critical designs, where it is better to stop the system than giving a false result.

## 4.5 Power and Energy Savings in NOCs

Power and energy savings in NOCs is another challenge that needs to be considered toward the goal of designing efficient interconnection structures for MPSoC architectures. Power savings obtained by only scaling down supply voltage levels are not going to be sufficient to compensate for a higher complexity, a larger interconnect capacitance and resistance, a higher operating frequency and an increased gate leakage [28].

Up to now, there are approaches targeting to provide energy management of NOCs mainly focused on controlling the power consumption of interconnects (network-centric approach) [68], or managing power of the cores (node-centric approach). Additional effort was paid in order to support both of these goals simultaneously [54]. This paragraph discusses at different layers of abstraction a number of well-established design approaches targeting to alleviate the impact of reliability degradation at NoC-based systems by providing power and energy savings.

### *4.5.1 Physical Layer*

At the physical layer, low-swing signaling is actively investigated to reduce communication energy on global interconnects [71]. Specifically, in the case of a simple CMOS driver, low-swing signaling is achieved by lowering the driver's supply voltage $V_{dd}$. This implies a quadratic dynamic power reduction, since the $P_{\text{dynamic}} \propto V_{dd}^2$. Unfortunately, swing reduction at the transmitter complicates the receiver's design.

In order to guarantee reliable data reception at NoC platforms, it is required increased sensitivity and noise immunity. Toward this goal, the usage of differential receivers has been proposed, which exhibit superior sensitivity and robustness. However, these receivers impose the doubling of the bus width. As an improvement of this overhead, pseudo-differential schemes have been proposed, where a reference signal is shared among several bus lines and receivers, and incoming data are compared against the reference in each receiver. Even though the usage of pseudo-differential signaling enables the reduction of the number of signal transitions; however, it has reduced noise margins with respect to fully differential signaling. Consequently, reduced switching activity is counterbalanced by higher swings and determining the minimum-energy solution requires careful circuit-level analysis.

Another key physical-layer issue affects the synchronization. Traditional on-chip communication schemes are based on the synchronous assumption. This selection implies the presence of global synchronization signals (i.e., global clocks) that define data sampling instants throughout the chip. Even though the usage of clocks improves considerably the design-time for performing system synchronization, they are extremely energy inefficient, whereas it is a well-known fact that a significant fraction of the total power budget in digital integrated systems is consumed by clock networks. Thus, postulating global synchronization when designing on-chip networks is not an optimal choice from the energy viewpoint. Alternative on-chip synchronization protocols that do not require the presence of a global clock have been proposed in the past [4, 25] but their effectiveness has not been studied in detail from the energy viewpoint.

## 4.5.2 Data-Link Layer

At the data-link layer, a key challenge is to achieve the specified communication reliability and/or fault coverage level with minimum mitigation cost in terms of power/energy consumption. Toward this goal, a number of alternative error recovery mechanisms can be applied to NoCs, but their energy efficiency should be carefully assessed in this context. For instance, we can consider two reliability-enhancement techniques: (i) error-correcting codes and (ii) error-detecting codes with retransmission. Both approaches are based on transmitting redundant information over the data link, but the error-correction imposes additional hardware redundancy and computational complexity. Consequently, it is expected that error-correcting transmission demands additional power consumption, as compared to the error-free case. Apart from this flexibility, in case an error arises, systems that incorporate error detecting schemes require that corrupted data will be retransmitted. Hence, depending on the network architecture, retransmission can be very costly in terms of energy (and performance).

This highlights that whenever an energy-efficient NoC has to be designed, there should be a careful trade-off between the increased cost of error correction and the energy penalty due to data retransmission. Either of these schemes may be optimal, depending on system constraints and on physical channel characteristics. Since such kind of decisions impose mentionable overheads in execution run time (they have to find a viable solution at a large search space), the usage of automatic design space exploration tools is almost a prerequest.

In case of shared-medium network links (such as busses), the media-access-control function of the data link layer is also critical for energy efficiency. Toward this goal, it is widely accepted the usage of arbitration (also known as time-division multiplexing) schemes [3, 16, 64]. These approaches assume that a single arbiter decides which transmitter accesses to the bus for every time slot. Unfortunately, such a centralized arbitration exhibits a limited scalability, which indicates that this approach is likely to be energy inefficient as the size of NoC scales up. As an improvement to this limitation, there are available approaches dealing with distributed arbitration schemes, as well as alternative multiplexing approaches, such as code division multiplexing [69].

### 4.5.3 Network Layer

The architecture of network infrastructure heavily influences the communication energy. As it was already mentioned in previous chapters, the shared-medium networks (e.g., busses) are inefficient in terms of energy dissipation, whereas this problem became far more savage as the network size scales up [22]. This mainly occurs because the data transmission at a bus-based communication scheme is performed by broadcasting from one transmitter to all possible receivers, ignoring if the message usually is destined to only one receiver, or a small group. In addition to that, the bus contention, with the related arbitration overhead further contributes to the energy overhead.

Preliminary studies on energy-efficient on-chip communication indicate that hierarchical and heterogeneous architectures are much more energy efficient than busses [25]. In their work, Zhang et al. [25] evaluate a hierarchical mesh network, where nodes with high communication bandwidth requirement are clustered and connected through a programmable generalized mesh consisting of several short communication channels joined by programmable switches. Clusters are then connected through a generalized mesh of global long communication channel.

Apart from network architecture, the network control is also critical for supporting efficient data communication with a NoC. For this purpose, careful study should be applied for selecting the best candidate switching schemes for indirect network architecture. From the energy viewpoint, the tradeoff is between the cost of setting up a circuit-switched connection once and for all, and the overhead of switching packets throughout the entire communication time on a packet-based connection. At the first case, the control overhead is constant and it is applied once (at the setup time), whereas in the latter case, it is distributed over many small contributions, one for each on-chip router.

In case, the communication flow between network nodes is extremely persistent and stationary, circuit-switched schemes are likely to be preferable. In contrast, if we have to design a network for an irregular architecture, or for an architecture with nonstationary communication pattern, then the packet switched scheme leads to better performance and energy metrics. Needless to say, circuit switching and packet switching are just two extremes of a spectrum, with many hybrid solutions in between [61].

### 4.5.4 Transport Layer

The goal of transport layer is concerned with optimizing the usage of network resources and providing a requested quality of service. An example of transport-layer design issue that affects the energy efficiency of a NoC is the choice between connection-oriented and connectionless protocols. In particular, the connection-oriented protocols can become energy inefficient under heavy traffic conditions, as they tend to increase the number of retransmissions. On the other hand, out-of-order delivery of data may imply additional work at the receiver, which causes additional energy consumption. This highlights the balance that should be applied between the communication energy and the computation energy at destination nodes.

Apart from the selected communication protocols, the flow control also contributes significantly to the energy consumption. More specifically, if many transmitters compete for limited communication resources, the network becomes congested and the cost per transmitted bit increases due to the increased contention and contention resolution overhead. The flow control can alleviate the consequences of this problem by regulating the amount of data that enters the network. Even though such an approach mitigates the congestion problem, it introduces a penalty at architecture's throughput.

## 4.6 On Designing Reliable NoCs

This section describes in more detail a number of selections aiming to alleviate the consequences of increased power and temperature values at reliability degradation. Toward this goal, we discuss some existing design methodologies for supporting efficient application realization onto NoC-based platforms under reliability constraints.

### *4.6.1 Reliability Improvement Through Mapping Algorithms*

One of the popular architecture level fault-tolerance techniques is to use redundancy-based design [36, 67]. In such a system, critical design components are replicated and results are voted to produce the output. However, replication techniques like DMR, TMR etc., come with high area penalty. Stringent cost budget is increasingly prohibiting the use of redundancy-based designs for MPSoCs.

A cost-effective solution for fault-tolerance involves migration of tasks from faulty cores. The decisions about task mapping and/or scheduling can be precomputed at design-time, or can be done at run-time. Accordingly, task mapping can be categorized as static, dynamic, and quasi-static. More specifically, static task mapping involves analysis at design-time to maximize system reliability [18, 38] but does not address task migration. On the other hand, dynamic approaches monitor system-status and decide on task migration at run-time to minimize migration overhead [17, 30, 58], or balance processor load [70]. However, the limitation of this approach affects the throughput, which is not always guaranteed. Moreover, migration algorithms need to be simple in order to minimize as much as possible the computation. Finally, quasi-static task migration techniques compute task mapping decisions at design-time for different fault scenarios [10, 12, 32]. As faults occur, these mappings are looked up at run-time to carry out task migration. The advantage of this technique is that any sophisticated algorithm can be used at design-time despite the associated mapping storage overhead.

A fixed order Band and Band reconfiguration technique is studied in [12]. Based on this approach, the cores of target architecture are partitioned into two bands. When one or more cores become faulty, tasks on these core(s) are migrated to other functional core(s) determined by the band in which the tasks belong. The core partitioning strategy is fixed at design-time and is independent of the application throughput requirement. Consequently, throughput cannot be guaranteed by this technique.

A re-execution slot based-reconfiguration mechanism is studied in [32]. Normal and re-execution slots of a task are scheduled at design-time using evolutionary algorithm to minimize certain parameters like throughput degradation. At run-time, tasks on a faulty core migrate to their re-execution slot on a different core. However, schedule length can become unbounded for high fault-tolerance systems. Moreover, analysis is based on task graphs and therefore cannot be applied to streaming applications with cyclic task dependencies.

Task re-mapping technique based on off-line computation and virtual mapping is proposed in [10]. Here, task mapping is performed in two steps—determining the highest throughput mapping followed by generation of a virtual mapping to minimize the cost of task migration to achieve this highest throughput mapping. These virtual mappings are computed at design-time based on different fault scenarios. A limitation of this technique is that the migration overhead significantly increases as this is not considered in the initial optimization process. Moreover, throughput constrained streaming applications do not benefit from a throughput higher than required and can increase buffer requirements at output.

### 4.6.2 Reliability Improvement Through Routing Algorithms

The software level can also improve reliability related issues of NoC architectures. The prevention of faults is mostly important for adaptive routing algorithms that are sensitive to topology changes. For instance, a faulty switch or link will degrade the topology into an irregular one and then the algorithms will fail. A candidate solution to this is the usage of agnostic routing algorithms (i.e. LASH [55], TOR [50], LASH-TOR [56], DL [52], multiple virtual networks [13], up*/down* [34], lturn [35], smart-routing [51] and FX [19]) in combination with static reconfiguration. In such a static fault-model, the network enters a reconfiguration phase when a fault is discovered, the system is halted and drained of packets, then all routing tables are recomputed. In case it is not possible to halt the network, then a possible solution is the usage of a dynamic fault model. Whenever a fault occurs in this approach, the faulty link or switch will be marked and another path configured. A limitation of methods based on reconfiguration is the appropriately handling the reconfiguration time in order to avoid possible packet loss, deadlock, and live-lock.

Apart from these, fault-tolerant routing algorithms were proposed in [53], while [41, 46] describes a routing algorithm that aims to decrease the probability of system failures. A technique that disables certain nodes for packet processing but not for packet routing is presented in [23]. Even though such a technique can provide acceptable results; however, it may disable a significant number of healthy processing elements. Another solution is to alter the routing tables in case of failure in order to adapt them to the new topology after the failure [9]. This technique is extremely flexible but it might result to significant performance degradation. In [20], a design methodology for enabling fault tolerant routing for mesh and tori topologies under a static fault model is discussed. The idea behind this fault tolerant approach is based on using intermediate nodes for routing. More specifically, for some source-destination pairs, packets are first forwarded to an intermediate node, and then from this node to the destination node, without being ejected. This approach allows the use of fully adaptive routing in the absence of failures, does not sacrifice any healthy nodes, and only requires the use of one additional virtual channel (note that two virtual channels are already required to provide fully adaptive routing [47]).

## Questions

**4.1** Why faults due to reliability degradation becoming so important in current process technologies?

**4.2** What are the main drawbacks of existing fault tolerant techniques?

**4.3** Do you think that total power consumption is more important for device architects compared to the power density?

**4.4** Why thermal stress, and consequently aging phenomena, is more crucial in 3-D NoCs, as compared to the conventional 2-D NoCs?

**4.5** Can you describe the "Bit Energy" model?

**4.6** Can you describe the energy model based on "packets" and "hops"?

**4.7** What are the main reasons for wear-out phenomenon?

**4.8** Mention the most important reliability degradation mechanisms for IC. Is there any common parameter for these mechanisms that highly affect reliability degradation?

**4.9** What is the fault, error and failure?

**4.10** Describe the three classical fault models for digital circuits and busses.

**4.11** What are the differences between transient, intermittent and permanent faults?

**4.12** What is a fault tolerant system?

**4.13** Is it possible to "transform" a conventional system to fault tolerant? If yes, please mention how is this possible.

**4.14** What are the advantages and disadvantages of applying fault tolerant techniques to software and hardware level?

**4.15** Can you define the Mean Time Between Failure (MTBF) rate?

**4.16** Can you define the Mean Time To Repair (MTTR) rate?

**4.17** How is it possible to define the availability of a system?

**4.18** Can you enumerate and describe the five key elements in a comprehensive approach to fault-tolerant design?

**4.19** How is it possible to apply fault tolerance at different layers of abstraction? Give an example.

**4.20** Where is applicable the error control coding?

**4.21** How is it possible to apply power and energy savings at different layers of abstraction of NoC-based architectures?

**4.22** Is it feasible to improve reliability of NoC-based architectures through careful mapping? If yes, please describe at least two different approaches.

**4.23** Is it feasible to improve reliability of NoC-based architectures by employing different routing algorithm? If yes, please describe at least two different approaches.

## References

1. J. Abraham, A. Krishnamachary, R. Tupuri, A comprehensive fault model for deep submicron digital circuits, in Internationl Workshop on Electronic Design, Test and Applications, 2002, pp. 360–364
2. R. Aitken, Nanometer technology effects on fault models for IC testing. IEEE Comput. **32**(11), 46–51 (1999)
3. P. Aldworth, System-on-a-chip bus architecture for embedded applications, in International Conference on, Computer Design, 1999, pp. 297–298
4. W. Bainbridge, S. Furber, Delay insensitive system-on-chip interconnect using 1-of-4 data encoding, in International Symposium on Asynchronus Circuits and Systems, 2001, pp. 118–126
5. R. Baumann, Soft errors in advanced computer systems. IEEE Des. Test Comput. **22**(3), 258–266 (2005)
6. R. Blahut, *Algebraic Codes for Data Transmission* (Cambridge University Press, Cambridge, 2003)
7. S. Borkar, Designing reliable systems from unreliable components: the challenges of transistor variability and degradation. IEEE Micro 25(6), 10–16 (2005)
8. D. Brooks, M. Martonosi, Dynamic thermal management for high-performance microprocessors, in International Symposium on High-Performance Computer, Architecture, 2001, pp. 171–182
9. R. Casado, A. Bermudez, J. Duato, F. Quiles, J. Sanchez, A protocol for deadlock-free dynamic reconfiguration in high-speed local area networks. IEEE Trans. Parallel Distrib. Syst. **12**(2), 115–132 (2001)

10. L. Chanhee, K. Hokeun Kim, P. Hae-woo, K. Sungchan, O. Hyunok Oh, H. Soonhoi, A task remapping technique for reliable multi-core embedded systems, in International Conference on Hardware/Software Codesign and System, Synthesis, 2010, pp. 307–316
11. L.H. Chee, W, Daasch, G. Cai, A thermal-aware superscalar microprocessor, in International Symposium on Quality, Electronic Design, 2002, pp. 517–522
12. Y. Chengmo, A. Orailoglu, Predictable execution adaptivity through embedding dynamic reconfigurability into static MPSoC schedules, in International Conference on Hardware/Software Codesign and System, Synthesis, 2007, pp. 15–20
13. L. Cherkasova, V. Kotov, T. Rokicki, Fibre channel fabrics: evaluation and design, in International Conference on System Sciences, 1996, pp. 53–62
14. J. Cong, Z. Yan Zhang, Thermal via planning for 3-D ICs, in International Conference on, Computer-Aided Design, 2005, pp. 745–752
15. C. Constantinescu, Trends and challenges in VLSI circuit reliability. IEEE Micro **23**(4), 14–19 (2003)
16. B. Cordan, An efficient bus architecture for system-on-chip design, in Custom Integrated Circuits, 1999, pp. 623–626
17. O. Derin, D. Kabakci, L. Fiorin, Online task remapping strategies for fault-tolerant Network-on-Chip multiprocessors, in International Symposium on Networks on Chip, 2011, pp. 129–136
18. A. Dogan, F. Ozguner, Matching and scheduling algorithms for minimizing execution time and failure probability of applications in heterogeneous computing. IEEE Trans. Parallel Distrib. Syst. **13**(3), 308–323 (2002)
19. J. Flich, P. Lopez, C. Sancho, A. Robles, J. Duato, Improving infiniBand routing through multiple virtual networks, in International Symposium on High, Performance Computing, 2002, pp. 49–63
20. M. Gomez, J. Duato, J. Flich, P. Lopez, A. Robles, N. Nordbotten, O. Lysne, T. Skeie, An efficient fault-tolerant routing methodology for meshes and tori. Comput. Archit. Lett. **3**(1), 3 (2004)
21. C. Grecu, A. Ivanov, R. Saleh, E. Sogomonyan, P. Partha Pratim, On-line fault detection and location for NoC interconnects, in International On-Line Testing, Symposium, 2006, p. 6
22. P. Guerrier, A. Greiner, A generic architecture for on-chip packet-switched interconnections, in Design, Automation and Test in Europe Conference and Exhibition, 2000, pp. 250–256
23. C. Ho, L. Stockmeyer, A new approach to fault-tolerant wormhole routing for mesh-connected parallel computers. IEEE Trans. Comput. **53**(4), 427–438 (2004)
24. http://www.xilinx.com/ise/optional_prod/tmrtool.htm
25. Z. Hui, W. Marlene, V. George, J. Rabaey, Interconnect architecture exploration for low-energy reconfigurable single-chip DSPs, in IEEE Workshop On VLSI, 1999, pp. 2–8
26. W. Hung, C. Addo-Quaye, T. Theocharides, Y. Xie, N. Vijakrishnan, M. Irwin, Thermal-aware IP virtualization and placement for networks-on-chip architecture, in International Conference on, Computer Design, 2004, pp. 430–437
27. International Standards Organization, Open Systems Interconnection (OSI) Standard 35.100 (available at http://www.iso.org)
28. ITRS 2011 (availabe at http://www.itrs.net)
29. A. Iwata, M. Sasaki, T. Kikkawa, S. Kameda, H. Ando, K. Kimoto, D. Arizono, H. Sunami, A 3D integration scheme utilizing wireless interconnections for implementing hyper brains, in International Solid-State Circuits Conference, 2005, pp. 262–597
30. V. Izosimov, P. Pop, P. Eles, Z. Peng, Design optimization of time- and cost-constrained fault-tolerant distributed embedded systems, in Design, Automation and Test in, Europe, 2005, pp. 864–869
31. S. Jayanth, S. Adve, B. Pradip, J. Rivers, Lifetime reliability: toward an architectural solution. IEEE Micro **25**(3), 70–80 (2005)
32. H. Jia, J. Blech, A. Raabe, C. Buckl, A. Knoll, Analysis and optimization of fault-tolerant task scheduling on multiprocessor embedded systems, in International Conference on Hardware/Software Codesign and System, Synthesis, 2011, pp. 247–256

33. B. Johnson, *Design and Analysis of Fault-Tolerant Digital Systems* (Addison-Wesley, MA, 1989)
34. M. Koibuchi, A. Funahashi, A. Jouraku, H. Amano, L-turn routing: an adaptive routing in irregular networks, in International Conference on Parallel Processing, 2001, pp. 383–392
35. M. Koibuchi, A. Jouraku, K. Watanabe, H. Amano, Descending layers routing: a deadlock-free deterministic routing using virtual channels in system area networks with irregular topologies, in International Conference on Parallel Processing, 2003, pp. 527–536
36. I. Koren, C. Krishna, *Fault-tolerant systems* (Morgan Kaufmann, CA, 2007)
37. P. Lala, *Self-Checking and Fault-Tolerant Digital Design* (Morgan Kaufmann Publishers, CA, 2001)
38. H. Lin, Y. Feng, X. Qiang, Lifetime reliability-aware task allocation and scheduling for MPSoC platforms, in Design, Automation and Test in Europe Conference and Exhibition, 2009, pp. 51–56
39. L. Lin, N. Vijaykrishnan, M. Kandemir, M. Irwin, Adaptive error protection for energy efficiency, in International Conference on, Computer Aided Design, 2003, pp. 2–7
40. A. Maheshwari, W. Burleson, R. Tessier, Trading off transient fault tolerance and power consumption in deep submicron VLSI circuits. IEEE Trans. Very Large Scale Integr. VLSI Syst. **12**(3), 299–311 (2004)
41. S. Murali, M. Coenen, A. Radulescu, K. Goossens, G. De Micheli, A methodology for mapping multiple use-cases onto networks on chips, in Design, Automation and Test in, Europe, 2006, pp. 1–6
42. S. Murali, T. Theocharides, N. Vijaykrishnan, M. Irwin, L. Benini, G. De Micheli, Analysis of error recovery schemes for networks on chips. IEEE Des. Test Comput. **22**(5), 434–442 (2005)
43. E. Nilsson, J. Oberg, PANACEA - a case study on the PANACEA NoC - a nostrum network on chip prototype, Royal Institute of Technology, Tech. Report. 229, 2006
44. P. Partha Pratim, C. Grecu, M. Jones, A. Ivanov, R. Saleh, Performance evaluation and design trade-offs for network-on-chip interconnect architectures. IEEE Trans. Comput. **54**(8), 1025–1040 (2005)
45. C. Patel, S. Chai, S. Yalamanchili, D. Schimmel, Power constrained design of multiprocessor interconnection networks, in International Conference on, Computer Design, 1997, pp. 408–416
46. M. Pirretti, G. Link, R. Brooks, N. Vijaykrishnan, M. Kandemir, M. Irwin, Fault tolerant algorithms for network-on-chip interconnect, in IEEE Annual Symposium on VLSI, 2004, pp. 46–51
47. V. Puente, R. Beivide, J. Gregorio, J. Prellezo, J. Duato, C. Izu, Adaptive bubble router: a design to improve performance in torus networks, in International Conference on Parallel Processing, 1999, pp. 58–67
48. J.M. Rabaey, *Low Power Design Essentials, Series on Integrated Circuits and Systems* (Springer, New York, 2009)
49. F. Ridruejo, J. Miguel-Alonso, INSEE: an interconnection network simulation and evaluation environment. Euro-Par Parallel Process. **3648**, 1014–1023 (2005)
50. J. Sancho, A. Robles, J. Flich, P. Lopez, J. Duato, Effective methodology for deadlock-free minimal routing in infiniBand networks, in International Conference on Parallel Processing, 2002, pp. 409–418
51. M. Schroeder, A. Birrell, M. Burrows, H. Murray, R. Needham, T. Rodeheffer, E. Satterthwaite, C. Thacker, Autonet: a high-speed, self-configuring local area network using point-to-point links. IEEE J. Sel. Areas Commun. **9**(8), 1318–1335 (1991)
52. R. Seifert, *Gigabit Ethernet* (Addison-Wesley, MA, 1998). ISBN 0-201-18553-9
53. N. Shanbhag, A mathematical basis for power-reduction in digital VLSI systems. IEEE Trans. Circuits Syst. II Analog Digital SSignal Proc. 44(11), 935–951 (1997)
54. T. Simunic, S. Boyd, P. Glynn, Managing power consumption in networks on chips. IEEE Trans. Very Large Scale Integr. VLSI Syst. **12**(1), 96–107 (2004)

55. K. Skadron, M. Stan, W. Huang, V. Sivakumar, S. Karthik, D. Tarjan, Temperature-aware microarchitecture, in International Symposium on Computer, Architecture, 2003, pp. 2–13
56. T. Skeie, O. Lysne, I. Theiss, Layered shortest path (LASH) routing in irregular system area networks, in International Symposium on Parallel and Distributed Processing, Symposium, 2002, pp. 162–169
57. J. Smolens, B. Gold, J. Hoe, B. Falsafi, K. Mai, Detecting emerging wearout faults, in Workshop on Silicon Errors in Logic - System Effects, 2007
58. T. Streichert, C. Strengert, C. Haubelt, J. Teich, Dynamic task binding for hardware/software reconfigurable networks, in Symposium on Integrated circuits and systems design, 2006, pp. 38–43
59. I. Sungjun, K. Banerjee, Full chip thermal analysis of planar (2-D) and vertically integrated (3-D) high performance ICs, in International Electron Devices Meeting, 2000, pp. 727–730
60. S. Tosun, N. Mansouri, E. Arvas, M. Kandemir, X. Yuan, Reliability-centric high-level synthesis, in Design, Automation and Test in, Europe, 2005, pp. 1258–1263
61. J. Walrand, P. Varaiya, *High-Performance Communication Networks* (Morgan Kaufman, CA, 2000)
62. H. Wei, K. Sankaranarayanan, K. Skadron, R. Ribando, M. Stan, Accurate, pre-RTL temperature-aware design using a parameterized, geometric thermal model. IEEE Trans. Comput. **57**(9), 1277–1288 (2008)
63. R. Wells, *Applied Coding and Information Theory for Engineers* (Prentice Hall, Inc., NJ, 1999)
64. S. Winegarden, A bus architecture centric configurable processor system, in Custom Integrated Circuits, 1999, pp. 627–630
65. D. Wingard, MicroNetwork-based integration for SOCs, in Design Automation Conference, 2001, pp. 673–677
66. F. Worm, P. Ienne, P. Thiran, G. De Micheli, A robust self-calibrating transmission scheme for on-chip networks. IEEE Trans. Very Large Scale Integr. VLSI Syst. **13**(1), 126–139 (2005)
67. Y. Xie, L. Li, M. Kandemir, N. Vijaykrishnan, M. Irwin, Reliability-aware co-synthesis for embedded systems, in International Conference on Application-Specific Systems, Architectures and Processors, 2004, pp. 41–50
68. T. Ye, L. Benini, G. De Micheli, Analysis of power consumption on switch fabrics in network routers, in Design Automation Conference, 2002, pp. 524–529
69. R. Yoshimura, K. Tan Boon, T. Ogawa, S. Hatanaka, T. Matsuoka, K. Taniguchi, DS-CDMA wired bus with simple interconnection topology for parallel processing system LSIs, in International Solid-State Circuits Conference, 2000, pp. 370–371
70. Z. Yuping, H. Zimian, X. Xianbin, Z. Wuqing, W. Zhuowei, Workload-balancing schedule with adaptive architecture of MPSoCs for fault tolerance, in International Conference on Biomedical Engineering and Informatics, 2010, pp. 2775–2779
71. H. Zhang, V. George, J. Rabaey, Low-swing on-chip signaling techniques: effectiveness and robustness. IEEE Trans. Very Large Scale Integr. VLSI Syst. **8**(3), 264–272 (2000)
72. H. Zimmer, A. Jantsch, A fault model notation and error-control scheme for switch-to-switch buses in a network-on-chip, in InternationAl Conference on Hardware/Software Codesign and System, Synthesis, 2003, pp. 188–193

# Chapter 5
# NoC-Based System Integration

**Abstract** Integrating the NoC subsystem with the core in an MPSoC is hardly trivial. It requires the design of complex NIs as well as tackle the problem of clock distribution, which requires appropriate synchronizers if the system clock cannot be distributed without skew as is often the case. The synchronizers must be reliable and yet minimize additional latency. Moreover, careful floor-planning is required to achieve good performance. Finally, any multi- and many-core environment must also efficiently solve the problem of cache coherence and therefore the NoC must support and even facilitate cache coherence mechanisms.

## 5.1 NoC Interface Design

The Network Interface (NI) is the logic required to connect the nodes to the NoC. Unlike routers which can be similar or even identical, NIs can differ significantly depending on the nature of the node, which can be a processor, memory, custom logic dedicated hardware etc. Using a NI allows IPs and communication infrastructure to be designed independently [22, 29]. One end of a NI is connected to a router using the selected flow control protocol (see Chap. 3) while the other end is connected to the node IP. Since most IPs are designed to communicate through a bus, the NI uses a bus interface. There is a number of standard bus specifications to this end, such as OCP (Open Core Protocol) [27], VCI (Virtual Component Interface) [34], the AMBA family of buses such as AMBA AHB [2], AMBA AXI (Advanced eXtensible Interface) [3] for performance and low power, AMBA ACE for full cache coherency between processors and AMBA ACE-Lite for I/O coherency as well as custom buses [33].

However, the NI is not merely a protocol adapter from a processor bus to a router port. It must provide services at the transport layer in the ISO-OSI reference model [29], because this is the first layer where offered services are independent of the network implementation. This is critical in order to decouple communication from computation [22, 30], allowing independent design and reuse of node IP

K. Tatas et al., *Designing 2D and 3D Network-on-Chip Architectures*,
DOI: 10.1007/978-1-4614-4274-5_5, © Springer Science+Business Media New York 2014

modules and communication infrastructure [19]. The NI must offer processing cores the view of a shared memory system, and the network itself should be transparent.

The services that the NI must provide can be classified as adaptation services, transaction reordering services, error and flow control services, route computation services and upper layer services.

Adaptation services include packetization/depacketization, protocol conversion and clock domain crossing. They are the absolute minimum services required of the NI so that data can be sent and received through the NoC subsystem.

Transaction reordering services are needed in the case protocols allow out-of-order reception of packets or even flits. The NI must place flits and/or packets in proper order before forwarding them to the node. This requires increased buffering resources and incurs a latency overhead.

Error and flow control services are used in the case of an unreliable medium. On the receive path the NIs must perform error detection and/or correction and request retransmission when required. On the transmit path, the NI must append error control information. Furthermore, the NI must be able to handle cases when the flow control protocol indicates that injected packets cannot be accepted by the router due to congestion.

Route computation is required in the case of source routing. Unlike distributed routing where the NI must only append the packet destination address, in source routing the NI must also calculate the route according to the selected routing algorithm and append the routing information to the packet header.

Upper layer services include cache coherence, bus transaction conversion to messages, etc. They can be implemented in software, but there is increasing tendency of migrating them to NIs for increased performance.

Supporting all these services can lead to large area NIs. In Steenhof et al. [31], a TV companion chip was redesigned with a NoC as the interconnect fabric, and 78 % of increase in chip area was proved to come from the NIs. Furthermore, the NoC-based version featured an increase of 4 % in total area compared to the original implementation. In the xpipes based system in [1], more than half of the NoC area is due to NIs. After extensive research focusing mostly on router design and optimization, attention has turned to NI design fairly recently. In order to save area, NI sharing among nodes was proposed in [14] and [4]. The basic concept is illustrated in Fig. 5.1. An arbiter is inserted between the PEs and the common NI, appropriately multiplexing transmit data and demultiplexing receive data.

Other aspects of NI design have also been addressed such as latency [23] and power consumption [8].

### *5.1.1 Message, Packet and Flit Format*

The most essential function of the NI is packetization before transmission and de-packetization after reception. Before NI design can begin message types, packet and flit format and size must be determined according to the system requirements.

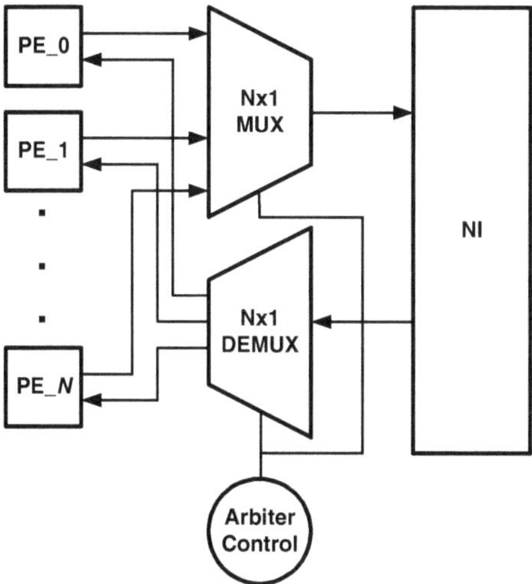


**Fig. 5.1** NI sharing between PEs

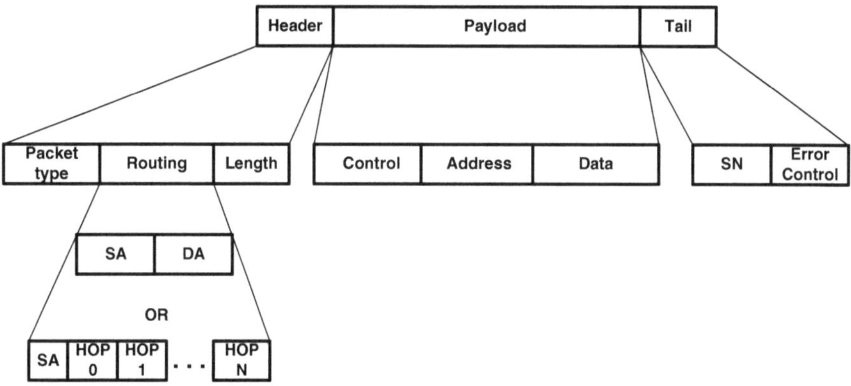


**Fig. 5.2** Typical NoC Packet Format

A typical packet format is shown in Fig. 5.2. It is generally composed of three parts, namely *header*, *payload* and *tail*. The header contains the necessary routing and network control information. In the case of distributed routing the information required is the destination and source addresses while in the case of source routing the complete routing information is written in a *routing field*. In the case of variable packet size a length field is also common.

The payload data comes from the IP. It includes a control field, which encapsulates control information such as bursts, and the data field which contains the actual address and/or data.

The tail typically includes the sequence number and error control fields such as hamming code or CRC fields are appended.

Studies on packet and flit size show that generally, increasing packet size reduces the cache miss rate, since more cache locations can be updated with fewer packets [12]. On the other hand, it increases miss penalty since more data must be fetched from the main memory in order to update the cache block [12].

In general the types of packets in an an MPSoC are [12]:

1. *Memory access request packet*: The packet is induced by an L2 cache miss that requests data fetch from memories. The header of this packet must contain the destination address of the target memory (node ID and memory address) as well as the type of memory operation requested (memory READ, for example). The address of the L2 cache is in the header as well, as it is needed to construct the data fetch packet (in case of a READ). Since there is no data being transported, the payload is empty.
2. *Data fetch packet*: This is the reply packet from memory, containing the requested data. The packet header contains the destination address of the packet (in this case the node ID of the cache that requested the data). The data is contained in the packet payload.
3. *Data update packet*: This packet contains the data that will be written back to the memory. It comes from L2 cache that requests the memory write operation. The header of the packet contains the destination memory address, and the payload contains the data.
4. *Cache coherence synchronization packet*: This type of packet is induced by the cache coherence operation from the memory. Packets come from the updated memory, and are sent to all caches, each cache will then update its content if it contains a copy of the data. The packet header contains the memory tag and block address of the data. If the synchronization uses the "update" method, the packet contains updated data as payload. If the synchronization uses the "invalidate" method, the packet header contains the operation type (INVALIDATE, in this case), and the payload is empty.
5. *I/O and interrupt packet*: This packet is used by I/O operations or interrupt operations. The header contains the destination address or node ID. If data exchange is involved, the payload contains the data.

Messages roughly correspond to bus transactions such as request and response. For example the data in a burst bus transaction together with the bus control signals can be packed into a message. (See AXI overview in subsection 5.1.2).

## 5.1.2 Basic Network Interface Design

Nodes in MPSoCs can be broadly classified as master and slave IPs. Master IPs such as processor cores generate request transactions and receive responses, while slave IPs such as memories receive requests and send back appropriate responses. Master IPs are connected to *initiator NIs*. Initiator NIs convert IP request transactions such

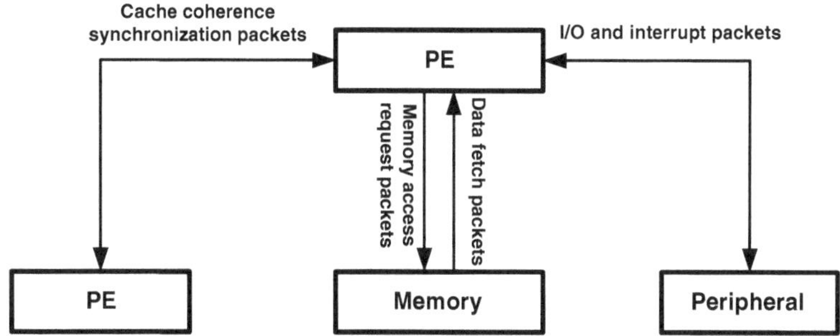


**Fig. 5.3** Relationship between nodes and packets

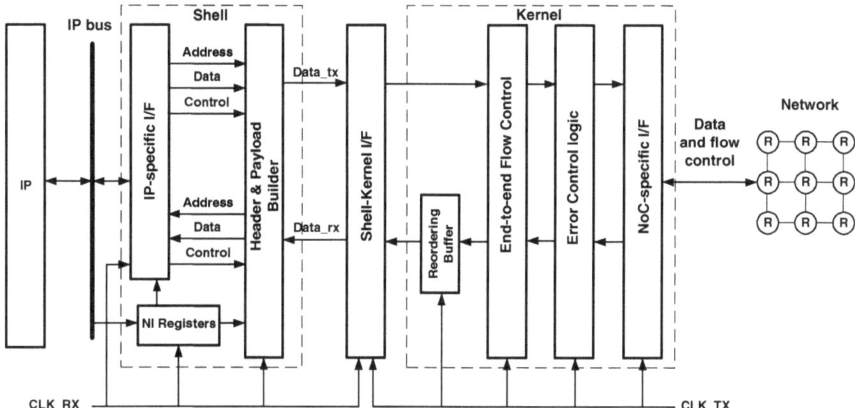


**Fig. 5.4** NI block diagram

as memory fetches into NoC traffic, and translate the packets received from the NoC into IP response transactions. On the other hand, *target NIs* are connected to slave IPs and must translate requests from the NoC and generate responses to the NoC. More specifically the relationships between nodes, protocols and packets is shown in Fig. 5.3.

The basic block diagram of a NI is shown in Fig. 5.4. For both initiator and target NIs two domains can be distinguished "vertically" namely the *Shell* which is IP-specific, and the *Kernel* which is NoC-specific, in order to allow independent design and reuse of IPs and routers. The kernel is responsible for packetization/depacketization, clock domain crossing, flow and error control. The shell is responsible for transaction reordering and higher-layer services. It is possible for multiple shells to share a single NI kernel [14].

Similarly, the NI can be "horizontally" split into a transmit and a receive path. The transmit path has the local node as source and the network as destination, while the receive path has the network as source and the local node as destination.

Along the transmit path, the IP-specific interface receives data from the node IP, splits it into packets and flits appropriately, creates the packet header appending the destination address or routing information, and sends packets to the shell-kernel interface. The kernel reads pending packets, calculates and appends error control information and sends it to the NoC-specific interface which implements the flow control protocol to send data to the attached port of the local router.

On the receive path, the NI accepts a packet after exchanging flow control information, reorders packets if necessary, checks the packet for errors, strips header and tail information and forwards the payload to the node according to the local bus protocol.

IP-Specific Interface

The IP-specific interface is responsible for exchanging data with the IP. This usually means generating and responding to bus transactions according to a specific protocol. It is commonly implemented as a wrapper in the HDL file hierarchy. As already mentioned, common standard bus protocols are OCP and the AMBA family. They are reviewed here briefly in order to discuss certain considerations in the NI design.

#### 5.1.2.1 OCP Overview

OCP features three different burst models, namely precise burst, imprecise burst and single request multiple data burst. Specifically, in a precise burst the burst length is known when the burst is sent. Each data word is transferred as a normal single transfer, where the address and command are given for each data-word, which has been written or read. In an imprecise burst the burst length can change within the transaction. A signal called MBurstLength shows estimation on the remaining data words that will be transferred. Each data word is transferred as in the precise burst model, with the command and address sent for every data word. Finally, in a single request multiple data burst the command and address fields are only sent once. That is in the beginning of the transaction. This means that the destination core must be able to reconstruct the whole address sequence, based on the first address and a MBurstSeq signal.

#### 5.1.2.2 AMBA AXI Overview

AXI is the third generation of AMBA interface, first defined in the AMBA three specification. In the latest AMBA four specification [3] it features three types of interfaces:

1. AXI4 for high-performance memory-mapped requirements.
2. AXI4-Lite for simple, low-throughput memory-mapped communication (for example, to and from control and status registers).
3. AXI4-Stream for high-speed streaming data.

The AXI and AXI-Lite standards specify a five-channel interface. A read address and command channel (ARADDR), a read data channel (RDATA), a write address and command channel (AWADDR), a write data channel (WDATA) and a write response channel (BRESP)

AXI4 allows a burst transaction of up to 256 data transfers, while AXI4-Lite allows only 1 data transfer per transaction (bursting is not supported).

The AXI4-Stream protocol defines a single channel for transmission of streaming data. The AXI4-Stream channel is modeled after the Write Data channel of the AXI4. Unlike AXI4, AXI4-Stream interfaces can burst an unlimited amount of data. There are additional, optional capabilities described in the AXI4-Stream Protocol Specification. The specification describes how AXI4-Stream-compliant interfaces can be split, merged, interleaved, upsized, and downsized. Unlike AXI4, AXI4-Stream transfers cannot be reordered. The AXI bus interface has been adopted by FPGA vendors such as Xilinx [18].

Supporting protocols with long and even dynamic bursts such as OCP and AXI4-stream, the NI requires large FIFOs, especially the receive FIFO which may have to support packet reordering.

#### 5.1.2.3 AMBA ACE Overview

The AMBA four ACE protocol [3] aims at supporting cache coherence (see subsection 5.1.3.2. It is based on a five-state cache model. Each cache line is either Valid or Invalid, meaning it either contains cached data or not. If it's Valid then it can be in one of four states defined by two properties. Either the line is Unique or Shared, meaning it's either non-shared data, or potentially shared data (the address space is marked as shared). And either the line is Clean or Dirty, generally meaning either memory contains the latest, most-up-to-date data and the cache line is merely a copy of memory, or if it's Dirty then the cache line is the latest most up-to-date data and it must be written back to memory at some stage. The one exception to the above description is when multiple caches share a line and it's dirty. In this case, all caches must contain the latest data value at all times, but only one may be in the SharedDirty state, the others being held in the SharedClean state. The SharedDirty state is thus used to indicate which cache has responsibility for writing the data back to memory, and SharedClean is more accurately described as meaning data is shared but there is no need to write it back to memory.

The ACE states can be mapped directly onto the MOESI cache coherency model states. However ACE is designed to support components that use a variety of internal cache state models, including MESI, MOESI, MEI and others.

ACE adds three more channels to the AXI bus standard for coherency transactions and also adds some additional signals to existing channels. The ACADDR channel is a snoop address input to the master. The CRRESP channel is used by the master to signal the response to snoops to the interconnect. The CDATA channel is output from the master to transfer snoop data to the originating master and/or external memory. The ARTERIS FlexNoC [24] supports the AMBA ACE interface.

#### 5.1.2.4 Packetizer/depacketizer

Packetization includes converting a bus transaction into message(s), splitting a message into packets as well as creating and appending the header and tail fields of each packet. The basic block diagram of the packetizer/depacketizer is shown in Fig. 5.5.

One of the most important functions of the NI is to calculate the network address or node identifier when sending a packet from the bus transaction. In the case of source routing the NI also calculates the exact route and encodes the information for each hop in the header. To the software running on a processor node, the memory is seen as a single contiguous address range starting at a specific offset. In a NoC environment, the physical memory may be distributed to several nodes. When a program requests data from a specific physical address, the NI must identify the network address of the physical memory as well as the address within that memory.

The header calculator must translate the address requested by a processor to memory transaction (read or write) to the network address of the memory node, which must be included in the header. In the case of source routing the entire routing path is included, while in the case of distributed routing the destination address is sufficient. The payload includes the memory physical address, the data to be written in the case of a memory write operation and additional control information such as burst mode or byte enable. In the case of out-of-order reception of packets or flits, a sequence number must be included. The tail includes the error control information that must be calculated in the destination NI and checked. In the case of end-to-end error control the transmitter NI must keep a copy of the sent packet and delete it only upon reception of an acknowledge message from the receiver.

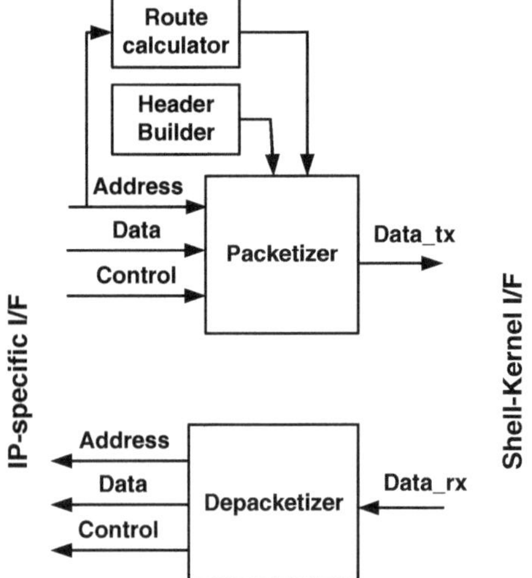


**Fig. 5.5** Packetizer/depacketizer diagram

Depacketization requires checking the incoming packet for errors, reordering packets if necessary, stripping the header and tail fields, and assembling messages from packets. Finally, the message must be converted to a valid bus transaction.

Alternatively, packetization/depacketization can be done in software. The authors in Bhojwani and Mahapatra [5] studied the impact of the packetization implementation on area, latency, complexity, and flexibility. Three packetization schemes were examined: (1) in software and (2) in hardware on core, and (3) in a wrapper. Software implementation results in high latencies and an increase of code size. On-core hardware packetization has a 13 K gates area overhead and a moderate latency of 10.8 ns. A hardware wrapper implementation has the lowest area overhead, equal to 4 K gates, and the lowest latency of 3.02 ns.

#### 5.1.2.5 Shell-Kernel Interface

The interface between the shell and the kernel is typically one or more dual-port FIFOs for each path. Since commonly the shell and kernel are in different clock domains, the FIFO, besides buffering, is used also for reliable clock domain crossing [15]. For that purpose the each pointer used for reading/writing must be read by the other clock domain as well. To prevent metastability problems two flip-flops are typically used for each pointer and the pointers often count in Gray code [25] so that only one bit flips at every pointer increment. This dual port FIFO synchronization scheme is illustrated in Fig. 5.6. When designing a shell-kernel interface with separate clock domains it should be a separate module in the design hierarchy according to synthesis guidelines [21]. This applies to any other possible clock domain crossing points. Furthermore, the flip-flops used for synchronization of the pointers should be placed physically close to reduce the possibility of metastability occurring in the second flip-flop [15].

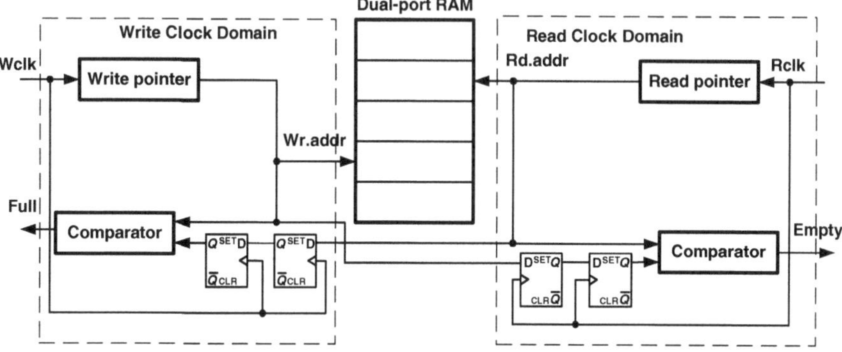


**Fig. 5.6** Synchronization FIFO

#### 5.1.2.6 NoC-Specific Interface

The NoC-specific interface is an implementation of the flow control protocol to transmit to (inject) and receive data (eject) from the router attached to the NI as discussed in Chap. 3.

### *5.1.3 Advanced Network Interface Features*

Another feature commonly required of NIs is support for error control. With devices becoming increasingly unreliable, error detection and/or correction is featured in modern NoCs while it was not considered in original NoC implementations. On the transmit path, the NI must calculate and append the error control information to a packet, while on the receive side the NI must calculate the error control information, compare it to the information received and detect or correct errors accordingly.

#### 5.1.3.1 End-to-End Flow Control

End-to-end flow control is desirable in order to regulate traffic*globally*, while link-level flow control regulates traffic*locally* between routers as examined in Chap. 3. Similarly to link-level flow control, end-to-end flow control ensures that no data is sent from a source NI unless there is enough space in the destination NI buffer to store it. it. Not using an end-to-end flow control scheme would lead either to dropping packets at the destination or en route due to congestion. However, using exclusively end-to-end flow control with no link-level flow control is not feasible.

End-to-end flow control can be implemented using credits similarly to credit-based flow control discussed in Chap. 3. The source NI uses a credit counter to track the empty buffer space of the remote destination queue for each channel which is initialized with the remote buffer size. When data is sent from the source queue, the counter is decremented. When data is consumed by the IP module at the other side, credits are produced in a counter in the destination NI (credit) to indicate that more empty space is available. These credits must also be sent to the producer of data to be added to its credit counter. This approach was used in [17] and is illustrated in Fig. 5.7.

In Tang and Lin [32], a flow control strategy aiming at regulating the sending packet rates using several injection levels according to network status (modelled as the number of data flows sharing the channel)and not just destination NI buffer space is presented. The proposed method requires obtaining the network status for the whole source to destination path and transmitting it from the destination to the source (instead of credits) using a dedicated control network.

In Lai et al. [23] instead of constantly tracking the empty buffer size in the target NI by the source NI, the target NI periodically sends a control packet that reports the available space to the producer of data. The source does not transmit pending data if

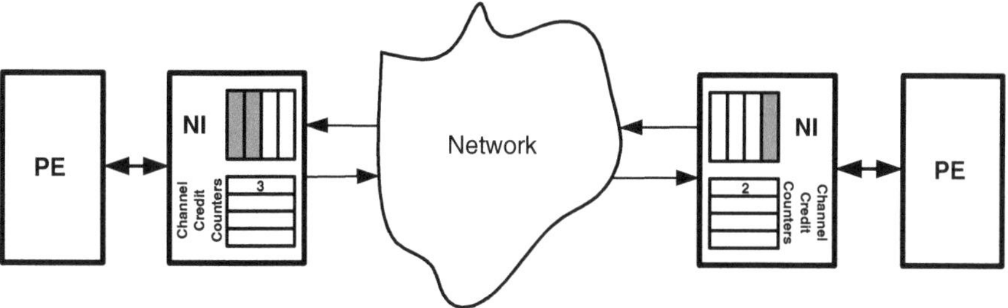


**Fig. 5.7** End-to-end flow control

the available buffer size received from the destination is insufficient to store the data pending transmission.

#### 5.1.3.2 Cache Coherence

An important consideration for all multi-core architectures is cache coherence. The cache coherency problem occurs when different processors in the SoC share data stored in the main memory. When processor P1 modifies data shared with processor P2, the copy of that data in the cache of processor P2 must be either updated or invalidated, before a new access to it.

Maintaining cache coherence in a multi-processor environment at all times can be difficult. Two basic cache coherence approaches exist, namely snoop-based and directory-based. The former relies on processors "snooping" on each other's address requests, while the latter relies on maintaining a common directory of shared data. Both approaches require additional states for the cache line than simply valid and invalid. A popular cache coherence protocol for write-back caches called MESI was originally proposed in Papamarcos and Patel [28]. It replaces the valid/dirty bit in a cache line with four possible states that the data in the block can have in an MPSoC:

1. *Modified*: The cache has the only valid copy that is in the whole system. The data which are in the main memory are invalid (out-of-date). A write-back operation will change this state to Exclusive.
2. *Exclusive*: The cache has the only valid copy of the block, but it has not been modified. The data in the main memory are valid. A read operation from another processor will change the state to Shared.
3. *Shared*: Another processor can have the data into its cache memory and both copies are updated.
4. *Invalid*: The data in the cache is not valid. Either the data the processor requests are not in the cache (miss), or the local copy of these data is not valid because another processor has updated the corresponding memory position.

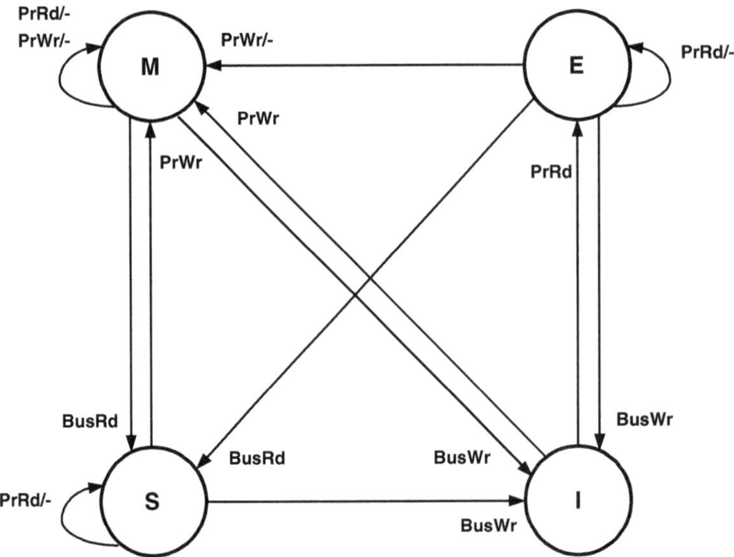


**Fig. 5.8** MESI cache coherence protocol

Depending on a write/read hit or miss and the state of the data in the cache (modified, exclusive, shared or invalid) the state of the data must be updated according to the state diagram of Fig. 5.8 and the appropriate action must be taken.

Variations of the MESI protocol also exist, such as the MOESI cache coherence protocol which adds an additional state to the MESI protocol called the "owned" state, which signifies that the processor is responsible for updating the main memory. Only one processor may have data in the owned state.

Implementing a snoop-based cache coherence protocol is a significant challenge in the NoC environment compared to bus-based communication which allows snooping between processors. Furthermore, transactions such as invalidate and update are naturally multicast, which many NoCs handle as unicast transactions. In Chaves et al. [7] the authors proposed the use of multicast messages to reduce the number of transactions to improve the performance of cache coherence protocols in NoC-based MPSoCs. Results show that performance of some transactions is improved up to 32 % when using multicast messages.

The exploration presented in Girao et al. [16] shows that the amount of data on the NoC for regular operations is much larger than the amount of data for cache coherence maintenance for almost all the cache sizes. However, as the increase in cache size leads to a decrease in the amount of data for regular operations, the amount of data for cache coherence becomes more significant. Further exploration shows that for small cache sizes, the amount of page replacement requests is the most responsible for the cache coherence injected load due to the amount of cache misses and replacements which increases as cache size decreases [16].

In Eisley et al. [13] implementation of the coherence protocol and directories within the network at each router node was proposed. Moving the coherence directories from the nodes to the routers opens up the possibility of optimizing a protocol with in-transit actions. Virtual trees, one for each cache line, are maintained within the network in place of coherence directories to keep track of sharers The virtual tree consists of one root node R which is the node that first loads a cache line from off-chip memory, all nodes that sharing this line and intermediate nodes between root and sharers Nodes of the tree are connected by virtual links Virtual trees are stored in virtual tree caches at each router within the network Reads and writes are routed towards the home node, if they encounter a virtual tree in-transit, the virtual tree takes over as the routing function and steers read requests and write invalidates appropriately towards the sharers instead.

A similar approach where, all directory caches are embedded within the router was proposed in [9].

In Walter et al. [36] a bus-enhanced NoC for supporting bus coherence was proposed, based on the concept of using a NoC for high-throughput communication of data and a bus for the low-latency, low-bandwidth communication of control information between pairs of modules such as acknowledgements and invalidates.

## 5.2 Clock Distribution

Ideally, it would be desirable to have a completely *synchronous* NoC, i.e. all routers clocked by the same clock, with no phase difference (skew) between routers. Unfortunately, one of the main problems in today's ASICs is to distribute efficiently a skew-free synchronous clock over the whole chip, while a large part of the power (about 70 %) is also consumed by the clock-tree [10].

The usage of fully *asynchronous* NoC implementations would eliminate the clock distribution concern and would make designs more modular since timing assumptions are explicit in the hand-shaking protocols. However, such an approach is not attractive yet, since the existing design tools and IP libraries heavily rely on the synchronous paradigm instead. As a result, middle ground solutions become more affordable in the short run. The proliferation of clock domains in a single chip imposes the design of reliable synchronizers to cross clock domains.

### *5.2.1 Globally Asynchronous Locally Synchronous (GALS)*

This approach is based on the fact that while it may not be feasible to distribute a global clock signal with acceptable skew, it is still possible to have a clock within a limited area (or region). A GALS architecture consists of multiple regions, each with the same clock, called *clock domains*, asynchronous to one another [20]. Usually, the GALS approach in NoC architectures is implemented by assuming that each switch/router is clocked with a different clock, all the same frequency, but not with the same phase. Passing data from one router to the other requires an appropriate

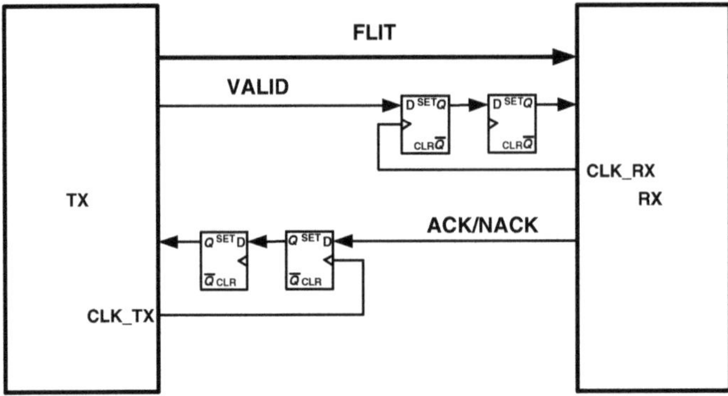


**Fig. 5.9** Flow control with two flip-flop synchronizer

synchronizer in order to avoid metastability. Instead of using the FIFO synchronizer shown in Fig. 5.6, if the clock frequency is the same, a simple two-flip-flop synchronizer of Fig. 5.9 can be used reliably [15] together with an ACK/NACK control flow scheme. Unfortunately, even this synhchronizer will introduce a two-cycle latency.

## *5.2.2 Mesochronous*

An improvement to synchronous communication scheme is the usage of mesochronous synchronization. Based on this approach, a single clock signal can be distributed to the various macro-cells in the design with an arbitrary amount of space-dependent time-invariant phase offset (i.e., the skew). Due to the fact that NoCs intricacies of skew-controlled chip-wide clock tree distribution, there is an increasing interest for mesochronous synchronizers. Consequently, the mesochronous synchronization can be viewed not just as an enabler for architecture scalability, but also as a means of eliminating (or relaxing) the skew constraints in the clock tree synthesis process, thus resulting in frequency speed-ups, power consumption reductions and fast back-end turnarounds [35].

Unfortunately, mesochronous synchronizers come with their own set of problems. For instance, an implementation of mesochronous communication is based on delaying either data or the clock signal so to sample data only when it is stable. However, the implementation of such a solution requires components often not available in a standard-cell design flow (e.g., configurable digital delay lines) or explicitly targeting full-custom design (e.g., Muller C-elements).

In addition to that, sometimes the synchronization latency, which impacts not just communication performance but has architecture-level implications as well, becomes a serious problem. In order to overcome this limitation a latency-intensive receiver architecture has to be designed, resulting in a more costly system overall.

A common mesochronous synchronizer is presented in Ginosar [15] and shown in Fig. 5.10. Alternatively, a FIFO synchronizer can always be used.

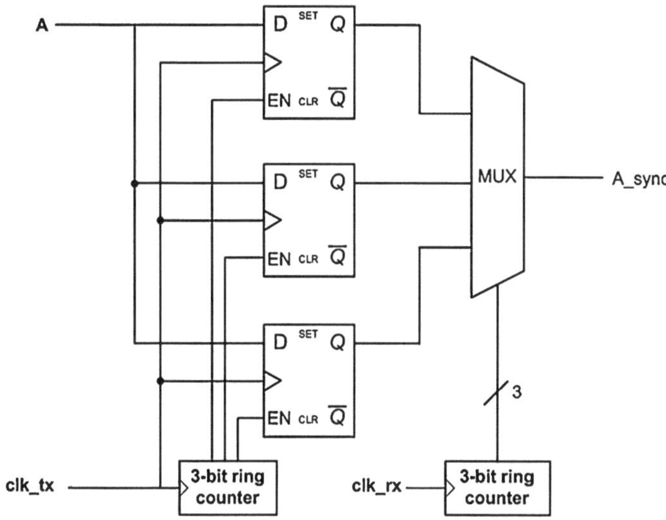


**Fig. 5.10** Mesochronous synchronizer

### *5.2.3 Multisynchronous*

Another consideration concerning mesochronous synchronizers is that often mesochronous clock domains can become *multisynchronous* due to temperature and voltage changes. This means that the relative phases of the two clocks drift slowly, and therefore the phase difference is no longer constant. Obviously, the general two-clock FIFO approach is valid in this case also. Alternatively, a mesonchronous synchronizer must be modified to monitor and adapt to the clock phase change [15].

### *5.2.4 Asynchronous*

An asynchronous NoC would completely eliminate clock skew considerations. Such an approach is the Mango clockless NoC [6]. However, since the IPs are synchronous, the NI must perform reliable crossing from the synchronous to the asynchronous domain and vice-versa. Another consideration is the fact that, as mentioned, EDA tools don't fully support asynchronous design practices.

## 5.3 NoC-based System Floorplanning

Floorplanning a NoC-based system must be done efficiently in order to reap the benefits of network communication. Otherwise, long wires (links) may lead to unacceptable delays. In systems with few cores, floorplanning can be done manually. However, as the number of nodes rapidly increases, EDA tools are required to achieve a good solution. Generic placement and routing tools that consider the NoC merely another component in the system may lead to poor results.

Traditional NoC design methodology that consists of generation of a symbolic mesh, followed by system-level floor-planning to generate the final SoC architecture. Since the symbolic mesh is generated without any knowledge of floor-planning, the final architecture can result in large link length and corresponding link power consumption. For the example shown in the figure, mapping and routing were performed by assuming a 4mm inter-router distance, and there was a deviation of about 50 % in link power consumption, and 80 % in wire length between the symbolic mesh, and the mesh after floor-planning. With further shrinking of technology, and increased contribution of link power to the overall power consumption, this deviation will result in a significant error between the estimated, and actual power consumption of the NoC.

One approach is to use floor-plan information in the topology synthesis phase. In Murali et al. [26] This paper obtains a more accurate estimation of the link lengths by integrating system-level floor-planning in the NoC design flow. A similar approach is presented in Yu et al. [37], where In this paper, we integrate partition into floor-planning to make use of physical information such as the length of interconnects among cores. At post-floor-planning optimization, a heuristic method is used to insert switches and a min-cost max-flow algorithm is used to insert network interfaces. Finally, we allocate paths to minimize power consumption.

In 3-D NoCs, a 3-D 2-layer and 3-layer NoC architectures that utilize homogeneous regular mesh networks on a separate layer and one or two heterogeneous floor-planning layers was proposed in de Paulo and Ababei [11].

## Questions

**5.1** What are the absolute minimum functions a NI must perform?

**5.2** Name the types of packets in a NoC-based MPSoC.

**5.3** What are the typical fields and sub-fields in a NoC packet?

**5.4** Name the typical NI shell services.

**5.5** Name the typical NI kernel services.

**5.6** What are the relative advantages and disadvantages of mesochronous and GALS architectures?

**5.7** Which types of clocking schemes are used in NoCs? What are their relative advantages and disadvantages?

## Laboratory Exercises

**5.8** Design a wrapper between a NI and an OCP bus in VHDL or Verilog.

**5.9** Design a wrapper between a NI and an ARM AMBA AXI bus in VHDL or Verilog.

**5.10** Design the FIFO synchronizer of Fig. 5.4.

**5.11** Design the mesochronous synchronizer of Fig. 5.6.

**5.12** Design a packetizer for an AMBA AXI burst node in a $4 \times 4$ mesh with distributed routing. The tail should use parity for error detection.

**5.13** Design a packetizer for an OCP burst node in a $4 \times 4$ mesh with distributed routing. The tail should use CRC-32 for error detection.

**5.14** Design an appropriate depacketizer for laboratory exercise 5.10.

**5.15** Likewise for laboratory exercise 5.11.

## References

1. F. Angiolini, P. Meloni, S. Carta, L. Benini, L. Raffo, Contrasting a NoC and a traditional interconnect fabric with layout awareness, in *Design, Automation and Test in Europe*, 2006, pp. 1–6
2. ARM, AMBA AHB Protocol Specification, version 2.0 ARM, 1999 (available at: http://infocenter.arm.com/help/index.jsp?topic=/com.arm.doc.ihi0011a/index.html)
3. ARM, Amba AXI and ACE protocol specification (available at: http://infocenter.arm.com/help/index.jsp?topic=/com.arm.doc.ihi0022d/index.html)
4. B. Attia, W. Chouchene, A. Zitouni, R. Tourki, Network interface sharing for SoCs based NoC, in *International Conference on Communications, Computing and Control Applications*, 2011, pp. 1–6
5. P. Bhojwani, R. Mahapatra, Interfacing cores with on-chip packet-switched networks, in *International Conference on, VLSI Design*, 2003, pp. 382–387
6. T. Bjerregaard, J. Sparso, Implementation of guaranteed services in the MANGO clockless network-on-chip. IEE Proc. Comput. Digital Tech. **153**(4), 217–229 (2006)
7. T. Chaves, E. Carara, F. Moraes, Exploiting multicast messages in cache-coherence protocols for NoC-based MPSoCs, in *International Workshop on Reconfigurable Communication-centric Systems-on-Chip*, 2011, pp. 1–6
8. W. Chouchene, B. Attia, A. Zitouni, N. Abid, R. Tourki, A low power network interface for network on chip, in *International Multi-Conference on Systems, Signals and Devices*, 2011, pp. 1–6
9. K. Daewook, K. Manho, G. Sobelman, DCOS: cache embedded switch architecture for distributed shared memory multiprocessor SoCs, in *International Symposium on Circuits and Systems*, 2006, p. 4
10. W. Dally, J. Poulton, *Digital Systems Engineering* (Cambridge university press, New York, 1998)
11. V. de Paulo,C. Ababei, A framework for 2.5D NoC exploration using homogeneous networks over heterogeneous floorplans, in *International Conference on Reconfigurable Computing and FPGAs*, 2009, pp. 267–272

12. G. DeMicheli, L. Benini, *Networks on Chips: Technology and Tools* (Morgan Kaufmann, Waltham, 2006)
13. N. Eisley, P. Li-Shiuan and S. Li, In-network cache coherence, in *International Symposium on Microarchitecture*, 2006, pp. 321–332
14. A. Ferrante, S. Medardoni and D. Bertozzi, Network interface sharing techniques for area optimized NoC architectures, in *EUROMICRO Conference on Digital System Design Architectures, Methods and Tools*, 2008, pp. 10–17
15. R. Ginosar, IEEE Des. Test Comput. Metastability and synchronizers: a tutorial. **28**(5), 23–35 (2011)
16. G. Girao, B.C. de Oliveira, R. Soares, I. Saraiva Silva, Cache coherency communication cost in a NoC-based MPSoC platform, in *Annual conference on Integrated circuits and systems design*, 2007, pp. 288–293
17. K. Goossens, J. Dielisse, A. Radulescu, AEthereal network on chip: Concepts, architectures, and implementations. IEEE Des. Test Comput. **22**(5), 21–31 (2005)
18. Introducing AXI for Xilinx System Development (available at: http://www.xilinx.com/support/documentation/ip_documentation/ug761_axi_reference_guide.pdf)
19. ITRS 2011 (availabe at http://www.itrs.net)
20. A. Jantsch, System modeling, models of computation and their applications, Compendium for the KTH course 2B1429, 2001
21. M. Keating, p Bricaud, *Reuse Methodology Manual for System-on-a-Chip Designs* (Springer, New York, 2007)
22. K. Keutzer, A. Newton, J. Rabaey, A. Sangiovanni-Vincentelli, System-level design: orthogonalization of concerns and platform-based design. IEEE Trans. Comput. Aided Des. Integr. Circuits Syst. **19**(12), 1523–1543 (2000)
23. Y. Lai, S. Yang, M. Sheu, Y. Hwang, H. Tang, P. Huang, A high-speed network interface design for packet-based NoC, in *International Conference on Communications, Circuits and Systems*, 2006, pp. 2667–2671
24. J.-J. Lecler, G. Baillieu, Application driven network-on-chip architecture exploration & refinement for a complex SoC. Des. Autom. Embed. Syst. **15**(2), 133–158 (2011)
25. M. Mano, M. Ciletti, *Digital Design* (Prentice Hall, Englewood Cliffs, 2006)
26. S. Murali, P. Meloni, F. Angiolini, D. Atienza, S. Carta, L. Benini, G. De Micheli, L. Raffo, Designing application-specific networks on chips with floorplan information, in *International Conference on Computer-Aided Design*, 2006, pp. 355–362
27. Open Core Protocol Specification, OCPIP Association, Release 2.0, 2003 (available at http://www.ocpip.org)
28. M. Papamarcos, J. Patel, A low-overhead coherence solution for multiprocessors with private cache memories, in *International Symposium on Computer, Architecture*, 1984, pp. 348–354
29. M. Rose, *The Open Book: A Practical Perspective on OSI* (Prentice Hall, Englewood Cliffs, 1990)
30. M. Sgroi, M. Sheets, A. Mihal, K. Keutzer, S. Malik, J. Rabaey, A. Sangiovanni-Vincentelli, Addressing the system-on-a-chip interconnect woes through communication-based design, in *Design Automation Conference*, 2001, pp. 667–672
31. F. Steenhof, H. Duque, B. Nilsson, K. Goossens, R. Llopis, Networks on chips for high-end consumer-electronics TV system architectures, in *Design, Automation and Test in Europe*, 2006, pp. 1–6
32. M. Tang, X. Lin, Injection level flow control for network-on-chip (NoC). J. Inf. Sci. Eng. **27**, 527–544 (2011)
33. T. Tayachi, P. Martinez, Integration of an STBus type 3 protocol custom component into a HLS tool, in *International Conference on Design and Technology of Integrated Systems in Nanoscale Era*, 2008, pp. 1–4
34. Virtual component interface standard - draft specification, v. 2.2.0 (available at : http://www.vsia.com), August 1997
35. F. Vitullo, N. L'Insalata, E. Petri, L. Fanucci, M. Casula, R. Locatelli, M. Coppola, Low-complexity link microarchitecture for mesochronous communication in networks-on-chip. IEEE Trans. Comput. **57**(9), 1196–1201 (2008)

36. I. Walter, I. Cidon, A. Kolodny, ENoC: a bus-enhanced network on-chip for a power efficient CMP. Comput. Archit. Lett. **7**(2), 61–64 (2008)
37. B. Yu, S. Dong, S. Chen, S. Goto, Floorplanning and topology generation for application-specific network-on-chip, in *Asia and South Pacific Design Automation Conference*, 2010, pp. 535–540

# Chapter 6
# NoC Verification and Testing

**Abstract** Verification and testing are tremendously costly steps in the design flow. In today's multi-million gate ICs the lion's share of design time is spent verifying the design against its specification. An inadequately verified design will lead to re-spins that could make the difference between success and failure for a product. On the other hand, manufacturing test must prevent defective parts from being shipped to customers.

## 6.1 Introduction

The term verification is used to describe the process of *functional testing*, in other words checking the design against its specification. The term testing, on the other hand, usually refers to the process of the *manufacturing test*, assuming a correct design [1]. Verification is performed against a *model* of the final implementation (Table 6.1).

An important difference between verification and testing is that the model that is being verified provides inherent observability and controllability, while the actual circuit under test is usually difficult to control and observe. For example, an RTL simulator allows forcing and observing all intermediate signals in the design hierarchy, while the actual circuit is only controllable and observable through its I/O ports.

## 6.2 NoC Verification

Verification is the process of demonstrating that a design meets its specification. Since system specification starts at a high level of abstraction and is refined as the design process goes down the abstraction levels, verification takes place in all

K. Tatas et al., *Designing 2D and 3D Network-on-Chip Architectures*, 
DOI: 10.1007/978-1-4614-4274-5_6, 

**Table 6.1** Verification and testing terminology

| Concept | Verification | Testing |
|---|---|---|
| Design | Design under verification (DUV) | Design under test (DUT) |
| Equipment | Testbench | Tester/BIST |
| Input | Stimuli | Test vectors |
| Output | Response | Response |
| Success metric | Coverage | Yield |

abstraction levels, verifying that each model of the system is correct. Therefore, there is algorithm-level verification, system-level verification, RTL verification and post-layout verification. It is also common to develop a working prototype of the system in FPGA in order to use real inputs. Furthermore, verification is done across design hierarchy levels. Typically a NoC-based MPSoC would be verified at the basic component hierarchy level (router and NI verification), at the NoC hierarchy level, and at the MPSoC hierarchy level. Given the distributed nature of the NoC as well as its unique position in the MPSoC and its potential for design reuse, it must be verified to a very high degree of confidence.

Since the register-transfer level is the highest abstraction level from which generally a circuit implementation can be automatically generated using EDA tools, a large part of the verification effort is spent at the RTL level.

RTL Verification of a complex NoC-based SoC requires the development of elaborate object-oriented self-checking test-benches and represents a significant amount of the design effort [2]. This chapter only covers RTL verification.

### 6.2.1 Verification Fundamentals

As already mentioned, verification is the process by which the designer demonstrates that a *Design Under Verification* (DUV) conforms to its specification. Verification is done by applying stimulus to the DUV and checking its response against its specification. It is most often practically impossible to prove that a design meets its specification requirements, due to the most often infeasibly large number of possible *input stimuli* that must be applied for exhaustive verification. However, a single counterexample is enough to disprove that the DUV is functionally correct.

Therefore, the verification engineer is not aiming at exhaustively testing a complex DUV but to achieve *functional coverage*, meaning to test all DUV features through the testbench. Since, exhaustive testing is most likely infeasible even for specific features of the DUV instead of all possible input combinations, the challenge in the verification process is to achieve coverage of all design features with the minimum set of input stimulus and simulation runs.

The structure used to apply the stimulus and check the result is called *testbench*, and is essentially a wrapper around the DUV as shown in Fig. 6.1. A testbench can

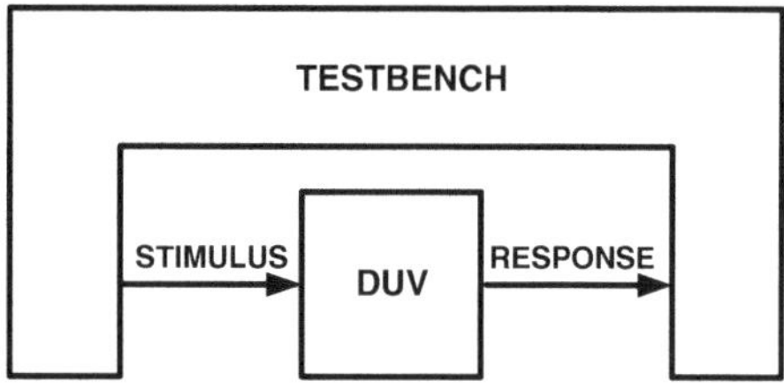


**Fig. 6.1** Testbench

be as simple as applying the truth table combinations to some simple glue logic to as complex as $4-5\times$ the lines of RTL code of the DUV of object-oriented code.

Testbenches can be classified according to the stimulus provided as:

- *Directed*: The stimulus is a set of test vectors required to test specific functional requirements. Directed stimulus is commonly used to verify initial system response. However, functional coverage with directed stimulus is too time-consuming, if at all feasible. Furthermore, since directed testing relies on testing specific requirements, it will not catch the bugs the designer has not thought of. However, directed testing is useful for simple low-level components in the hierarchy, or in order to identify a bug that has been caught by constrained-random verification.
- *Constrained-random*: The stimulus is a set of valid transactions with random data. For example, valid control signals and timing bus transaction, with random addresses and data, for example valid NoC headers with random source and destination addresses and random payloads. Constrained-random verification is very useful because it can catch bugs the designer has not thought of, but it should be mentioned that by walking the space of possible inputs at random may never reach some specific inputs even after long simulation times.
- *Realistic*: Realistic stimulus in NoC testing requires traces of actual packets and messages. The advantage of this stimulus is obvious, however it is often difficult to generate.

An important concept and metric for successful test-benches is coverage. The test-bench must cover all DUV features as documented in its specification, based on the requirements in the verification plan. Modern simulators provide certain coverage metrics such as statement coverage, branch coverage and state coverage.

An example of statement coverage output using ModelSim is shown in Fig. 6.2. The output shows both the percentage of statements that were covered and the actual statements that were missed to help guiding the verification effort.

Obviously, high coverage does not ensure a bug-free DUV. It is, however, a metric of how successful the test-bench was in sensitizing the DUV model. Low coverage means that a large portion of the design functionality has not be tested at all.

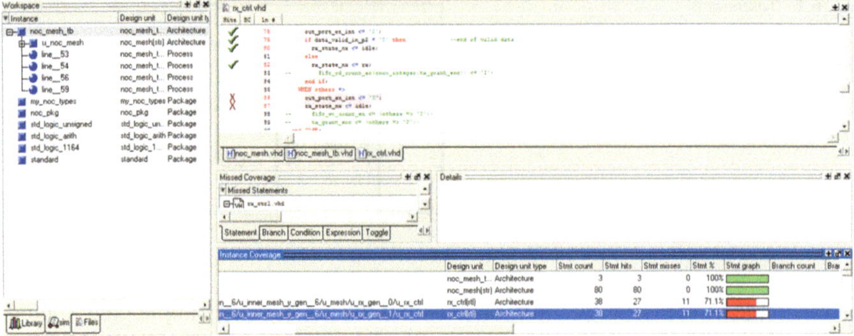

**Fig. 6.2** Test-bench coverage

## 6.2.2 Verification Plan

The verification plan is a specification document used for verification. Verification is usually performed using a bottom-up approach. The verification process should be split to requirements and each requirement should be numbered and prioritized. First-time success should be also documented in the verification plan. The conditions which are considered outside the scope of the system should also be explicitly stated in the specification document and the verification plan. Sample requirements for a NoC system verification plan are shown in Fig. 6.3.

```
Router verification requirements

R1. Packets are transmitted and received from router MUST
   R1.0. Packets are transmitted from all router ports
        R1.1.0. Single packet TX, RX disabled
        R1.1.1. Multiple packet TX to the same port, in appropriate order, RX
                disabled
   R1.1. Packets are received from all router ports
        R1.1.0. Packets are correctly stored in appropriate direction buffer
                Likewise with buffer connected to RX state machine
        R1.1.1. Single packet RX, TX disabled
        R1.1.2. Multiple packet RX from various ports, TX disabled Constrained-
                random test with RX requests from various ports at random with
                random data
   R1.2. Multiple packet RX and TX from/to various ports Constrained-random
         test with RX and TX requests from various ports at random intervals
         with random data

Network verification requirements

R2.0. All traffic in NoC reaching destination MUST
R2.1. Single packet routed from source to correct destination node MUST
R2.2. Multiple unicast packets routed through the network)
R3.1. Multicast packets reach their destinations SHOULD
R4.1. Flow control between neighboring routers works properly MUST
R4.2. Deadlock is resolved SHOULD
```

**Fig. 6.3** Sample NoC test-bench requirements

### 6.2.3 Test-Bench Creation for NoC

With verification becoming increasingly challenging, research efforts have concentrated on developing reusable, self-checking test-benches based on object-oriented programming. Open Verification Methodology (OVM) and Advanced Verification Methodology (AVM) are methodology examples.

The AVM [3] developed by Mentor Graphics comprises an open-source library of modular, reusable verification components implemented in both System Verilog and SystemC.

The OVM [4] is a methodology backed-up by an open-source library for functional verification using System-Verilog from Mentor and Cadence. OVM combines the features from mentor's AVM and Cadence ERM Methodology. Its scalable to system level verification. It is written in IEEE 1800 System-Verilog and therefore runs on any simulator supporting the IEEE 1800 standard.

#### 6.2.3.1 Testbench Structure

The traffic models used in the exploration step can be used in the verification process also. They are classified as either *realistic* or *synthetic*. Specifically, realistic traffic models are traces of application execution onto NoCs. On the other hand, the synthetic traffic patterns corresponds to abstract models of packets exchanged between nodes of the NoCs. In contrast to realistic traffic, which is representative of a more specific class of applications, synthetic traffic is generated based on mathematical models. Hence, it covers a broad class of applications executed onto NoC platforms. The competitive advantage of incorporating synthetic models is that it allows a network to be stressed with a regular and predictable pattern. However, since they do not represent traffic from real-life applications, they cannot be employed for accurate design-space exploration, whenever an application-specific NoC platform has to be designed.

A general test-bench structure used in modern verification methodologies such as AVM and OVM is shown in Fig. 6.4. This structure is applicable to NoC verification as well. The test-bench comprises a number of verification components, namely the generator, the agent, the driver the monitor and the checker or scorecard.

The generator is the highest layer of the test-bench. It is composed of classes which aim at generating transactions for the DUV. In the case of a constrained-random NoC test-bench, it should generate NoC transactions at random time intervals, from/to random (but valid) nodes with random payloads.

The agent is the next layer down in the hierarchy. It should receive pending transactions from the generator and create packets and flits according to each input transaction. Those are then forwarded to the driver.

The driver is responsible of generating the bit-level stimulus for the DUV according to the input received from the agent. It essentially generates appropriate

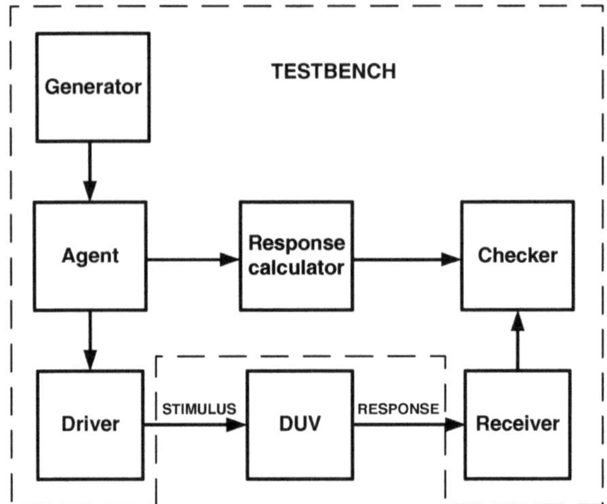


**Fig. 6.4** NoC test-bench structure

waveforms for the NoC routers flow control signals in order to convert the packets and flits received from the agent into valid inter-router transactions.

The receiver is responsible for receiving the flits from each router according to the flow control outputs in the NoC and encapsulating them back into flits, then packets and, finally, transactions.

The response calculator generally calculates the DUV response to the provided stimulus at a higher abstraction level. In the NoC environment this is fairly straightforward since the payload and destination addresses are known to the agent. However, it is harder to calculate the packet flight time for the traffic conditions generated by the test-bench. A solution to this would be to also create a test-bench with exactly the same scenario as the one run in the high-level NoC simulator (same injection rate and traffic type).

The checker is responsible for comparing the NoC input and output transactions and check if the data were received from the expected node and within the expected time interval.

Obviously creating such elaborate test-bench structures requires significant design effort, however the coverage that can be achieved in relatively short simulation time is well worth it.

### *6.2.4 NoC Prototyping*

Another step in the verification process after simulation is building a working prototype of the system or subsystem. The prototype is only required to satisfy functional requirements (specifications) and not performance and power constraints. It is

**Table 6.2** Verification framework comparison

| Framework | NoC generation | IP support | Simulation support | Prototyping | Case study |
|---|---|---|---|---|---|
| [5] | Yes | Partial | Yes | Yes | 4×4 NoC |
| [6] | Yes | No | Yes | Yes | |
| [7] | Yes | Yes | Yes | Yes | 3 PEs with ring, 2 × 2 mesh, single router topologies |
| [10] | Yes | Yes | Yes | Yes | 3 PEs with single router |

typically built using FPGA technology, often requiring multiple devices to emulate the final multi-million gate SoC.

Due to the time-consuming and error-prone process of NoC verification and prototyping, developing a framework of EDA tools that automates NoC generation, testbench creation, simulation and analysis is important in order to reduce system development time. A few attempts at developing a more or less complete NoC generation/verification/prototyping framework have been made. Specifically, The MAIA framework [5] provides parametric NoC generation for three NoC topologies (mesh, torus and star) together with appropriate testbenches and simulation/analysis scripts. The ATLAS framework [6] for NoC generation and evaluation contains tools for NoC generation, traffic generation, simulation, traffic and power evaluation. An integrated flow to automatically generate a configurable NoC-based multi-processor SoC has been proposed in [7].

Other research efforts have concentrated exclusively on the prototyping framework/platform part of the NoC verification problem. In [8] a 4 × 4 mesh network without processing cores is implemented. In [9] an FPGA NoC prototyping framework that utilizes the partial reconfiguration capabilities of an FPGA to reduce re-synthesis time for accelerating the emulation process was introduced and a 2 × 2 mesh was used as case study.

A hardware-software emulation framework implemented on an FPGA has been presented [10]. It is shown that the FPGA-based emulation framework achieves four orders of magnitude of speedup over a software simulator. The complete NoC architecture with three cores, a single router and two network interfaces was emulated on an FPGA. In [11], a multi-FPGA NoC emulation framework was proposed.

Table 6.2 compares the features of some of the frameworks discussed above.

## 6.3 Testing Fundamentals

The objective of the testing process is to cover possible defects while minimizing the testing time and area overhead (see Sect. 6.3.4). As in verification, the test vectors must be chose wisely, since exhaustive testing is not feasible. Defects themselves cannot be tested, therefore fault models are used, so that the test can capture the *effect* of the defect on the circuit behavior.

### 6.3.1 2-D IC Defects

Defects can be broadly classified as bridging defects (short circuit) and open circuit defects [12]. A bridging defect is an unintentional connection between two or more circuit nodes. The bridging effect introduces a bridging resistance $R_{short}$ between the shorted nodes, which if it is lower than a *critical* resistance, the circuit will fail either in terms of logic levels or sensitivity to noise.

Bridging defects can be intra-transistor (gate oxide shorts), inter-transistor (combinational or sequential circuit shorts) and power rail shorts. Combinational circuit shorts may or may not create feedback.

An open circuit defect is an unintended break in interconnect lines. The open circuit can occur in either metal, polysilicon or diffusion regions. Their main effect is that one circuit node ceases to be driven by any gate and may be left floating. The final voltage of the floating node depends on the size of the crack and the amount of charge present in the floating node [12].

### 6.3.2 3-D IC Defects

As more dies are integrated into a 3-D IC, the yield of the final product falls off exponentially, since a single faulty die can render the entire IC defective, even without defects in the vertical links. For that reason, testing of each die before bonding was proposed in [13]. This can be very challenging since a functional unit may be partitioned between layers, making controllability and observability difficult.

Furthermore, the new 3-D processing steps cause new defects. TSV processing only allows for limited TSV heights and aspect ratios [14]. Wafers need to be thinned down before bonding to another die which can cause new defects such as degradation of some $I$–$V$ characteristics, shifts in device performances, and limited yield losses [15].

Thermal dissipation and thermo-mechanical stress are other possible causes of defects. The heat has no means of escape in densely packed stacks of thinned dies, with the danger of causing faults in various dies, especially heat-sensitive ones such as DRAMs.

Fortunately, these additional faults, unique to TSV-based interconnects are rather similar to the faults normally considered for wiring interconnects [16] such as opens, shorts, and delay faults, even though the nature of the underlying defects is different. Therefore, test engineers can use the existing large body of work with respect to test pattern generation for detecting hard open and short faults through a set of test patterns that grows only logarithmically with the number of interconnects [17–20].

### 6.3.3 Fault Modeling

Circuit defects such as shorts, opens, cracks etc. cannot be used directly for test generation and diagnosis. Therefore, the defects are represented by *fault models*. The most common fault models used in manufacturing test are the following:

- *Single Stuck-at Fault Models*: It is one of the first introduced fault models which is common even now. In this model, faults are represented as a node having a fixed logic value (stuck-at-0 or stuck-at-1). Such a kind of fault is permanent, while the basic functionality of the circuit is not altered. The main advantage of employing such model is its conceptual simplicity. Current research efforts still examine the single stuck-at fault model. However, this model does not provide an accurate representation of the physical defect's behavior [21].
- *Bridging Fault Models*: It models two signals shorted together. Such kind of fault may change the circuit's sequential behavior, while the voltage level at the end of two shorted wires can massively depend on the location of the bridge. A problem with this kind of fault is that the effect cannot be predicted in advanced. For instance, a strong driver will for sure overpower a smaller transistor if their outputs are shorted but two equally strong drivers will probably generate an unpredictable value as their common output.
- *Open Fault Models*: Similar to bridging faults, this kind of defect is common in CMOS processes, since the latest architectures incorporate more metal layers (and hence much more vias). Additionally, the copper interconnects make reliability issues even more critical [22]. This kind of fault fixes the gate of a transistor at the open value, and hence the transistor cannot be switched on.

### 6.3.4 Design for Testability

Tests can be applied either *offline* or *online*. The equipment that applies the test vectors and checks the response can be either external or internal. The second method is known as Built-In Self-Test (BIST). The advantage of BIST techniques is their speed, while their disadvantage is the generally significant area overhead, since the test structures are included in the chip. Obviously the BIST itself is not immune to defects.

## 6.4 NoC Testing

Like all other SoC components the NoC is susceptible to manufacturing defects and must be tested. Therefore, testing a NoC-based system includes testing of embedded cores and testing of the on-chip network. Testing of embedded cores is similar to conventional SoC testing, while the NoC can be considered as just another core of a SoC, but it is also special in two ways:

1. It is often composed of many identical sub-cores (routers and network interfaces), and
2. It occupies a privileged, central position in the SOC, by virtue of its interconnecting role. Testing of on-chip network includes testing of interconnects, switches/routers, input/output ports, and other mechanism other than the cores.

NoC-based SoCs require the application of a number of test methods from SoC, memory and FPGA domains, such as functional test, scan test, logic and memory BIST and testing of wires and switches [23].

## *6.4.1 BIST Structures for NoC*

A number of BIST structures for NoC architectures have been proposed [24–27]. In [24] a four-step composite test strategy and corresponding BIST structures are presented. The inter-router links are checked first, because they are used as Test Access Mechanism (TAM) for the following tests. The second part of the test checks all router FIFO buffers in the chip, the third checks the routing logic blocks in the chip and the fourth checks the multiplexers at the output ports and the intra-router links.

The BIST mechanism for testing the inter-switch links is shown in Fig. 6.5.

In [27] the authors proposed a built-in self-test/self-diagnosis procedure at start-up of an on-chip network (NoC) for bi-synchronous communication channels. Concurrent BIST operations are carried out after reset at each switch, thus resulting in scalable test application time with network size. The key principle consists of exploiting the inherent structural redundancy of the NoC architecture in a cooperative way for the effective diagnosis and error detection. At-speed testing of stuck-at faults can be performed in less than 4000 cycles regardless of their size, with an hardware overhead of less than 30 %.

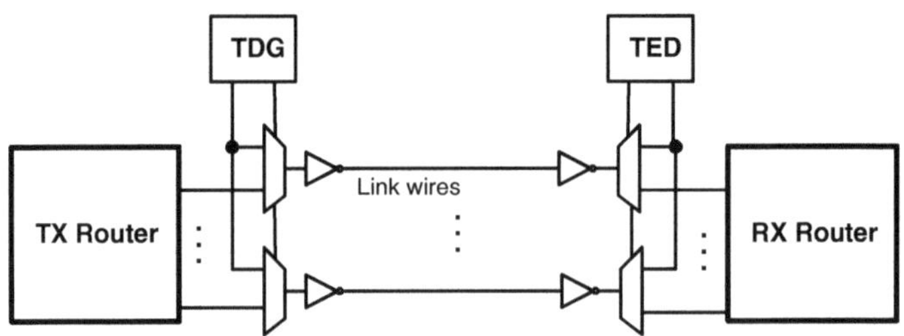


**Fig. 6.5** Link test BIST

### 6.4.2 *NoC as Test Access Mechanism*

On the other hand, the NoC can be used to feed test vectors to the cores also [25]. This has proven to be effective in reducing the routing overhead since NoC replaces test buses. However, this approach is not without disadvantages [28]. The DFT wrappers and test control are more complex due to the NoC protocols, testing time is higher and reliability is lower due to the NoC sensitivity to soft errors.

## Questions

**6.1.** What is the main difference between verification and testing?
**6.2.** What is the purpose of the verification plan?
**6.3.** List a few important verification requirements for a NoC design
**6.4.** What are the main types of defects in 2-D and 3-D ICs?
**6.5.** Describe the basic fault models used in testing?

## Problems

**6.1.** A 5-port NoC router with ACK/NACK flow control has a phit size of 64 bits. How many test vectors would be required to test it exhaustively?
**6.2.** Write a directed testbench in VHDL or Verilog for requirement R.

## Projects

**6.1.** Write a complete verification plan for the NoC router of Prob. 1.
**6.2.** Write directed testbenches in VHDL or Verilog for the verification plan of Prob.2.
**6.3.** Design in VHDL or Verilog the inter-router link test structure of Fig. 6.3.

## References

1. N. Weste, D. Harris, *Principles of CMOS VLSI Design*, 4th edn. (Addison Wesley, Boston, 2010)
2. C. Spear, *SystemVerilog for Verification* (Springer, New York, 2008)
3. http://www.mentor.com/products/fv/methodologies/uvm-ovm/cb_dll
4. https://verificationacademy.com/topics/verification-methodology

5. A. Mello, J. Palma, F. Moraes, N. Calazans, MAIA - a framework for networks on chip generation and verification. Proceedings of the Asia and South Pacific design automation conference (ASP-DAC), pp. 49–52, Jan 2005 (Shanghai, China)
6. A. Mello, A. Amory, N. Calazans, Fernando Moraes, ATLAS A framework for NoC generation and evaluation. Paper presented at the design automation and test in Europe (DATE), University Booth, Grenoble, 2011
7. A. Kumar, A. Hansson, J. Huisken, H. Corporaal, An FPGA design flow for reconfigurable network-based multi-processor systems on chip. Paper presented at the design automation and test in Europe and exhibition (DATE), p. 16, 2007
8. U. Ogras, R. Marculescu, H. Lee, P. Choudhary, D. Marculescu, M. Kaufman, P. Nelson, Challenges and promising results in NoC prototyping using FPGA. IEEE Micro. **27**(5), 8695 (2007)
9. Y. Krasteva, F. Criado, E. Torre, T. Riesgo, A fast emulation-based NoC prototyping framework, Paper presented at the international conference on reconfigurable computing and FPGAs (ReConFig), pp. 211–216, 2008
10. N. Genko, D. Atienza, G. De Micheli, L. Benini, Feature-NoC emulation: a tool and design flow for MPSoC. IEEE Circ. Syst. Mag. **7**(4), 42–51 (2007)
11. Y. Liu, P. Liu, Y. Jiang, M. Yang, K. Wua, W. Wang, Q. Yao, Building a multi-FPGA-based emulation framework to support networks on-chip design and verification. Int. J. Electron. **97**(10), 1241–1262 (2010)
12. J. Segura, C. Hawkins, *CMOS Electronics: How It Works, How It Fails* (Wiley-IEEE Press, 2004)
13. D. Lewis, H. Lee, Testing circuit-partitioned 3D IC designs, Paper presented at the annual symposium on VLSI (ISVLSI), pp. 139–144, 2009
14. B. Kim, C. Sharbono, T. Ritzdorf, D. Schmauch, Factors affecting copper filling process within high aspect ratio deep vias for 3D chip stacking. Paper presented at the electronic components and technology conference, p. 6, 2006
15. A. Ikeda, T. Kuwata, S. Kajiwara, T. Fujimura, H. Kuriyaki, R. Hattori, H. Ogi, K. Hamaguchi and Y. Kuroki, Design and measurements of test element group wafer thinned to 10$\mu$m for 3D system in package. Paper presented at the international conference on microelectronic test structures (ICMTS), pp. 161–164, 2004
16. E. Marinissen, Y. Zorian, Testing 3D chips containing through-silicon vias. Paper presented at the international conference (ICT), pp. 1–11, 2009
17. W. Kautz, Testing for faults in wiring networks. IEEE Trans. Comput. **C–23**(4), 358–363 (1974)
18. P. Goel, M. McMahon, Electronic chip-in-place test. Paper presented at the design automation conference (DAC), pp. 482–388, 1982
19. G. Robinson, J. Deshayes, Interconnect testing of boards with partial boundary scan. Paper presented at the international test conference, pp. 572–581, 1990
20. E. Marinissen, B. Vermeulen, H. Hollmann, R. Bennetts, Minimizing pattern count for interconnect test under a ground bounce constraint. IEEE Des. Test Comput. **20**(2), 8–18 (2003)
21. R. Aitken, Nanometer technology effects on fault models for IC testing. IEEE Comput. **32**(11), 46–51 (1999)
22. J. Abraham, A. Krishnamachary, R.S. Tupuri, A comprehensive fault model for deep submicron digital circuits. Paper presented at the international workshop on electronic design, test and applications (DELTA), pp. 360–364, 2002
23. A. Jantsch, H. Tenhunen, *Networks on Chip* (Springer, New York, 2003)
24. http://www.iti.uni-stuttgart.de/studies/ss09/seminar/pdfs/03_offline_test_and_diagnosis_of_noc_report.pdf
25. J. Lee, R. Mahapatra, In-field NoC-based SoC testing with distributed test vector storage. Paper presented at the international conference on computer design (ICCD), pp. 206–211, 2008
26. C. Grecu, P. Pande, A. Ivanov, R. Saleh, BIST for network-on-chip interconnect infrastructures. Paper presented at the VLSI test symposium, pp. 6–35, 2006
27. N. Caselli, A. Strano, D. Ludovici, D. Bertozzi, Cooperative built-in self-testing and self-diagnosis of NoC bisynchronous channels. Paper presented at the international symposium on embedded multicore SoCs (MCSoC), pp. 159–166, 2012

28. Y. Feng, H. Lin Huang, X. Qiang, Re-examining the use of network-on-chip as test access mechanism. Paper presented at the design automation and test in Europe (DATE), pp. 808–811, 2008

# Chapter 7
# The Spidergon STNoC

**Abstract** Spidergon STNoC is a state-of-the-art, low-cost on-chip interconnect that plays a vital role in enabling multiprocessor system-on-chip by providing structure, performance, and modularity. This chapter outlines topological and routing characteristics of the packet-switched Spidergon STNoC, focusing on its low diameter, vertex-symmetric, point-to-point chordal ring topology, and its low-cost, efficient deterministic, shortest-path routing algorithm. It also describes interesting design tools and discusses new Spidergon extensions toward fault tolerant routing.

## 7.1 Introduction

Multicore system-on-chip (MPSoC) technology has created exciting opportunities for developing new advanced engineering products and market scenarios. In the heart of current SoC technology, Moore's law expresses a continually increasing CMOS integration capability, thus challenging the electronic design automation community for delivering new design methodology and tools that address an ever-increasing system complexity with improved power-performance-price ratios and reduced time-to-market requirements.

By embedding multi-dimensional regularity mechanisms within the topology infrastructure, we enable full-scale integration of complex MPSoC. More specifically, MPSoC architectures can scale beyond traditional 32 nm processes using a packet-switched network-on-chip (NoC) design methodology with a pseudo-regular fabric that helps resolve manufacturing problems due to increased design complexity and growing variability and achieve higher yields [3, 5, 22, 28, 37, 38, 48].

Within this context, we explore structural regularity in architectural, topological, and algorithmic features to simplify NoC and network interface (NI) realization

Miltos Grammatikakis, Marcello Coppola, George Kornaros.

K. Tatas et al., *Designing 2D and 3D Network-on-Chip Architectures*, 
DOI: 10.1007/978-1-4614-4274-5_7, 

and improve MPSoC concurrency using efficient routing, intensive communication algorithms, and mapping mechanisms.

More specifically, in Sect. 7.2, we focus on the topological and routing issues of the Spidergon STNoC architecture describe its network topology characteristics, including design foundations, graph properties, and routing strategy. We also outline the Spidergon STNoC design methodology, indicating existing open source tools that target design space exploration. In Sect. 7.3, we examine recent extensions toward fault tolerant routing using network reconfiguration techniques at the network interface. Finally, in Sect. 7.4, we provide conclusions and outline directions for further work.

## 7.2 Spidergon STNoC Architecture

In the STNoC architecture, processing elements and routers are connected together through a (patented) novel, regular, point-to-point topology, called Spidergon. The architecture exploits IP reuse, modularity, and multiple levels of abstraction through OSI-like network layering. The Spidergon STNoC is configured using four generic building blocks appropriately interconnected to each other.

- The STNoC router is responsible for efficient and reliable data transfer of packet flits within the Spidergon STNoC topology, the elements into which packets (and messages) are logically divided. It implements the network and data link layers of the NoC protocol, offering best effort latency, throughput and quality of service (QoS).
- The network interface (NI) provides connectivity of individual IP by converting subsystem transactions into packets transported within the NoC. It is the building block to perform correct frequency and protocol conversions between the router and connected IP cores. These cores, which are generic devices such as embedded processors and memories, might act as initiators through a Master interface or targets by means of a Slave interface. In the former case, they are responsible for sending request transactions throughout the Spidergon STNoC and receiving responses from elsewhere, while in the latter case, they can only receive transactions and answer to them. Thus, in Spidergon STNoC, NI behavior depends on the type of IPs (initiators and targets) connected to it. Fig. 7.1 illustrates a Spidergon STNoC configuration with 12 routers, 8 initiators (noted as $I$), and 4 targets (noted as $T$) connected to corresponding NIs. A generic NI initiator provides conversion of initiator bus protocols into Spidergon STNoC packets which can be sent through the packet switched network of routers, splitting them into many flits as necessary. Once the flits from the initiator come to the NI relative to destination IP (Target), they are decoded at the NI target. Target and initiator might operate at different frequencies between them and the router and they can also have different bus sizes. The NIs provides support to adapt to such differences. NIs facing toward initiators or targets are almost similar in architecture except few details which do not change

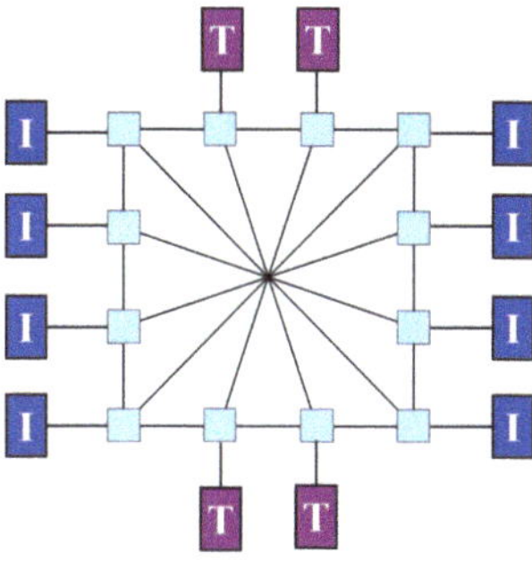


**Fig. 7.1** Spidergon STNoC configuration with 12 routers, 8 initiators and 4 targets

their global complexity and behavior. NI is also responsible to provide correctness (completion) information on packet routing, e.g., concerning an initiator/target transaction, and allows for IP reuse by hiding network-dependent aspects from the transport layer, a critical benefit for reducing multicore SoC design complexity.

- The Spidergon STNoC network plug switch (concentrator) is a block for connecting a generic number of IPs to a single Spidergon STNoC router. The multi-IP traffic injected in the network is concentrated (and arbitrated), while traffic extracted from the network is split to the proper IP.
- The physical link is responsible for the actual propagation of the signals among the network routers and also to/from external IPs and subsystems through the NIs. The choice of a physical link technology, e.g., serial versus parallel or synchronous versus asynchronous, involves tradeoffs between clock distribution area, on-chip wiring and required chip area.

The switching strategy, flow control, and routing algorithm are fundamental functions of the NoC topology used to establish reliable and efficient communication [14–16].

Spidergon STNoC supports wormhole switching, i.e., a packet is further subdivided into flow control units (flits), with each flit having unique flow control [2]. Thus, the rest of the packet flits follow the same path reserved for the header. This idea drastically reduces the amount of network buffering, while making the latency for congestion-free (cut-through) communication almost independent of the distance between the source and destination [12].

Flow control refers to the scheduling policy for resolving conflicts during packet traversal. Thus, flow control deals with fair allocation of buffer and channel resources, and packet buffering, dropping, or deflection. We distinguish between link-level flow control dealing with managing short-term imbalances, e.g., through Request/Grant protocol or threshold-based (watermark) link protocols, and end-to-end flow control dealing with packet scheduling, and admission control, i.e., acknowledgment-based, credit-based, or threshold-based protocols.

The current Spidergon STNoC link-level flow control protocol is very simple. The first flit is transmitted according to credit-based hop-by-hop flow control which completely avoids flit dropping and retransmission. A simplified description of the protocol is as follows. Each physical link of a router consists of two unidirectional data channels: upstream (US) and downstream (DS). Each transmitted flit on the US channel of any router is always accepted by the DS channel of the following router (or network interface), since there is always enough buffer space. The protocol works by automatically setting an initial number of credits in the US channel, equal to the input buffer size of the DS channel it communicates with. Since the US interface transmits a flit only if the connected DS channel can accept it, there are no pending flits on the link wires. This approach also allows virtual channel flit-level interleaving, so that separate virtual networks can share the same physical link. This fact together with the implementation of output queues on the US physical ports enhances performance by avoiding head-of-line blocking effects. The input buffer in the DS interface is a small FIFO, which stores the incoming flits of packets traveling on the link on the specific VN. Its depth is dimensioned according to the credit round-trip delay to guarantee maximum throughput. Finally, notice that crosspoint output queueing (separate output queue for each input port) and no output queueing are alternatives for high performance communication or low cost solutions, respectively. A quality metric for this scheme is the round-trip credit delay which is defined as the minimum time between two consecutive credits for the same buffer location. Only once the grant signal is set, the second flit may be sent; this allows the router to arbitrate the output queue requested by this packet. If space is available, the packet locks that queue and the second flit is accepted. From this point on, flow of subsequent flits is stopped only if the queue becomes full.

For *end-to-end flow control*, Spidergon STNoC supports two bus protocols at the network interface (NI): the Amba AXI4 protocol, as well as the STMicroelectronics' STBus Type3 protocol. This choice for STBus allows exploiting past user experience, e.g., connection to existing proprietary IPs, or integration of hierarchical on-chip communication architectures based on STBus clusters into a high-level NoC backbone. The NI implements "access" to the NoC, translating transactions into packets that are exchanged within the network. The NI hides network-dependent end-to-end aspects (such as message disassembly) within the transport layer of the connected module.

## 7.3 The Spidergon STNoC Topology

The choice of NoC topology has a significant impact on MPSoC price/performance. To reduce design and verification time for a wide range of applications, a simple, regular topology with efficient buffer management, flow control, and routing, mechanisms has been proposed. The topology also provides efficient solutions for resolving different network problems, e.g., balance of virtual channels, quality of service, protocol or message-dependent deadlock, and intensive communication.

Figure. 7.2 illustrates examples of previously proposed NoC topologies. These topologies are usually compared based on theoretical metrics that affect routing cost and performance, such as number of nodes, number of edges, vertex symmetry, net-

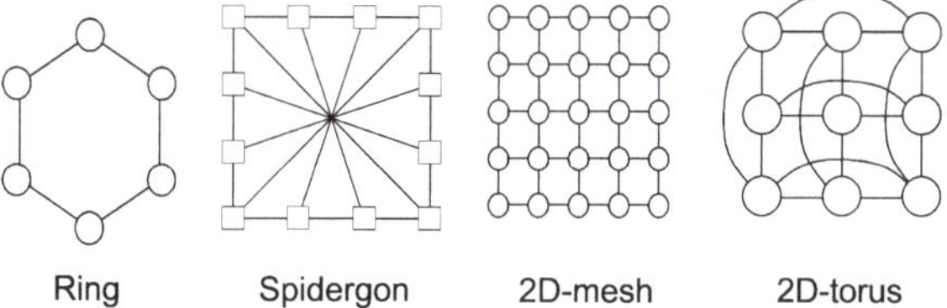


**Fig. 7.2** Examples of regular network topologies

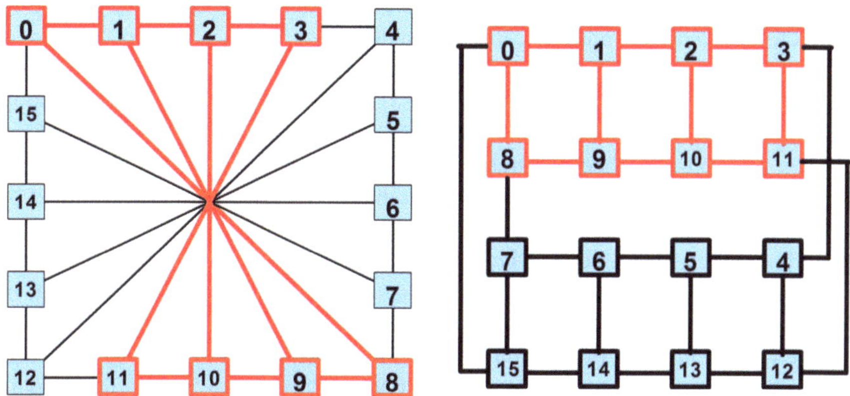


**Fig. 7.3** Spidergon topology and its planar layout

work degree, network size granularity, network diameter, average distance, network bisection width, as well as embedding properties for common communication patterns.

The main driving factor of the Spidergon STNoC design approach was to explore the complex network design space in order to achieve a low-cost, high performance hardware implementation. In this respect, Spidergon STNoC adopts two main topologies: a custom one used in the embedded systems (not discussed here), and a simple, regular topology that could be used for chip multiprocessor architectures. Spidergon STNoC promises to deliver the best tradeoff between theoretical metrics and final price/performance ratio, considering commercial realities of the SoC market. The Spidergon topology is based on a bidirectional ring, with extra cross links from each node to its diagonally opposite neighbor. Thus, each packet arriving at a non-final node is forwarded in a clockwise, anticlockwise or cross direction. As shown in Fig. 7.3, the Spidergon topology translates into a practical low-cost layout implementation (single crossing); notice that each square corresponds to a router connected to a processing element. In general, chip area relates to edge bisection, while the longest wire affects NoC latency.

The Spidergon topology is vertex-symmetric with a relatively small number of links and constant (equal to 2) size granularity (called network extendibility). The total number of edges in an N-node Spidergon is $\frac{3N}{2}$, while its diameter is $\frac{N}{4}$ . For currently realistic NoC configurations with up to 60 nodes, the proposed graph has a smaller number of edges, and a smaller diameter than fat-tree, 2D-mesh or 2D-torus topologies, leading to latency reduction for small packets, even if wormhole routing is employed [31, 47]. For example, the diameter of a $4 \times 5$ mesh with 31 bidirectional edges is 7, while that of a 20-node Spidergon STNoC topology with 30 bidirectional edges is only 5. Hence, higher degree topologies do not provide significant benefits due to reduced performance for small, non-square, irregular networks; this is especially true when mapping practical, non-random NoC application traffic, e.g., multimedia traffic.

The Spidergon STNoC topology is a *chordal ring* that belongs to the family of undirected $k$-circulant graphs, i.e., it is represented as $G(N; s_1; s_2; \ldots; s_k), 0 \leq s_i < N$, where $s_i$ is an undirected edge between network node l and node $(l + s_i)$ mod $N$. Chordal rings are circulant graphs, with $s_1 = 1$ [6, 24, 25], while double loop networks (also called double fixed-step graphs) are chordal rings, with $k = 2$. These families of graphs have been theoretically studied as competitors to meshes and tori in respect to graph optimality, i.e., minimum diameter graphs for a given number of nodes and constant degree; see Moore graphs [51].

Within the class of circulant graphs, the Spidergon STNoC network connects an even number of nodes $N = 2n$, where $n$ = 2, 3, ... as a bidirectional ring, in clockwise, node $i$ to node $(i + 1)$ mod $N$ and anticlockwise direction, i.e., node $i$ to node $(i - 1)$ mod $N$. In addition, for each node there is a cross connection, i.e., from node $i$, $0 \leq i < N$ to node $(i + n)$ mod $N$. The Spidergon STNoC topology is a fixed degree topology, thus the same type of router is used to compose the entire network of a generic $N$.

Spidergon specializes to a basic Octagon topology, for $n = 8$. In fact, the STMicroelectronic's Octagon network processor, based on a Cartesian product of basic Octagon topologies with a processing element at each node (see Fig. 7.4), has been proposed as an alternative to complex on-chip busses for high concurrency, low latency network processors, able to meet OC-768 network processing speeds [10].

We next outline the Spidergon STNoC design approach starting with its theoretical foundations.

### *7.3.1 Theoretical Foundations*

The Spidergon graph as a circulant graph has a non-optimal diameter. However, due to its constant bisection width (8), we have also examined chordal rings of constant degree, for alternative NoC topologies with the following constraints.

- small diameter for less than 100 nodes;
- large edge bisection width that scales well;
- efficient (wire balanced) point-to-point routing without preprocessing;
- efficient intensive communication algorithms, including broadcast, scatter and gather versions;
- good fault tolerance;
- efficient layout with short, mostly local (small chordal links) wires.

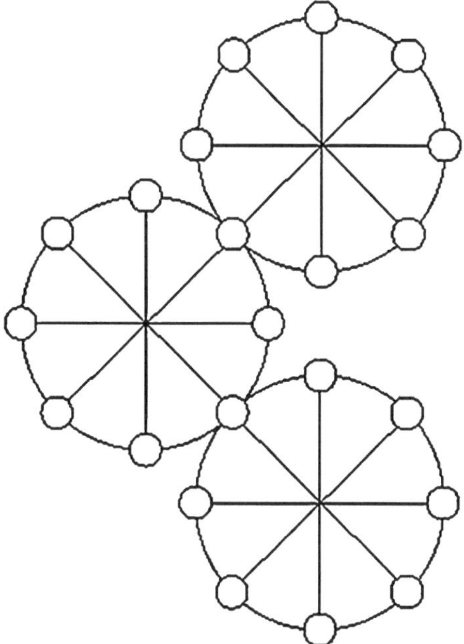

**Fig. 7.4** The ST Octagon topology

Searching for optimal chordal rings involves several tradeoffs and peculiarities. For example, diameter and average distance metrics of chordal rings are not always monotonically increasing and cannot be minimized together. Through a theoretical graph exploration phase, we have identified interesting families of chordal rings. For example, there exist degree 2 and 3 chordal rings with smaller diameter and higher edge bisection than Spidergon. These graphs have simple routing algorithms, but usually a non-constant (linearly increasing) network extendibility metric.

An example of these interesting NoC candidate topologies is the 3-circulant graph shown in Fig. 7.5. This graph uses chords with $s_1 = \pm 1$ and $s_2 = \lfloor \sqrt{N} \rfloor$ and achieves a $\lfloor \sqrt{N} \rfloor$ diameter (similar to torus) with a fairly simple Spidergon-like deterministic shortest-path routing algorithm. The algorithm first routes packets forward using long steps not overshooting their destination (by following $s_2$ links in either direction), and then uses backward or forward links to reach the target node. Since forward and backward steps can be taken in any order, this routing algorithm has a very high adaptivity factor. Moreover, assuming a synchronous communication model optimal permutation routing very similar to point-to-point routing can be obtained using $L$-shapes, i.e., a planar geometrical representation of 2-circulants [8, 12, 18]. More specifically, permutation routing a) first routes packets using a number of long steps, and then b) routes packets using short steps, always assigning higher priority to the packet that takes the longest distance; this can be implemented by extending the packet structure with hop counter or address comparison mechanisms. Since all

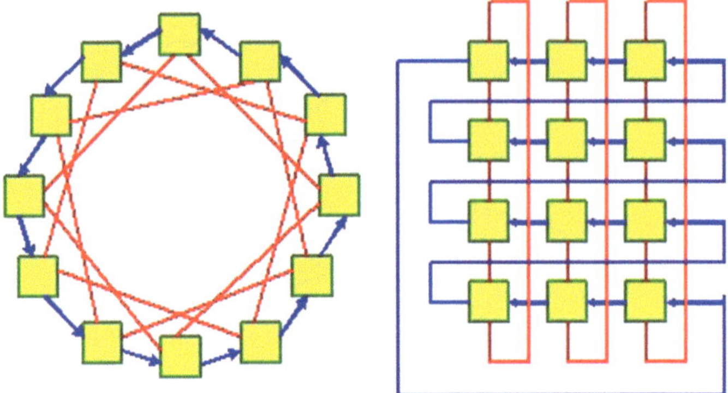

**Fig. 7.5** Equivalent representations of an undirected 3-circulant

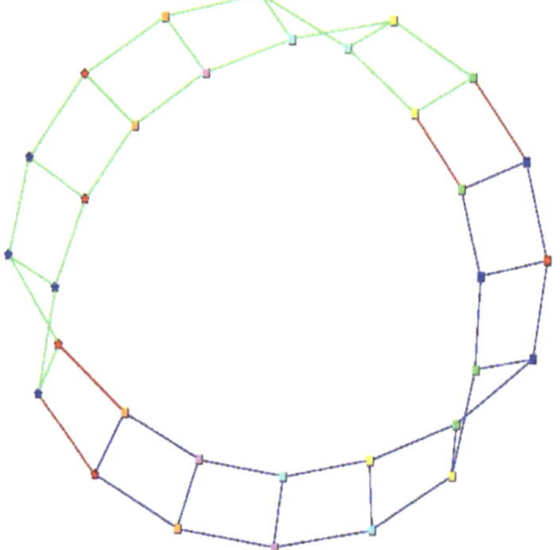

**Fig. 7.6** Neato layout for 32-node spidergon (min bisection is identified)

distances are different there is no extra queuing delay. Efficient broadcast, multicast and intensive communication algorithms, e.g., multinode gather and many-to-few routing can be designed using similar techniques, while also taking into account traffic balancing.

In order to explore topological properties in alternative constant degree NoC topologies (especially symmetric chordal rings) we have explored theoretical results through the use of several available open source software packages targeting network topology selection and visualization. Nauty [36] and metis [30] tools are able to analyze graph theoretical properties and automorphisms, while neato can visualize topologically equivalent graph vertices based on spring relaxation [41]; two vertices are equivalent (identical display attributes), if there is a vertex-to-vertex bijection preserving adjacency. For example, Fig. 7.6 shows the planar layout and limited bisection of a 32-node spidergon. Nauty also determines the orbits that partition graph vertices into equivalence classes, thus providing symmetry-related topological metrics. In addition, metis provides an extremely fast, multilevel graph partitioning embedding heuristic that can also extract topological and scalability-related metrics, e.g. diameter, average distance, in/out-degree, and bisection width. In fact, for small graphs ($N < 40$ nodes), a custom-coded version of Lukes's exponential-time dynamic programming approach to partitioning provides an exact bisection if one exists [34]. For larger graphs, metis partitioning uses fast heuristics to approximate a near-minimum bisection width.

Graph embedding and corresponding partitioning techniques form a binding element between parallel algorithms and architectural topology. In particular, graph embedding (called simulation, emulation, or mapping) targets efficient task allocation or IP mapping. Task graphs express the necessary computation, communication, and synchronization patterns for realizing a particular functionality. They provide a generic framework for developing efficient parallel algorithms for new topologies by porting existing strategies in existing topologies.

Embedding is also used for reconfiguring faulty networks, i.e., providing fault-free physical sub-graphs in "injured" physical system graphs. In this approach, the operating system hides the system graph and allows users to define virtual topologies. Thus, when hardware redundancy is implemented, it is possible to maintain network performance (bandwidth and latency) in the presence of a limited number of faults without requiring fault-tolerant routing. For a given node expansion, the quality of graph embedding is characterized by several metrics, such as dilation which measures latency overhead during point-to-point communication in the target graph, or congestion which indicates edge contention. An embedding of Spidergon onto an undirected chordal graph with node expansion and dilation one is shown in Fig. 7.7; this embedding applies to any Spidergon graph of size $n = 2k + 2$, where $k$ is an integer.

Small dilation, expansion, and congestion embedding have been examined on different topologies using Scotch [13, 54]. *Scotch* partitioning is faster than *metis*

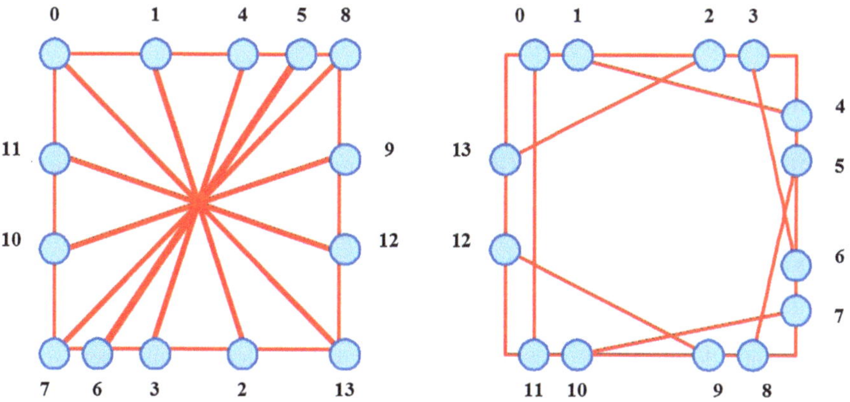


**Fig. 7.7** Isomorphism between Spidergon and an undirected chordal of size 14, with $s_1 = \pm 1$, and $s_2 = \pm 3$ or -3 for odd-/even-numbered nodes

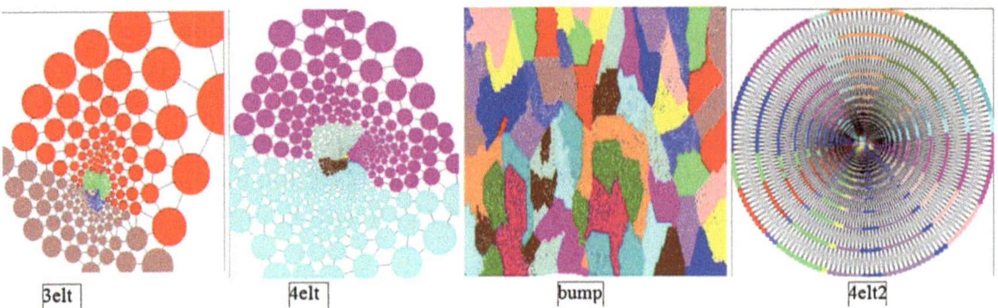


**Fig. 7.8** Various graphic representations for embedding finite elements onto different topologies (different colors specify the different node mapping)

(and usually far better than other similar tools); it works in linear time to the number of edges in the source graph and logarithmic time to the number of vertices in the target graph. Moreover, Scotch provides graphical visualization of the computed mappings based on a predefined geometry (see Fig. 7.8) and easily interfaces to other partitioning or theoretical graph analysis programs, e.g., *Metis* or *Nauty*, due to standardized vertex/edge labeling format.

## *7.3.2 Open Source System-Level Simulation in Spidergon STNoC*

System-level design methodology focuses on the functionality and relationships of the primary system components, separating system design from implementation, thus providing rapid, high quality, cost-effective design in a time-critical fashion by evaluating a vast number of complex architectural and technological communication alternatives.

System-level design provides increased productivity through innovative design methodology and tools focusing on higher levels of abstraction, where massive reuse and algorithmic or architectural optimizations are possible. The general consensus decision to elevate the abstraction level significantly above RTL by creating a system-level TLM model (e.g., IEEE's SystemC and TLM [53]) is becoming a real necessity for new product conception, platform validation, and design space exploration. SystemC increases the productivity of developing communication driver models through the definition of a universal API providing a new design pattern that enables creation and reuse of executable transaction level models across a variety of design environments and simulation platforms.

Executional specifications for the Spidergon STNoC router and network interface has been developed using SystemC and OMNeT frameworks [43].

SystemC development was also supported by developing the open source On-Chip Communication Network (OCCN) methodology and design framework [9]. OCCN is a flexible, highly parameterized, object-oriented, user-friendly C++-based framework consisting of an open-source, GNU GPL library, built on top of

SystemC, aiming at NoC modeling at different levels of abstraction. In addition, OCCN employs a multilayered NoC modeling methodology and defines a universal API for NoC specification, modeling, simulation, and design exploration.

OMNeT is a flexible, highly parameterized, object-oriented, user-friendly, state-of-the-art C++-based framework consisting of an open-source, GNU GPL library, built on top of SystemC, aiming at NoC modeling at various levels of abstraction, simulation, and design space exploration [43].

### 7.3.3 Custom Tools for Design Space Exploration

Besides the use of open source technology for topology exploration and embedding, it is also interesting to outline the main concepts surrounding many innovative industrial EDA tools designed to automate and enhance the design of a Spidergon STNoC design flow. This sequence of complementary and modular tools define STMicroelectronic's proprietary Interconnect Design Kit (IDK) ecosystem aimed to support specification and validation of a customized NoC architecture, while enabling the exploration of a huge design space. As shown in Fig. 7.9, IDK consists of several subtools that are combined into a single design product tool-chain.

- *I-Put* is used for optimally embedding application IPs described as task graphs to Spidergon STNoC tiles described as architecture graphs. The tool is based on

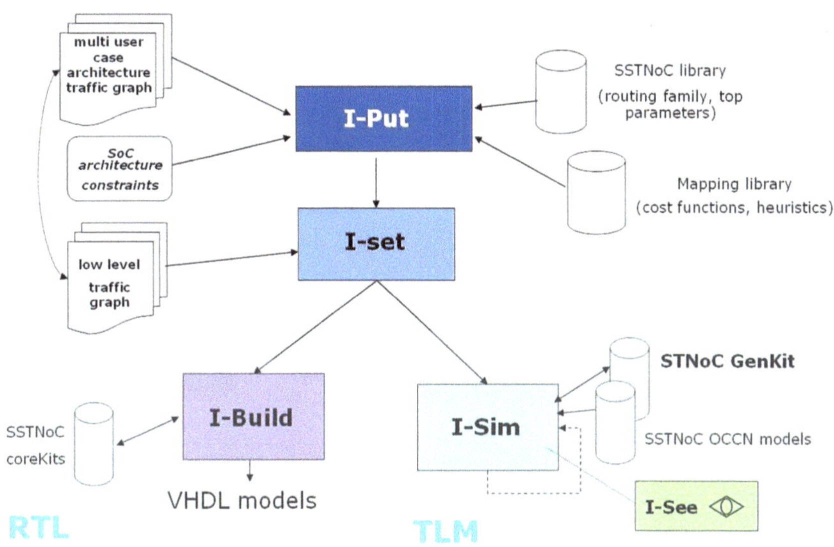


**Fig. 7.9** Custom Spidergon STNoC design tools

simulation and obtains the best mapping of the IPs, assuming a frozen architecture and use case scenario.

- Assuming a given NoC architecture (possibly output from *I-Put*), *I-Set* is used to automatically set subsystem parameters (or provide suggestions to the user regarding possible configuration), thereby providing cost-efficiency tradeoffs. For example, *I-Set* enables setting the buffer size at router or network interface and quality of service features. The output of *I-Set* is used as a configuration file for *I-Sim*, the architecture simulator.
- *I-Sim* takes as input an XML description of the platform and generates a SystemC simulation model at transaction-level, bus cycle-accurate level, or co-simulation VHDL/SystemC.
- *I-See* is used to interact with the simulator, validate, and explore the NoC architecture.
- *I-Build* takes as input an XML description of the platform and a library of Spidergon STNoC components from Synopsys CoreKit and generates a synthesizable VHDL model.

In order to further enhance STNoC design with high-level functions, including design customization, RTL generation and early software development, the *I-Build* design environment has been extended with two inter-related tools: *I-NoC* and *Meta-NoC* (see Fig. 7.10).

*I-NoC* is a tool integrated with an IDE environment that allows the architect to draw a NoC platform using a worksheet, parameterize each instantiated component, properly set-related system/network information (e.g., memory maps and packet routing), identify and resolve design errors or mismatches, generate the so-called

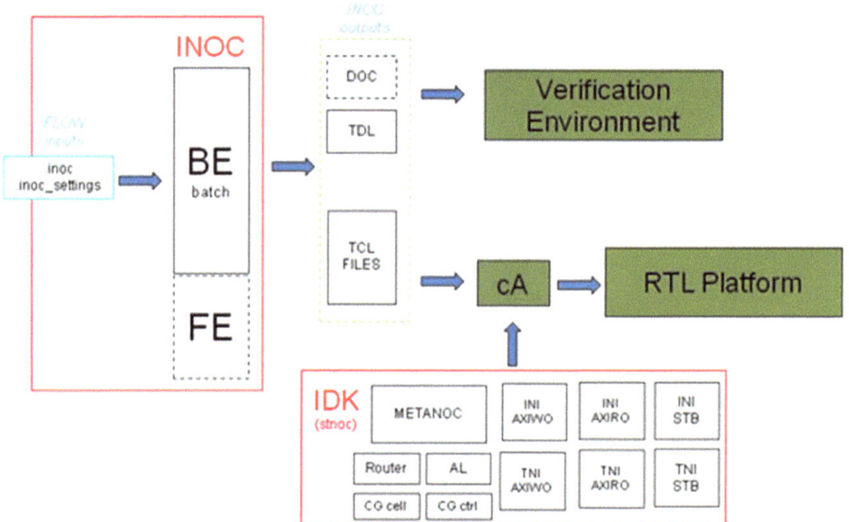


Fig. 7.10 STNoC design flow

RTL architecture generation scripts, and eventually support software development activities. More specifically, *I-NoC* design flow is based on two parts.

- The front end consists of graphical STNoC platform creation; this design stage produces a series of TCL files that accurately describe the platform.
- The backend is the "hidden" part of *I-NoC*; this is responsible for subsequently translating the TCL files using the *CoreAssembler* (cA) Synopsys technology to automatically generate the RTL corresponding to the parameterized IP created using the graphical front end. Notice that another key input to the *coreAssembler* tools is the *coreKit* library composed of highly configurable STNoC RTL meta-components implemented using ad hoc code generators on top of basic VHDL units. For example, corekit conveniently packages STNoC router, AXI NI, and STBus components using the Synopsys *coreKit* technology. The backend also produces a set of input files to aid STMicroelectronics' verification flow and regression activity. Finally, *I-NoC* can be used to generate a software driver for a Linux kernel which provides a user space interface to the programmable services provided by the NoC design. In addition, a portable C library uses the driver interfaces to provide a homogeneous C interface to applications and higher level functionality, such as determining appropriate QoS parameters based on specific bandwidth requirements.

*Meta-NoC* technology enhances STNoC design flow by supporting well-defined NoC customization. Since Meta-NoC technology is currently patent pending, we cannot disclose all the fine details. As shown in Fig. 7.10, *Meta-NoC*'s entry point in the design flow is through *I-NoC*, while adding powerful NoC configuration capabilities based on the notion of a metacompiler. This metacompiler is essentially a well-structured software layer which generates configurable IP blocks in RTL, overcoming limits of HDL-based, tool-dependent configuration statements. The metacompiler is implemented as a set of high-level XOTCL instructions; XOTCL is an object/aspect-oriented scripting language based on TCL.

The main concept of *Meta-NoC* is to maintain the global HDL description similar to a traditional design flow, but to also implement certain configurable parts of the design (that require complex configuration) using specialized instructions of the metacompiler. During the configuration phase, the metacompiler substitutes these instructions with the configured RTL block, by providing a "hook" to *coreAssembler* through which the "metacompiler" is called, allowing it to perform the necessary substitution. The obtained RTL is optimized HDL code that prevents generation of unnecessary code and allows instantiation of different types of components according to the user-defined configuration, hence resulting in new features and product improvements, while shortening the design cycle; in this process, unconnected signals and ports are removed. An example where *Meta-NoC* has already been applied effectively is NoC reconfigurability (see Sect. 7.4).

NoC modeling and evaluation is an essential topic in multicore SoC design. In this scope, researchers from both academy and industry have also developed open design methodologies and tools which are briefly outlined in the following section.

### 7.3.4 Related Open Tools for NoC Design Space Exploration

Today's modern system-level design focuses on methodologies, tools, and frameworks for exploring multicore SoC and NoC infrastructures. In particular, NoC design flow involves different layers from algorithmic description of system functionality to representation and modeling in a hardware description language (HDL). Driven by the advances and greater capability of electronic system-level automation, several system-level tools have been proposed for exploring the design of multicore systems, including on-chip interconnects.

Employing a NoC involves several challenges in SoC design flow, such as selecting the most suitable combination of NoC topology, routing algorithm, buffering strategy, flow control scheme in terms of performance or power-efficiency. Because of the wide spectrum of alternatives that these challenges may impose to the required performance of the final application, the evaluation of NoCs turns into a mandatory practice in the design flow of modern SoCs. Exploration at design time of these alternatives is usually the primary methodology that researchers propose; even though network calculus and runtime schemes have emerged as an attempt to evaluate the best tradeoff in terms of NoC architecture characteristics, the system constraints and requirements of a given application are usually examined through extensive simulation.

Open source research simulators for NoC are usually built to facilitate dissemination. In this subsection, we elaborate on open source system- down to RT-level design tools and methodologies for NoC design, while related VLSI simulation, synthesis, and physical design tools actually fall beyond the scope of this work.

The computational effort of a simulation framework usually indicates a level of accuracy. Careful consideration of this tradeoff is necessary in order to avoid significant deviation from real system operation. In fact, simulation results beyond an error margin can be rendered meaningless; on the other hand, excessive computational effort could make simulation time prohibitive. While most network simulation tools lack the required flexibility to allow dynamic adaptation of their computational effort to different operating scenarios, systems such as GEM5 can dynamically adjust this tradeoff when moving from initial exploration of the design space to the final product [7]. A different solution, employed by GEMS [35], relies on simulating different parts of the system with varying levels of detail, in order to optimize either for speed or accuracy.

As NoC development draws a lot of attention due to enabling scalable and power-efficient multicore SoC solutions, many researchers have proposed specialized modeling frameworks, in addition to general-purpose network simulators (e.g., *ns-2*) that were previously used to estimate NoC performance [46].

Orion is a NoC simulator developed to evaluate network performance and power estimation for different traffic patterns and flow control parameters [29, 50]. Orion 2.0 models support any topology that uses wormhole routers, while developers may provide more accurate system-level power and area models for NoC routers.

SICOSYS [45] is a general purpose interconnection network simulator that captures essential details of low-level simulation. It was originally conceived to obtain results very close to those obtained by modeling network components in HDL hardware simulators, albeit at a lower computational cost.

RSIM simulates shared memory multiprocessors and uniprocessors built from processors that aggressively exploit instruction-level parallelism (ILP) [44]. RSIM is execution-driven, modeling state-of-the-art ILP processors, an aggressive memory system, and a multiprocessor coherence protocol and interconnect, including contention at all resources.

On top of SICOSYS, TOPAZ [1] is proposed as a general-purpose interconnection network simulator that enables the modeling of a wide variety of message routers, with different tradeoffs between speed and precision. The design of TOPAZ is object-oriented and has been implemented in C++.

The Nostrum NoC simulation environment (NNSE) explores the design space for Nostrum NoC [32, 33]. Furthermore, HNOCS is a modular open-source OMNeT++ based NoC simulator supporting heterogeneous NoCs with variable link capacities and number of VCs per each unidirectional port [27]. The simulator features parallelism and arbitrary topologies and includes synchronous virtual output queues and asynchronous NoC routers.

The Noxim NoC simulator supports various routing algorithms, NoC sizes, and arbitrary traffic injection rates [42]. NiRGAM NoC interconnect modeling tool [40] is developed giving the user the ability to investigate different QoS levels and parameters for network traffic. These tools are useful to explore design space and power estimation and are capable of obtaining performance metrics. These simulators are based on SystemC and are not synthesizable, providing limited support for exploring error control features in NoCs.

The Atlas framework is a set of tools developed in Java that automate the various processes related to the design flow of mesh-based NoCs [21]. In particular, the Atlas environment enables the designer to evaluate the performance and power consumption of different NoC configurations. The toolkit includes several tools.

- A NoC generation tool that creates a user-parameterizable network which is described in VHDL and its testbench described in SystemC.
- A traffic generator that produces traffic following different temporal traffic distributions (e.g., normal, uniform, and exponential).
- A simulation tool, which actually invokes an external VHDL/SystemC simulator, Mentor Graphics' ModelSim with the generated traffic files. At the end, this simulation driver collects the generated traffic output files for evaluation.
- A traffic evaluation tool, which uses the simulator output traffic to provide statistical analysis results, such as: (i) total number of received packets, (ii) average time to deliver the packets, (iii) total time to deliver all packets, and (iv) the average, minimal, maximal, and standard deviation time to deliver a packet.
- A power evaluation tool that integrates an internal power estimation model annotated using a commercial power estimation tool (Synopsys PrimePower).

**Table 7.1** Summary of open NoC simulation frameworks

| Simulator | Framework | Topologies | Comments |
|---|---|---|---|
| Atlas | Java, SystemC, HDL | Hermes | Synthesizable |
| HNOCS | OMNeT++ | All | Parallelism |
| Netmaker | SystemVerilog | - | Synthesizable |
| NiRGAM | SystemC | All | |
| NNSE | SystemC | Mesh, Torus | |
| Noxim | SystemC | Mesh | |
| OCCN | C++, SystemC | - | |
| ORION | C | - | power, area |
| SICOSYS | C++ | Mesh with constant transfer time | |
| TOPAZ | C++ | 2-d/3-d mesh/torus, midimew, ring | |

In addition to this open source framework, Netmaker is a library of fully synthesizable parameterized NoC implementations [39]. These networks are designed to provide packet-based communication for complex multiprocessor SoCs and multi/many-core processors. This library is developed in order to aid the accurate characterization of potential router microarchitectures, routing algorithms, and network topologies. The ability to synthesize gate-level implementations also allows the library to be used when exploiting FPGA-based environments for many-core systems research.

More recently, FPGA-based NoC emulators have been proposed that reduce simulation time by several orders of magnitude compared to software [19, 52]. However, resource restrictions of the FPGA device and costly redesign cycles may render these approaches impractical. A middle-ground proposal is the DART architecture [49] that can be parameterized by software at runtime to simulate different NoCs without modifying the hardware simulator on the FPGA. It has been shown that this infrastructure can achieve over 100× speedup relative to a cycle-based software simulator, while maintaining the same level of simulation accuracy.

A summary of the open source tools presented above is provided in Table 7.1. Besides open modeling frameworks which are too many to mention here, commercial design kits (e.g., NoCexplorer and NoCcomplier proposed by Arteris [4]) allow defining NoC interfaces, quality of service requirements and topologies, and are capable of estimating area and performance.

## 7.4 The Spidergon STNoC Routing Algorithm

Routing algorithms are responsible for the selection of a path from a source node to a destination node in a particular network topology. A good routing algorithm balances load across the various network channels, even in the presence of non-uniform or intensive traffic patterns. A well-designed routing algorithm also keeps path lengths as short as possible, thus reducing the overall packet latency.

Another important aspect of a routing algorithm is its ability to work in the presence of faults in the network. If a particular algorithm is hardwired into the router and a network link or node fails, then the entire network communication fails. Alternatively, if the routing algorithm can be reprogrammed (adapted to the failure), then the system can continue to operate with only a slight performance loss.

Routing algorithms are classified depending on how they select between all possible paths from a source node to a destination node. Three main categories are specified:

- deterministic, where always the same path is chosen between a source and a destination node, even if multiple paths exist;
- oblivious, where the path is chosen without taking into account the present state of the network; oblivious routing algorithms include deterministic routing algorithms as a subset;
- adaptive, where the current state of the network is used to select the path.

The routing algorithms adopted in the SSTNoC belong to the class of deterministic algorithms. The choice of deterministic routing guarantees an ordered end-to-end communication, thus avoiding costly flit reordering at packet reception. Moreover, adaptive routing has an implementation cost that is generally considered unacceptable in the NoC domain for the near term future.

After discussing the fault-free routing algorithms, we will also focus on SSTNoC methods for changing the routing paths at runtime, thus maintaining fault tolerance in the case of network router or link faults. These methods have already been implemented in the corresponding SSTNoC RTL IP.

### *7.4.1 Across-First Routing*

Across-First is a deterministic, shortest-path routing algorithm. The algorithm either moves packets along the ring, in the proper direction, to reach nodes which are close to the source node (not farther than $d = ceiling(N/4)$ hops), or otherwise, it uses the cross link to send packets to the far away subnetwork, and then move them along the ring.

Fig. 7.11 shows the routing paths in the SSTNoC with $N$=12, considering node 0 as the starting point; the clockwise direction is indicated by $R$ (right), the counter-clockwise by $L$ (left) and the cross connection by $A$ (across). Since the routing algorithm is local, and the NoC topology is vertex-symmetric, we may describe the routing algorithm at any node, and only for half of all possible destinations.

The Across-First routing scheme is based on the following principles:

- the $A$ port is selected at most once, always in the beginning of each packet's route;
- when the packet moves in the ring, it follows the same direction, either right ($R$) or left ($L$) for its entire path.

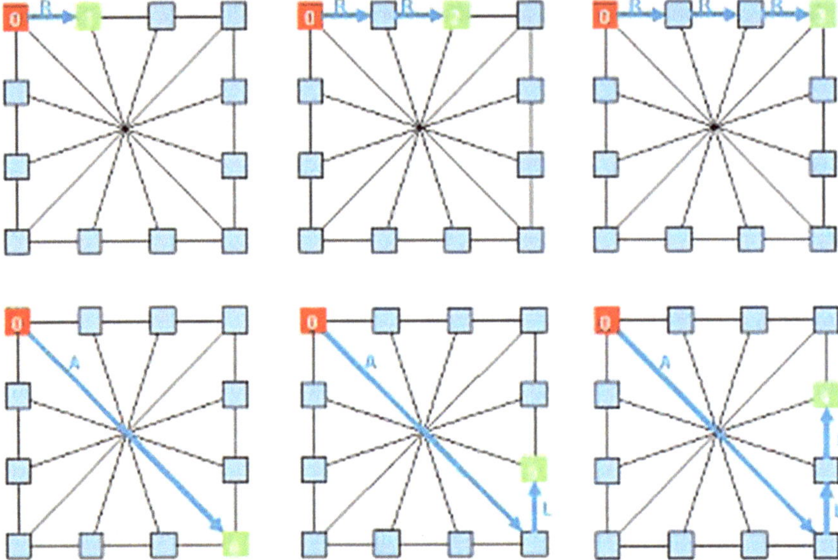


**Fig. 7.11** Across-First routing paths in SSTNoC with $N = 12$

According to the previous two properties, the Across-First scheme can be implemented as simple source-based routing, i.e., the entire path is encoded in the packet header, so each router can extract the forward information without any need of complex computation or any lookup table, implying fast routing decisions at each router. A common drawback of source-based routing is that the header path field has a variable size depending on the network size. However, this does not occur in SSTNoC because of the symmetry of the topology. Indeed the Across-First scheme describes the packet route in the packet header through a fixed size field, called dir (2 bits). These two bits specify the direction of the packet in the network with four choices (right, left, across, and then left, or across and then right).

Moreover, the routing scheme is extremely simple and fast, consisting of two steps, one at packet generation in the NI and the other along the packet path on each input port of the router.

Formally, let us assume that each router has a unique network address $i$, where i is from 0 to $N$-1 and $N$ is the network size. We define the following node identifiers: $dest$ as the address of the destination node and current as the address of the node along the path crossed by the packet. We also assume that a sense of direction field ($dir$) indicates the possible directions: right ($R$), left ($L$), left and finally across ($LA$), or right and finally across ($RA$).

At packet generation, given the destination node, the two fields in the packet header ($dest$ and $dir$) are set. The dir bits provide information on the direction of the packet route in the ring, so it is $R$ or $L$ if the first hop is respectively right or left in the ring, or $AR$ and $AL$ if the first hop is across. When the destination is the

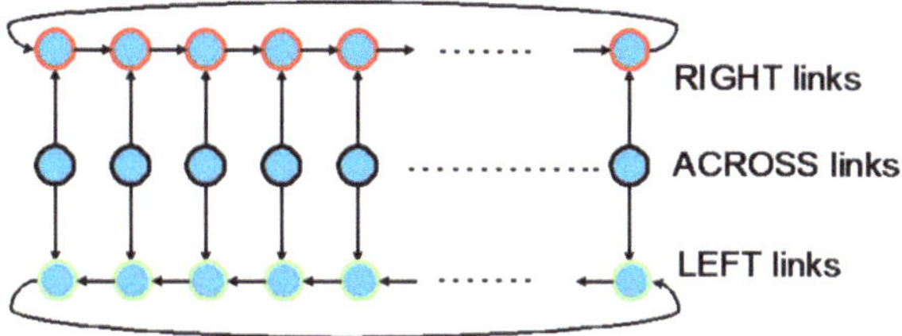


**Fig. 7.12** Channel dependency graph for Across-First routing on Spidergon STNoC

one directly connected by the cross link, the $L$ or $R$ information of the $dir$ field is meaningless; this case can be included in the $AL$ direction field, and then the $L$ value is not considered. For instance, in Fig. 7.11, when the destination is node 5, the first hop is $A$ and the $dir$ bit is set to $AL$; when the destination is node 6, the $dir$ bit is set to $AL$ as well, but actually $L$ is not used.

After this first step at packet generation, a simple algorithm is used to forward the packet at each router; the $dest$ and $dir$ fields are exploited.

```
if (current = destination)
  route packet to the NI port;
else
  route packet using dir field direction;
```

Across-First routing is appropriate not only for one-to-one (point-to-point), but also for one-to-many (broadcast, scatter), and many-to-many (total exchange) traffic configurations. However, it is not optimal for many-to-one (gather) traffic scenarios. For this scenario (and related some-to-many scenarios), the cross communication port should be selected at the end of each packet's route (called $Across - Last$, defined later in this section), thus requiring a slight modification of the above algorithm. Although, we next focus only on the point-to-point routing algorithm, necessary modifications for many-to-one (or some-to-many) scenarios are immediate and will be omitted.

If router arbitration is fair, Across-First routing is livelock and starvation free, but not deadlock free. To provide a deadlock free routing algorithm, we may restrict the routing paths that a packet may take, e.g., as in the 2-D mesh turn model [20], or alternatively use virtual channels (VCs) to break cycles in the channel dependency graph [11, 13, 17]. By examining the corresponding cyclic dependency graph (see Fig. 7.12), deadlock avoidance the Across-First scheme requires 2 virtual channels ($VC_0$ and $VC_1$) for each link on the clockwise and anticlockwise ($R$ and $L$) links. Notice that the use of FIFO virtual channel buffers at each output port implies guaranteed ordering of transmissions from packets originating at the same node. Hence, with a simple modification, Across-First algorithm can be made deadlock free: "During routing, $VC_0$ is always used, if and only if $currdestin$ '?'0, otherwise $VC_1$ is selected".

### 7.4.2 Across-First Zero VC Routing

Across-First routing can be slightly modified to become deadlock free without the use of any virtual channels. We call this strategy Across-First_zero, since it behaves like Across-First, but forbids taking a path, while traveling in the ring, that crosses node 0. The choice of node 0 (called a physical dateline) is arbitrary due to network symmetry.

With this scheme, the initial routing step has to be modified with respect to the Across-First scheme: packets crossing node 0 must be routed in the opposite direction of the ring, thus possibly not using the cross connection as a first hop. In general, this routing scheme is not shortest-path, but despite performance degradation, Across-First_zero has been introduced, since

- it can be used when low-cost is priority and performance requirements are satisfied, and
- it is foreseen that many SoC applications can be mapped in the network minimizing or avoiding traffic patterns that cross node 0, thus canceling the performance penalty of Across-First_zero with respect to Across-First.

Hybrid routing is also possible, e.g., by combining Across-Last_zero routing for the request packets and Across-First_zero routing for the response. This type of routing is critical in request/response flows, whereby many initiators communicate with few targets.

Furthermore, notice that the physical dateline assignment (e.g., node 0) for the naive algorithm leaves considerable flexibility for a more balanced virtual channel allocation. For example, if a packet route does not cross the dateline, any virtual channel can be used. These logical datelines can be exploited further to achieve a more balanced assignment of packets to virtual channels, thereby making more efficient use of network buffer space, reducing network contention, lowering communication latency and increasing network bandwidth.

Based on an application task graph, a balanced virtual channel allocation must be *fair* to all IPs, so that all accesses complete the same amount of work in approximately the same amount of time. Otherwise, system performance may be determined by the speed of the slowest IP. Balancing virtual channel allocations depends on different issues, such as:

- exact placement of datelines,
- assignment of unconstrained communication paths,
- qualitative and quantitative aspects of the communication protocol, e.g., latency hiding, e.g., multithreading, data prefetching, cache coherency [8], multipoint versus point-to-point relative bandwidth allocations, priority classes, relative size of each packet type, e.g., requests versus replies,
- multiprogramming, e.g., relative importance of system partitions, and
- application traffic, esp. dynamic (or stochastic) and static communication/ computation patterns. Dynamic patterns include models for continuous routing,

random permutations/destinations, memory/network hot spots, locality of reference, periodic data packet bursts, and latency hiding [23, 31].
- intensive communications, e.g., single and multinode broadcast (or accumulate), single node scatter (or gather), and total exchange,
- linear permutations, e.g., affine, Omega, Inverse Omega, BPC permutations, and Ascend/Descend computation patterns, and
- generalized communication patterns common in parallel algorithms, such as isotonic routing (contraction, or expansions), parallel scan (prefix), and sorting. Dynamic patterns include routing according to random permutations or destinations.

Virtual channel balancing can also help avoid head-of-line blocking, at least partially, since we usually have a limited number of virtual channels compared to $N^2$ possible source-destination pairs. In this sense, a blocked packet flit on a congested virtual channel can no longer block a packet on a different non-congested virtual channel over a common physical channel.

We have also explored related models in order to optimize virtual channel allocations for Spidergon STNoC.

### 7.4.3 Across-Last Routing

Across-Last is also a deterministic, shortest-path routing algorithm. For nodes closer than $d = ceiling(N/4)$ hops, it moves packets in the proper direction along the ring to reach the destination node. Otherwise, for nodes further, the packet moves in the opposite position with respect to the destination; the final target is reached by jumping through the cross link at the last hop. The algorithm is similar to the Across-First scheme, but the cross link, used to reach the far away nodes from the source one, is now selected as last hop, Fig. 7.13 shows the routing paths in the SSTNoC with $N = 12$, considering node 0 as starting point.

Across-Last routing is based on following principles:

- the $A$ port is selected at most once, always at the end of each packet's route;
- when the packet moves in the ring, it follows the same direction (right or left) for its entire path.

The Across-Last scheme can be implemented in a similar way to the Across-First scheme. In this case, the dir field indicates the following possible directions: right ($R$), left ($L$), left and finally across ($LA$), or right and finally across ($RA$).

At packet generation, given the destination node, the two fields in the packet header ($dest$ and $dir$) are set. The dir bits provide information on the direction of the packet route in the ring, so it is $R$ or $L$ if the first hop is right or left respectively, or $LA$ and $RA$ if the last hop is across. When the destination is the one directly connected by the cross link, the $LA$ or $RA$ cases can be used, although the $L$ or $R$ information are really meaningless. For instance, in Fig. 7.13, when the destination

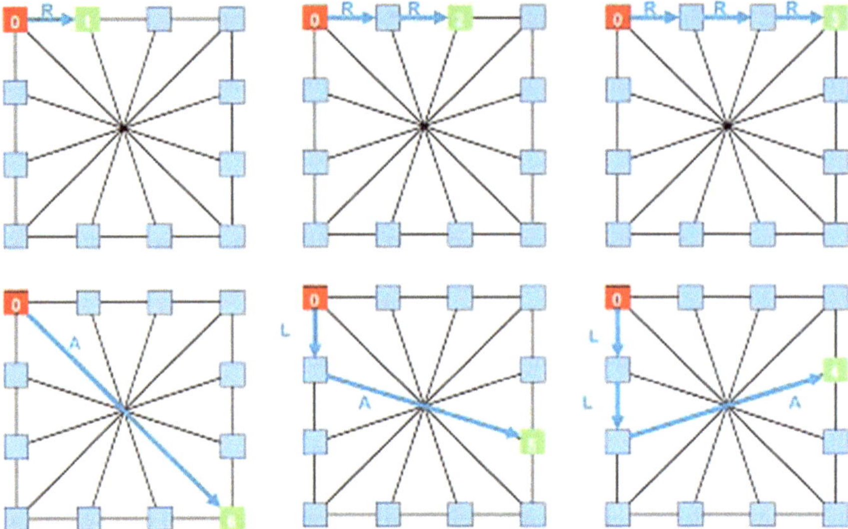


**Fig. 7.13** Across-Last routing paths in SSTNoC with $N = 12$

is node 5, the path is $L$ then $A$, and the dir bit is set to $LA$; when the destination is node 6, the dir bit is set to $LA$ as well, but actually $L$ is not used.

After this first step at packet generation, during packet routing, a simpler algorithm is used to forward the packet at each router; the destination and dir fields are exploited.

```
if (current = destination)
  route packet to the NI port;
else
  if ((current+N/2) MOD N = destination)
      route packet to the A direction;
  else
      route packet using dir field direction;
```

### 7.4.4 Across-Last Zero VC Routing

A deadlock-free Across-last algorithm would require two virtual channels along the Spidergon STNoC ring links. Hence, Across-Last Zero VC routing modifies the initial routing step with respect to the Across-First scheme: packets crossing node 0 must be routed in the opposite direction of the ring, thus possibly not using the cross connection as a last hop. Alike Across-First_zero, the physical dateline at node 0 is arbitrary and general thanks to the symmetry of the network. The concepts regarding VC balancing developed for Across-First_zero also apply here.

While traditional Spidergon STNoC routing algorithms rely on a simple form of source routing, thus allowing routers to operate at very high frequencies without complex processing or access to internal routing tables, they do not allow the SoC architect to design a platform-specific network instance. By considering source routing technology, we can enhance the interconnection capability of Spidergon STNoC technology, allowing the implementation of more general graph topologies of a maximum degree 4 (excluding network interface connection) that are fundamentally different from Spidergon interconnect and derivatives. With only a small overhead in terms of router complexity and consequently its performance, this concept provides much more freedom in terms of building an application-specific network topology. The new routing technology requires unification of the *dir* and *dest* header subfields to a single routing header and the introduction of a specialized Routing Computing Unit (RCU) which makes decisions based on a preset number of bits in the header field. Spidergon STNoC currently allows selecting any of these two routing strategies for a given network, but does not allow both spidergon and source routing to operate simultaneously on the same network. The header is created within the network interface, where routing information is computed and stored during address/source decoding. Further details fall beyond the scope of this chapter and have been omitted.

## 7.5 Fault Tolerant Routing on Spidergon STNoC

In the previous section we have introduced the SSTNoC routing algorithms. These algorithms allow routers to run at very high frequencies since the steps for routing packets are very simple and it has been possible to implement them in a very efficient way. In this section, we focus on a novel routing and destination reprogramming capability that leads to a simple, robust fault tolerant routing mechanism.

A fault within a NoC can happen at two levels: link or router node. If the design does not provide a fault tolerant routing scheme, both faults can lead to a complete failure of the multicore SoC implemented around this interconnect. Fault tolerant routing instead allows maintaining all SoC capabilities and services running. Performance can be deteriorated and therefore some high-end capabilities can be lost, but the complete system continues to run.

We now consider an example of a failure within SSTNoC and explain how to realize a fault tolerant system through the use of direction and destination reprogramming at the network interface. If a router is faulty, the network interface attached to this router cannot communicate online with the rest of the system anymore, if there is no fault tolerant routing scheme available. In case of target nodes, such as memory elements, we can build a robust system using destination reprogramming which allows the most important targets to continue to be connected even if a router connected to them stops operating.

In Fig. 7.14, if the router with ID 3 becomes faulty, then we can imagine that as a direct consequence all packets that were previously routed through this router (with

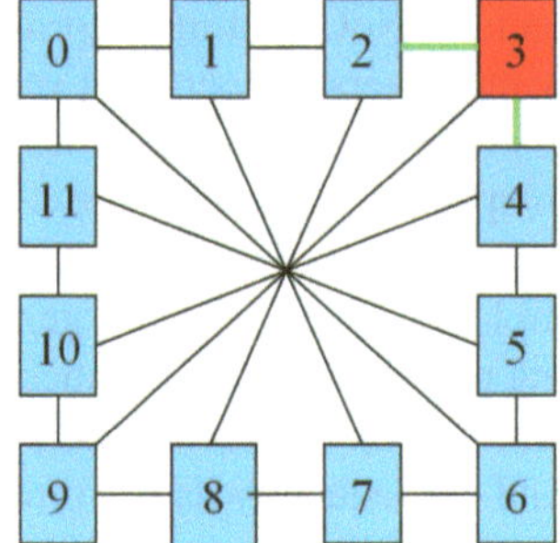


**Fig. 7.14** SSTNoC router fault

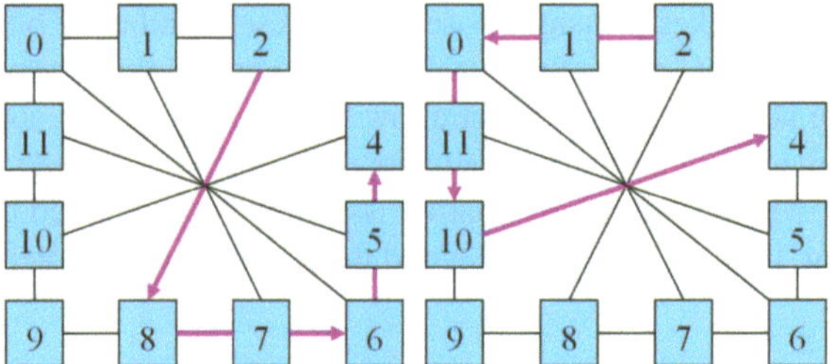


**Fig. 7.15** Two possible alternative routes for SSTNoC with router fault

any of the previously explained algorithms) would be lost, unless an appropriate fault tolerant routing scheme is established. Thus, in this case, packets from router 2 and going to router 4 would continue to follow the ring (green bold line in Fig. 7.14), and eventually they would be lost since router 3 has faulted.

Based on this example, observe that if the association between each packet destination and direction field was kept fixed, then the NoC would not be fault tolerant. Hence, for this purpose, we propose that each SSTNoC network interface connected to an Initiator or Target device can provide a sorted table containing for each possible destination the corresponding direction ($dir$) field to be used.

In this manner, SSTNoC can be made fault tolerant by reprogramming the destination and direction fields at runtime. For example, by reprogramming the table contained in the network interface attached to router 2 (see Fig. 7.14), the packet destined to router 4 can still reach its destination. While some paths in Fig. 7.14 are shortest path (a preferable path), there are several other non faulty (but possibly longer paths) that can be used, if this path becomes unreliable.

Two alternatives paths (of non-shortest length) are shown in Fig. 7.15. If we want to implement the one on the right, the direction field associated to destination 4 will be reprogrammed, taking the new value $RA$ (right and across as last hop). If the path on the left is preferred the direction field would take the value $AR$ (across first then right).

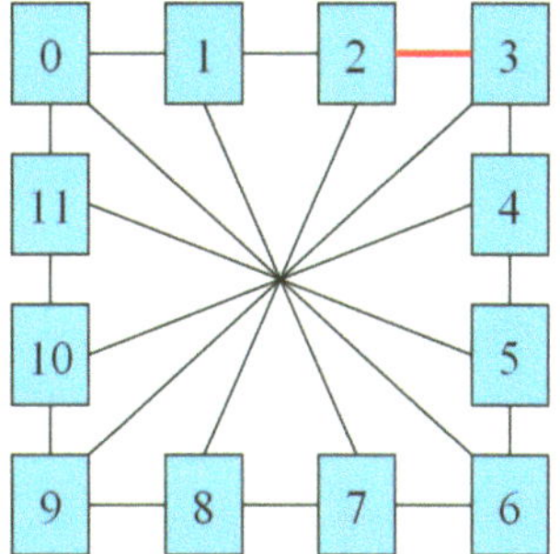


**Fig. 7.16** SSTNoC link fault

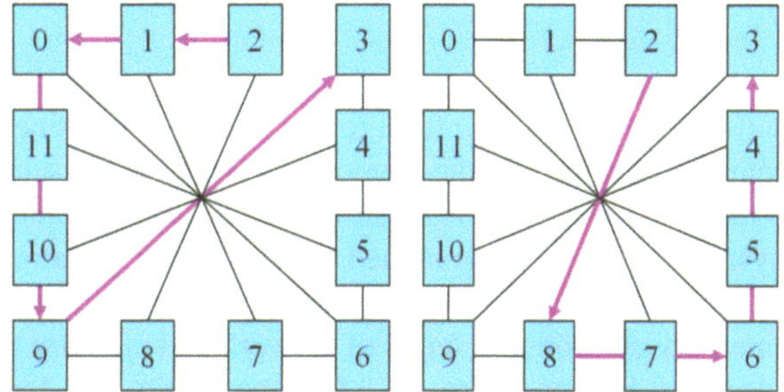


**Fig. 7.17** Two possible alternatives of SSTNoC routing (link fault)

Another cause of SSTNoC failure can be a link breakdown, as shown in Fig. 7.16. In this case, consequences of the link fault are less important than in the case of router breakdown. As before, this fault would make all packets normally traveling around the ring to stop when trying to pass through the faulty link. Similar to the previous case of a router fault, by reprogramming the direction field associated to the traffic impacted by this failure, we can still support a working system.

Fig. 7.17 shows two alternative paths (of non-shortest length) than can be exploited to overcome the link failure. Both schemes allow the traffic coming from router 2 still to reach router 3. If we want to implement the one on the right, the direction field associated to destination 4 would be reprogrammed to $AR$ (across first then right). If the path on the left is preferred the direction field would take the value $RA$ (right and across as last hop).

Thus far, we have discussed efforts toward making SSTNoC even more robust to permanent link or router faults through direction reprogramming.

This feature is often useful when attaching external components to the NoC, such as memory tiles. The following example provides motivation toward fully exploring the benefits of fault tolerant routing mechanisms based on reprogramming at the network interface.

In Fig. 7.18, we show such a system (notice that network interfaces are not drawn for sake of simplicity) where a dual port memory controller is connected to Spidergon

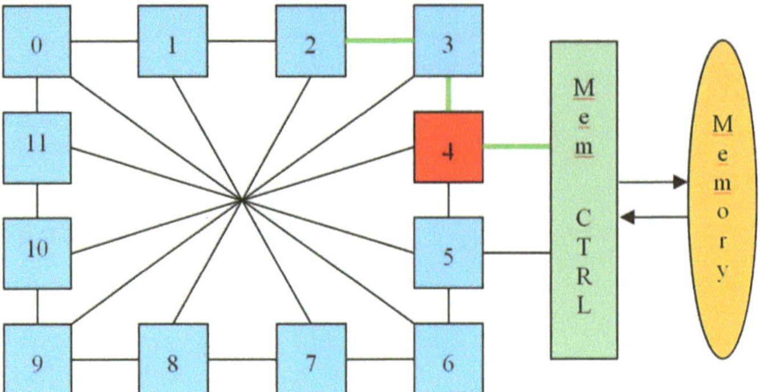


**Fig. 7.18** A dual port memory controller connected to SSTNoC

STNoC through two connections: one from router 4, and another one from router 5. The physical memory space behind the memory controller is the same. This means that the address range associated to destination router 4 and destination router 5 is the same. The system is usually programmed to allow some Initiators to access the memory through one port and some others through the second port, trying to optimize accesses to the memory controller in terms of network and memory bandwidth.

Let us consider what happens, if router 4 becomes faulty (see Fig. 7.18). Then, all masters that access the memory using this router as final destination (e.g., the one connected to router 2) cannot anymore access memory, hence in this case direction reprogramming at the network interface alone does not help. However, if in addition to direction reprogramming, we also offer destination reprogramming, then we can realize a fault tolerant routing solution.

In Fig. 7.19, we can clearly see how traffic coming from an Initiator connected to router 2 can still reach the memory (the initiator is omitted from the figure). This is achieved through reprogramming both the direction and destination: the direction field takes a new value *RA* "right and across as last" and also the destination filed takes a new value "5".

### *7.5.1 NoC Reconfigurability using Meta-NoC*

Meta-NoC has already been applied effectively in allowing the user to customize the transport header of the STNoC packet, as required in network interface reprogramming.

Fig. 7.20 shows a typical STNoC packet. Notice the distinction between transport header and network header. The first header contains all information that is end-to-end protocol-dependent, i.e., coded at the master NI and decoded at the target NI, while the second one is used by routers to transmit packets to their correct output port.

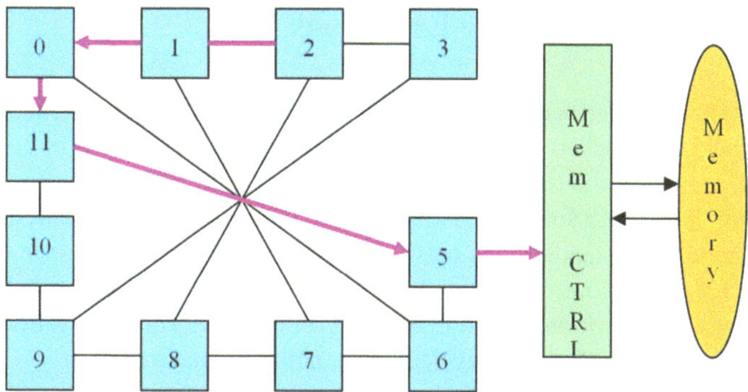


**Fig. 7.19** Fault tolerant routing with destination/direction reprogramming on SSTNoC

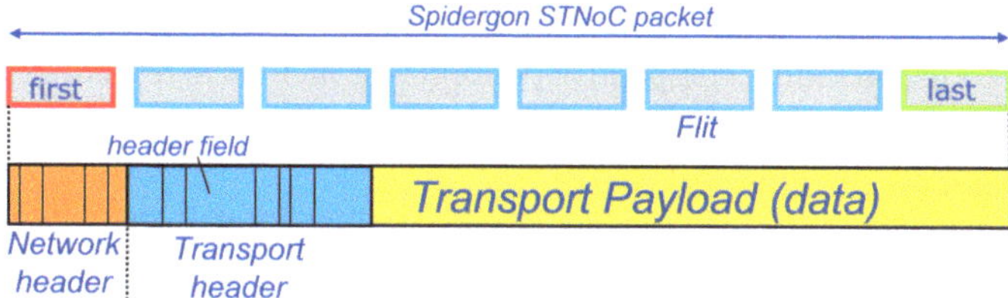


**Fig. 7.20** The STNoC packet

By fusing the necessary RTL mechanisms required to translate a user-defined transport header into the *Meta-NoC* approach, the user can introduce a TCL description of its "transport header" into the STNoC design flow which defines all required fields. These fields can be subsequently translated to appropriate RTL by applying the *Meta-NoC* design flow explained in Sect. 7.3.2.

Moreover, network header size is currently fixed to 16 bits, with only 12 bits related to source routing information. This is a potential limitation on large-scale systems realized by STNoC technology, with workarounds, such as connecting one master and one target NI back to back so that information that flows out of one "NoC port" reenters in another" one, leading to latency and resource utilization overheads. Hence, similar to the transport header, *Meta-NoC* technology can exploit the transition from a fixed 16-bit network header to a dynamically configurable $N$-bit network header specified and customized by the architect. The protocol would also enable the architect implement specialized fields in the STNoC packet header, such as error correction/detection, information dispersal, and secret sharing. Notice that, NI and router must also be instrumented to cope with dynamic processing of these specialized bits.

Notice that there are many related concepts that would be explored in the future, so this example only provides a glimpse into a powerful *Meta-NoC* approach designed for STNoC methodology.

## 7.6 Conclusion and Future Extensions

Market, application and technology developments impose new challenges for SoC design. A packet-switched NoC is foreseen as the natural evolution of current SoC buses, such as STBus and IBM CoreConnect [26], for achieving crucial cost-effective tradeoffs for future MPSoC applications. The Spidergon STNoC consists of three basic components (a standardized network interface, a high performance wormhole router and a physical communication link) configured in ST's proprietary Spidergon NoC topology. Furthermore, STNoC supports an OSI-like communication protocol stack composed of four layers: physical, data, network, and transport.

The Spidergon STNoC topology is a regular chordal ring with vertex-transitivity. Thus, all nodes have global knowledge of the network, providing for simple, local shortest-path routing, and scheduling based on virtual circuits. In addition, for practical network sizes, the Spidergon STNoC topology is a low-cost tradeoff compared to mesh or torus, providing competitive performance metrics in respect to number of links, diameter, average distance, size granularity (just 2), and embedding properties for mapping MPSoC application traffic.

The implementation of fault tolerant STNoC routing through reprogramming network interface routing registers avoids unreliable paths due to link and/or router faults failures, allowing the system to continue to work with slight degradation. The RTL has already been implemented by STMicroelectronics and integrated into the high-level *Meta-NoC* tool for industrial exploitation. Similar extensions toward a STNoC source-based routing algorithm that can overcome limitations in the number of network routers will lead toward eventual commercialization of topology-independent STNoC technology.

Due to STMicroelectronics's efforts toward the development of *Meta-NoC*, it is now possible to provide a dynamic approach toward customized design of many important ingredients of STNoC technology.

The Interconnect Design Kit outlined in Sect. 7.3.2 can be further generalized an array of tools targeting efficient design of any Interconnect Processing Unit (IPU). By definition, an IPU is any on-chip communication network (NoC) with hardware and software components which jointly implement key functions of different SoC programming models through a set of communication and synchronization primitives and provide low-level platform services to enable advanced features in modern heterogeneous applications on a single die. Thus, we anticipate that the Spidergon NoC topology will enable the transition from today's SoC integration to future generations of complex MPSoC applications, delivering competitive products with ever-increasing performance/price requirements without compromising time-to-market.

## References

1. P. Abad, P. Prieto, L.G. Menezo, A. Colaso, V. Puente, J.A. Gregorio, "TOPAZ: An open source interconnection network simulator for chip multi-processing and supercomputers", in

Proc. Symp. Networks-on-Chip, 2012, pp. 99–106.
2. Amba Bus, Arm, http://www.arm.com
3. A. Andriahantenaina, H. Charlery, A. Greiner, L. Mortiez and C. Zeferino, "SPIN: a scalable, packet switched, on-chip micro-network", in Proc. Design, Automation & Test in Europe Conf., 2003, pp. 70–73.
4. Arteris. http://www.arteris.com
5. L. Benini, G. De Micheli, Networks on Chips: A new SoC paradigm. IEEE Computer, vol. **35**(1), 70–781 (2002)
6. J.C. Bermond, F. Comellas, D.F. Hsu, Distributed loop computer networks: a survey. J. Parallel Distrib. Comput. **24**(1), 2–10 (1995)
7. N. Binkert, B.M. Beckmann, G. Black, S.K. Reinhardt et al., The GEM5 simulator. ACM SIGARCH Computer Architecture News **39**(2), 1 (2011)
8. M. Chaudhuri, M. Heinrich, Exploring virtual network selection algorithms in DSM cache coherence protocols. IEEE Transactions on Parallel and Distributed Systems **15**(8), 699–712 (2004)
9. M. Coppola, S. Curaba, M.D. Grammatikakis, R. Locatelli, G. Maruccia, F. Papariello, OCCN: a NoC modeling framework for design exploration. J. Systems Arch. **50**(2–3), 129–163 (2004)
10. M. Coppola, M. D. Grammatikakis, R. Locatelli, G. Maruccia, and L. Pieralisi, "Design of Cost-Efficient Interconnect Processing Units: Spidergon STNoC", CRC Press, 2008.
11. W.J. Dally, "Virtual-channel flow control", IEEE Trans. Parallel Distrib. Syst., C-2 (2), 1992, pp. 194–205.
12. W.J. Dally, "Performance analysis of k-ary n-cube interconnection networks", IEEE Trans. Computers, C-39 (6), 1990, pp. 775–785.
13. W.J. Dally and H. Aoki, "Deadlock-free adaptive routing in multi-computer networks using virtual channels", IEEE Trans. Parallel Distrib. Syst., C-4 (4), 1993, pp. 466–474
14. W.J. Dally and C. Seitz, "Deadlock free message routing in multiprocessor interconnection networks", IEEE Trans. Computers, C-36 (5), 1987, pp. 547–553
15. W.J. Dally and B. Towles, "Route packets, not wires: on-chip interconnection networks", In Proc. IEEE/ACM Design Automation Conf., 2001, pp. 684–689
16. J. Ding and L.N. Bhuyan, "Analysis of multi-queue buffer allocation schemes in multistage interconnection networks", Tech. Rep. - 053/93, Dept. Comput. Sci., Texas A & M University, 1993
17. J. Duato, S. Yalamanchili and L. Ni, "Interconnection networks: an engineering approach", Morgan Kaufmann, 2nd Edition, 2002
18. M. Forsell, A scalable high-performance computing solution for networks on chips. IEEE Micro **22**(5), 46–55 (2002)
19. N. Genko, D. Atienza, G. De Micheli, J. M. Mendias, R. Hermida, F. Catthoor, "A complete network-on-chip emulation framework", in Proc. Design Automation and Test in, Europe, 2005
20. C. Glass, L. Ni, "The turn model for adaptive routing", J. ACM, v. **41**(5), 874–902 (1994)
21. G. Guindani, C. Reinbrecht, T. Raupp, N. Calazans, F.G. Moraes, *"NoC power estimation at the RTL abstraction level", in Proc* (Symp, VLSI, 2008)
22. K. Goossens, J. Dielissen, J. van Meerbergen, P. Poplavko et al., *"Guaranteeing the quality of services in networks on chip", Networks on Chip* (Kluwer Academic Publishers, Eds. A. Jantsch and H. Tenhunen, 2003)
23. M.D. Grammatikakis, D.F. Hsu and M. Kraetzl, "Parallel System Interconnections and Communications", CRC press, 2000
24. F.K. Hwang, A complementary survey on double-loop networks. Theor. Comput. Sci. **263**(1–2), 211–229 (2001)
25. F.K. Hwang, "A survey on multi-loop networks", Theor. Comput. Sci., Elsevier (North Holland), 299(1–3), 2003, pp. 107–121
26. "IBM on-chip CoreConnect bus", Available from http://www.chips.ibm.com/products/coreconnect
27. Y. Ben-Itzhak, E. Zahavi, I. Cidon and A. Kolodny, "HNOCS: Modular open source simulator for heterogeneous NoCs", in Proc. Embedded Computer Systems: Architectures, Modeling, and Simulation (SAMOS), 2012

28. A. Jalabert, S. Murali, L. Benini, G. De Micheli, *"xpipesCompiler: A tool for instantiating application specific Networks on Chip", in Proc* (Automation & Test in Europe Conf, Design, 2004)
29. A. Kahng, B. Li, L.-S. Peh and K. Samadi, "ORION 2.0: A fast and accurate NoC power and area model for early-stage design space exploration", in Proc. Design Automation and Test in, Europe, 2009
30. G. Karypis and V. Kumar, "Metis: a software package for partitioning unstructured graphs, meshes, and computing fill-reducing orderings of sparse matrices (v 3.0.3)", Tech. Rep., University of Minnesota, Dept. Comp. Sci. and Army HPC Research Center, November 1997
31. F.T. Leighton, *Introduction to Parallel Algorithms and Architectures* (Academic Press, New York, 1992)
32. Z. Lu, R. Thid, M. Millberg, E. Nilsson and A. Jantsch, "NNSE: Nostrum NoC simulation environment", in Proc. Design Automation and Test in, Europe, 2005
33. Z. Lu, "A user introduction to NNSE: Nostrum NoC simulation environment", 2005, pp. 1–12, available from http://www.imit.kth.se/info/FOFU/Nostrum/NNSE/
34. J.A. Lukes, Combinatorial solution to the partitioning of general graphs. IBM Journal of Research and Development **19**, 170–180 (1975)
35. M.M.K. Martin, D.J. Sorin, B.M. Beckmann, M.R. Marty et al., Multifacet's general execution-driven multiprocessor simulator (GEMS) toolset. ACM SIGARCH Computer Architecture News **33**(4), 99 (2005)
36. B. McKay. Nauty users guide (v 1.5), Technical report, Australian National University, Dept. Comp. Sci., 2009. Available from http://cs.anu.edu.au/ bdm/nauty/nug.pdf
37. M. Millberg, E. Nilsson, R. Thid, A. Jantsch, *"Guaranteed bandwidth using looped containers in temporally disjoint networks within the Nostrum network on chip", in Proc* (Automation & Test in Europe Conf, Design, 2004)
38. F. Moraes, N. Calazans, A. Mello, L. Moller, L. Ost, Hermes: an infrastructure for low area overhead packet-switching networks on chip. Integr. VLSI J. **38**, 69–93 (2004)
39. http://www-dyn.cl.cam.ac.uk/ rdm34/wiki
40. Nirgam. http://nirgam.ecs.soton.ac.uk/
41. S.C. North. NEATO user's guide, Tech. Report, AT&T Bell Laboratories, Murray Hill, NJ, USA, October 2004. Available from http://www.graphviz.org/pdf/neatoguide.pdf
42. Noxim. http://sourceforge.net/projects/noxim
43. OMNET, available from http://www.omnetpp.org
44. V.S. Pai, P. Ranganathan, S. Adve, RSIM: Rice Simulator for ILP Multiprocessors. ACM SIGARCH Computer Architecture News **25**(5), 1 (1997)
45. V. Puente, J.A. Gregorio and R. Beivide, "Sicosys: An integrated framework for studying interconnection network performance in multiprocessor systems", in Proc. Euromicro Conf. on Parallel, Distributed, and Network-Based Processing, 2002
46. Y. R. Sun, S. Kumar and A. Jantsch, "Simulation and evaluation of a network on chip architecture using NS-2", in Proc. IEEE NorChip Conf., 2002, pp. 6
47. J. Upadhyay, V. Varavithya, and P. Mohapatra, "Routing algorithms for torus networks", Int. Conf. High Perf. Comput., 1995, pp. 743–748
48. M. Taylor, J. Kim, J. Miller, D. Wentzlaff, et al., "The Raw micro-processor: a computational fabric for software circuits and general-purpose programs", IEEE Micro, 22 (2), 20002, pp. 25–35. Also available from, http://www.cag.lcs.mit.edu/raw
49. D. Wang, N.E. Jerger, J.G. Steffan, *"DART: A programmable architecture for NoC simulation on FPGAs", in Proc* (Symp, Networks-on-Chip, 2011)
50. H. Wang, X. Zhu, L.-S. Peh and S. Malik, "Orion: A power-performance simulator for interconnection networks", in Proc. of MICRO 35, Istanbul, Turkey, November 2002
51. E.W. Weisstein, "Moore Graph", refer to http://mathworld.wolfram.com/MooreGraph.html
52. P. Wolkotte, P. Holzenspies, and G. Smit, "Fast, accurate and detailed NoC simulations", in Proc. Symp. on Networks-on-Chip, 2007
53. SystemC, http://www.systemc.org
54. C. Chevalier, F. Pellegrini, PT-Scotch. Parallel Computing **34**(6–8), 318–331 (2008)

# Chapter 8
# Middleware Memory Management in NoC

**Abstract** Since modern platforms have moved from single core to Multi-Processor Systems-on-Chip and Many-Core architectures, NoC prevails as the key solution to overcome on-chip communication problems. However, memory management has become a major challenge in improving applications performance on top of the services provided by the NoC infrastructure. Specifically, dynamic memory requests over distributed memory organizations appear to be a critical problem, since they can be unknown at design-time and unsuccessful management can lead to severe bottlenecks and excessive power consumption. In this chapter a middleware (microcode) approach to providing Dynamic Memory Management is presented in detail.

## 8.1 Introduction

The current trend in computing and embedded architectures is to replace complex superscalar architectures with many processing units connected by an on-chip network. Future integrated systems will contain billion of transistors [23], composing tens to hundreds of IP cores and the number of cores to be integrated in a single chip is expected to rapidly increase in the coming years, moving from multi- to many-core architectures. Modern embedded platforms take advantage of this manufacturing technology advancement and the Network-on-Chip (NoC) communication paradigm prevails as the solution for modern platforms. From an industrial point of view, the vision goes as far as thousand core chips [7]. Additionally, the development of such many-core architectures is driven also by the development of highly parallel/multi-threaded demanding applications.

Memory is an important contributor to the performance and power consumption of NoC platforms. As the number of on-chip cores increased, the memory content also increased from 20 % ten years ago to 85 % of the chip area today and will continue

Iraklis Anagnostopoulos, Sotirios Xydis, Alexandros Bartzas, Zhonghai Lu, Dimitrios Soudris, Axel Jantsch.

K. Tatas et al., *Designing 2D and 3D Network-on-Chip Architectures*,
DOI: 10.1007/978-1-4614-4274-5_8, © Springer Science+Business Media New York 2014

to increase in the future. Memories are preferably distributed for medium and large scale system sizes, since centralized memory has already become the bottleneck of performance, power and cost. Traditional memory optimization uses compile-time information and focuses on static allocation in respect to memory hierarchy [8]. For modern dynamic applications using many-core architectural templates, this is no longer possible since there is a lot of memory unpredictability, which cannot be captured by source code analysis alone and the increased dynamism in data storage leads to unexpected memory footprint variations unknown at design time. Insufficient memory management leads to overall performance degradation, big memory footprint and increased power consumption as shown in Fig. 8.1.

Dynamic Memory managers, also known as heap managers or allocators, are responsible for organizing the dynamically allocated data in memory and servicing the application memory requests at run-time [26]. The efficient implementation of dynamic memory managers, which can be implemented either in software, in hardware or in middleware, plays an important role to the application performance and platform's power consumption. As illustrated in [6] simple Dynamic Memory Management (DMM) implementations often form a performance and scalability bottleneck in the case of multi-threaded applications, affecting the memory and energy consumption of the overall system. Thus, customized DMM solutions are critical components during the design phase of modern NoC systems.

## 8.2 Categorization of Dynamic Memory Managers

There are three ways to offer memory management services on NoC modern Distributed Shared Memory (DSM) platforms: (a) software-only; (b) dedicated hardware; and (c) middleware-microcode.

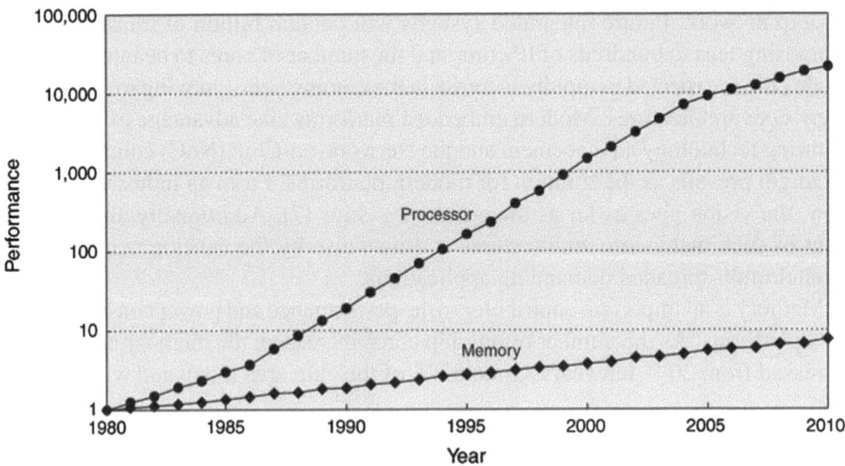


**Fig. 8.1** Performance gap between processors and memory [12]

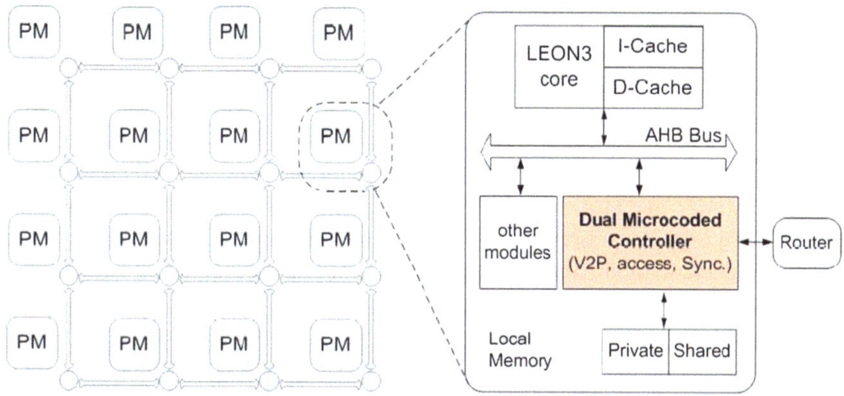


**Fig. 8.2** 16-node mesh McNoC; Processor-Memory (PM) node [10]

Historically, software-only solutions are the current practice, being flexible but consuming many processor cycles, limiting system performance. Extensive research has been conducted for general-purpose DMM, which target either the single processor [15, 26], or the multi-processor domain [6, 14, 17, 18, 25]. However, the amount of work that the DM managers perform is not always the same, it changes at run-time since the state of the heap and the number of applications accessing it varies, and it also depends on the type of calls and their parameters. Developing dynamic multi-threaded applications, using worst-case estimates for managing memory in a static manner, would impose severe overheads in memory footprint and power consumption. To avoid such type of costly over-estimations, developers are motivated to efficiently utilize dynamic memory.

Dedicated hardware solutions can achieve high performance, but any small change in functionality leads to re-design of the entire hardware module. A memory allocator which favors cache locality on specific SMP systems is proposed in S. Schneider [21]. Authors in [4] study heap management in the Cell processor, a relevant hardware architecture, but they do not handle shared memory; instead of this, the processing units have to handle their own, dedicated memory and they communicate with the system through explicit DMA calls, a limitation posed by the individual hardware platform. A hardware memory management unit (SoCDMMU), responsible for the dynamic allocation and de-allocation of memory is presented in M. Shalan and V.J. Mooney [22]. However, this is a centralized unit and could be a potential bottleneck in NoCs. Furthermore, SoCDMMU is able to allocate only complete global memory pages and the management of the data (de)allocation of the local (or private) memories is left out to the processors. A hardware MMU (HwMMU) offering dynamic allocation of data on the DSM space of an NoC is proposed in M. Monchiero et al. [19]. HwMMU supports dynamic allocation and de-allocation of shared memory with a granularity of complete memory pages, supported by new API calls.

Middleware-microcode approach is a good alternative to overcome the perfor mance-flexibility dilemma, offering a programmable and flexible solution to accelerate a wide range of applications [24]. The microcode approach has been used in previous multiprocessor systems to solve DSM related memory management issues. The Alewife [1] machine addresses the problem of providing a single address space machine with integrated message passing mechanism. However, it is a dedicated hardware solution and also does not support virtual memory. Both FLASH [16] and Typhoon [20] use a programmable co-processor for supporting flexible cache coherence policy and communication protocol. If two or more requests come concurrently, only one can compete to be handled while the others have to be delayed, resulting in contention delay. Furthermore, the FLASH and the Typhoon organize memory banks to form a cache-coherent shared memory. Memory accesses are handled by the programmable coprocessor. The SMTp [9] exploits SMT in conjunction with a standard integrated memory controller to enable a coherence protocol thread used to support DSM multiprocessors. The protocol programmability is offered by a system thread context rather than an extra programmable coprocessor. A distributed application-specific DMM for distributed shared-memory MPSoC platforms, implemented at the microcode level has been proposed in I. Anagnostopoulos et al. [2].

## 8.3 NoC Platform Example Providing Middleware Services

An NoC system able to provide mddleware services and customized microcoded DMMs is the one presented in X. Chen et al. [10]. The platform is composed of Processor-Memory (PM) nodes interconnected via a packet-switched mesh network (Fig. 8.2). A PM node is composed of a LEON3 processor with its own I-Cache and D-Cache, a Dual Microcoded Controller (DMC) and memory which can be shared among the nodes. The key module, on which the developed techniques are based for memory and data management, is the DMC, able to simultaneously serve various requests from the local core and the remote ones via the network. The platform offers base DSM services such as:

- virtual-to-physical (V2P) address translation
- synchronization
- cache coherency
- memory consistency
- shared memory access

To speed up frequent memory accesses as well as to maintain a single logical addressing space, the local memory of each node is partitioned into two parts: private and shared. Accordingly, two addressing schemes are introduced: physical addressing and virtual addressing. *The local core using physical addressing can only access the private memory. All shared memories are globally visible to all nodes and organized as a single virtual addressing space using virtual addressing and V2P translation.*

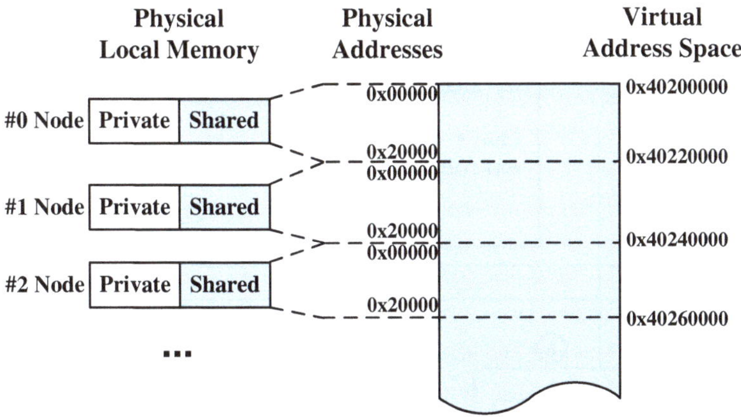


**Fig. 8.3** DSM organization and V2P translation

Such translation incurs overhead but makes the DSM organization transparent to the application and the other DSM services, thus facilitating programming.

Figure 8.3 shows the DSM organization and V2P translation. On the left, there are the platform's nodes each of which have their private and shared memory. The physical addresses of the shared part range from `0x00000` to `0x20000`. Under V2P translation in the DSM environment, all shared memories are organized as a single virtual addressing space. The application uses the virtual addresses (`0x40200000`, etc.) in order to access the shared memory and the triggering of the corresponding physical node is performed but the DMC after the V2P translation.

### *8.3.1 Microcoded DMM on NoC Platform*

DMM is a technique in which programs determine at run-time how and where dynamic data should be stored. It is needed when the amount of required memory or when the life cycle of memory usage depends on factors that are not known a-priori. For example, complex 3D multimedia applications greatly rely on DMM due to the unpredictability of the input data at compile-time. In DMM, memory is allocated from a large pool of unused memory area called the heap. A heap, is an area of memory used for dynamic data allocation and all blocks of memory are allocated and freed in an arbitrary order. The allocation pattern and the size of blocks is not known until run-time. Under the `C` programming language, dynamic memory management is performed by two basic functions, one for dynamically allocating a memory block (`malloc`), and one for returning a previously allocated block to the system (`free`). Other routines (such as `calloc` and `realloc`) are implemented on top of these two procedures.

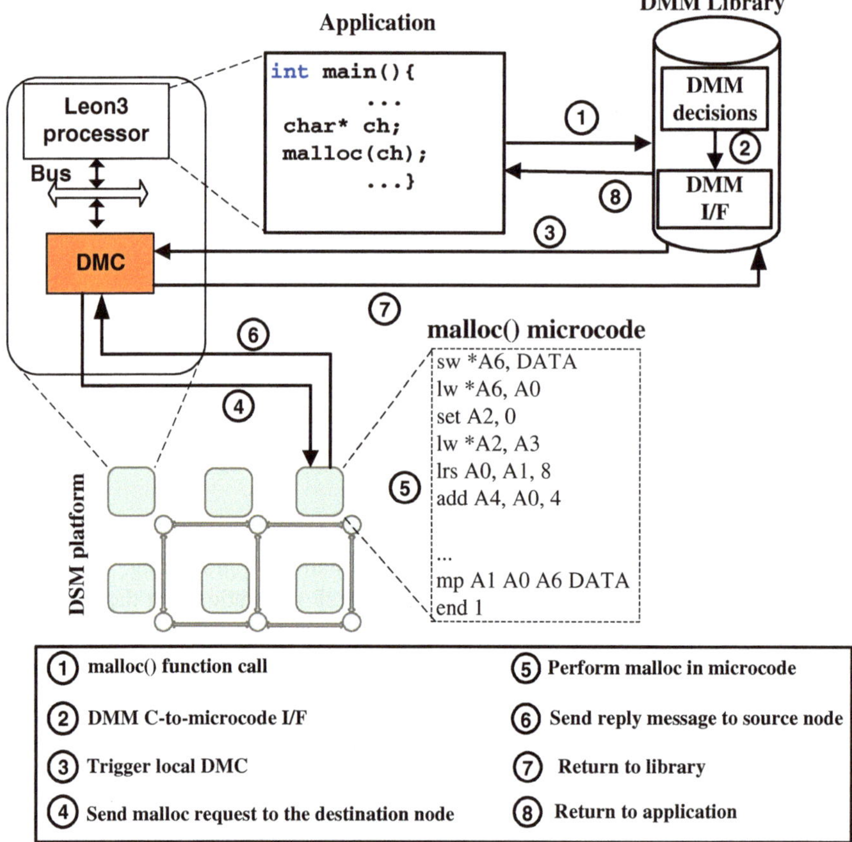


**Fig. 8.4** Overview of the MAD-DMM distributed allocation allocation procedure

DMM is based on "chunks", memory blocks that consist of application usable regions and additional in-band management information. A header is attached to each allocated object and contains its size and the size of its previous object connected them in a linked-list style. Heap chunks can be either allocated or freed. Allocated chunks are still in use by the application, whereas, freed chunks are chunks that were allocated by the application, used, and then freed. The header's metadata allows the efficient indexing of allocated/freed chunks on the appropriate linked lists and the performing of actions such as coalescing adjacent freed objects into a larger chunk when needed.

Figure 8.4 shows an overview of how an application interacts with the microcoded DM managers. The entry point is any `C` written application. When a DMM function call appears (`malloc()/free()`), the DMM library is triggered (STEP 1). The DM manager, based on the pre-selected policies handles allocated and freed lists by employing list searching, fitting and fragmentation handling techniques (Table 8.1).

**Table 8.1** DMM library decision building blocks

| Feature | Description |
|---|---|
| High-level architecture | It determines the way the dynamic memory allocator organizes and architects its heaps in order to exploit the available thread-level parallelism into memory management. Five different families exist: (1) single serial heap; (2) concurrent single heap; (3) pure private heaps; (4) private heaps with ownership; and (5) private heaps with thresholds |
| Data coherency | It deals with the synchronization mechanisms in order to ensure the data coherency in each heap |
| Inter-heap allocation | Where multiple heaps are present design parameters of this category manage the way where each thread allocate memory at the inter-heap level. Allocation in this level is strongly connected with decisions that consider both the thread clustering in order to share a heap and the thread to heap mapping. Allocation decisions of finer granularity i.e. fit policies etc. are included into the intra-heap design space |
| Inter-heap de-allocation | It includes the decisions concerning ownership-aware de-allocation of each memory block and the placement decisions for the de-allocated blocks (i.e., in which heap to be place) |
| Inter-heap fragmentation | It manages the potential memory blowup of the multi-threaded application and consider decisions in order to reduce or bound the worst memory blowup (i.e., free block movement among different heaps) |
| Block structure | It handles the data structures, which organize the memory blocks inside each heap of the allocator |
| Pool organization | It defines the pool organization inside each heap (i.e., single pool, one pool per size, traversing order etc.) |
| Block allocation and de-allocation | They deal with the operations that satisfy the allocation and de-allocation requests |
| Block splitting and coalescing | They formalize the decisions to handle the current coalescing and splitting blocks techniques, i.e. the threshold logic for coalescing and splitting the blocks |

These blocks contain information about the heap architecture, the data coherency, inter- and intra- thread (de)allocation and fragmentation, block structure and pool organization. Further details regarding these building blocks can be found in P.R. Wilson et al., D. Atienza et al., S. Xydis et al. [3, 26, 27]. Afterwards, the DM manager, through the corresponding C-to-microcode interfaces (STEP 2), triggers the local DMC (STEP 3) which sends a message to the selected, by the V2P mechanism, node (STEP 4). Then, that node performs the `malloc()/free()` functions using a microcoded implementation of the DMM which is stored in its local control store (STEP 5). Once STEP 5 is completed, a message is sent to the first node (STEP 6) and through the DMM library (STEP 7), the corresponding information is propagating to the application (STEP 8).

High-level (C/C++) implementations of DMM can be transformed to equivalent microcode functions. An example of such a tool, has been presented in [2], in which the authors built generic and fully configurable microcode templates based on the C/C++ dynamic data structures and exploit the platform's features. Figure 8.5 shows an example of the C++ to microcode translation for the First Fit [26] algorithm. Some of the parameters that can be customized by the aforementioned tool are:

- **Number and Type of Fixed List Heaps:** This type of Heaps serve fixed sized allocation requests and in a quick manner. They can be viewed as memory caching mechanisms for dynamic data thus having a great impact on DMM's performance and memory fragmentation.
- **Heap size:** The maximum assigned memory space that DMM can use for allocation request in a specific heap. In case heap size threshold is crossed, additional memory is required from the system, otherwise allocation requests fail.
- **Heap positioning:** Refers to the mapping of the overall heap organization onto the distributed memory of the platform. With microcode functions various heap organizations can be generated and mapped onto the NoC nodes. For example, heaps can be characterized according to their position either as local (heaps laying into processor nodes) or global (heaps laying into memory nodes). Heap size is closely connected with heap positioning due to different memory sizes across the NoC.

The C/C++ DMM implementations work at a high abstraction level, thus leaving to the host operating system or in the case that there is no OS to the selected policies the decision of which (part of) physical memory is accessed during allocation requests. However, the management of accessing physical memory becomes dominant in NoC architectures due to memory distribution over the platform. So, the question that arises here is which part of the distributed memory should be used for the allocation?

Authors in I. Anagnostopoulos et al. [2] increase the performance of microcoded DMM by instructing nodes to try allocation according to topology criteria. For this reason, at design time and based on *topology criteria*, priority tables $PT_{s,d}$, $(s, d \in N)$ are built for each node $N$ of the NoC platform. $P \in N$ represents the processing nodes of the MPSoC and $M \in N$ represents the memory ones. $PT_{s,d}$ describes the priority weight of source $s$ accessing destination $d$. $PT_{s,d}$ priorities are exploited at run-time guiding the allocation to (neighboring) nodes according to $PT_{s,d}$ table, starting from the node with the highest priority. The $PT_{s,d}$ value is defined in Eq. 8.1 [2].

$$PT_{s,d} = \frac{w_1 P_{s,d} + (1 - w_1)(w_2 ML_d + (1 - w_2)(w_3 MP_d + (1 - w_3 D_{s,d})))}{\sum_{\forall i}^{(i \neq d)} \{w_1 Ps.i + (1 - w_1)(w_2 ML_i + (1 - w_2)(w_3 MP_i + (1 - w_3 D_{s,i})))\}} \tag{8.1}$$

where $i \in M$, $P_{s,d}$ and $D_{s,d}$ are the power consumption and delay of the $(s, d)$ link respectively. $ML_d$ and $MP_d$ are the memory latency and memory power

C/C++ code

```
inline void * first_fit_free_list (size_t sz) {
      // Check the free list first.
      freeObject * prev = &head;
      while (prev->next != &tail) {
        if (Header::getSize(prev->next) >= sz) {
          freeObject * ptr = prev->next;
          prev->next = ptr->next;
          if (prev->next == &tail) {
            tail.next = prev;
          }
          return (void *) ptr;
        }
        prev = prev->next;
      }
      return 0
}
```

C/C++ to DMC Microcode

```
FIRSTFIT:
set A1 {N_FREES}            ; If there are no empty blocks perform an INSERT
set A6 {G_HEAP_START}
beqz A1 INSERT
LOOP:
set A4 0       ; Check the indexes of the single linked list,
               ;to check where is the      empty block
add A6 A6 4
lw *A6 A2
lrs A2 A4 1
beqz A4 FOUND_EMPTY_BLOCK    ; Empty block found?
add A6 A6 A2
jmp LOOP
FOUND_EMPTY_BLOCK:    ;As soon anempty block is found we check
               ;whether we can accommodate the allocation request

set A5 DATA
bleq A5 A2 ADD ;If so, INSERT (place) the element to that address
sub A1 A1 1
beqz A1 ADD           ;If not, check other free blocks (if any), else INSERT
                      ;as a new element to the end of the Allocated List
Add A6 A6 5
add A6 A6 A2
jmp LOOP
```

**Fig. 8.5** Code translation example. From C++ to microcode. First Fit algorithm

consumption per access of the $d$ memory respectively. Also, $\sum_{i=1}^{3} w_i = 1, w_i \geq 0$, are the weights for configuring the cost function.

Figure 8.6 shows an illustrated example of customized microcoded DMM according to memory distribution. We assume a $3 \times 3$ NoC architecture which consists of three memory nodes ($M_{(0,1)}$, $M_{(1,2)}$ and $M_{(2,0)}$). There are three processing nodes executing threads ($\mathrm{Th}^{d}_{(0,2)}$, $\mathrm{Th}^{d}_{(1,0)}$, $\mathrm{Th}^{d}_{(1,1)}$) with dynamic allocation operations and the rest execute code with static data ($\mathrm{Th}^{s}_{(0,0)}$, $\mathrm{Th}^{s}_{(2,1)}$, $\mathrm{Th}^{s}_{(2,2)}$). All processing nodes have their own Local Memory (LM). Topology aware DMM customization manages

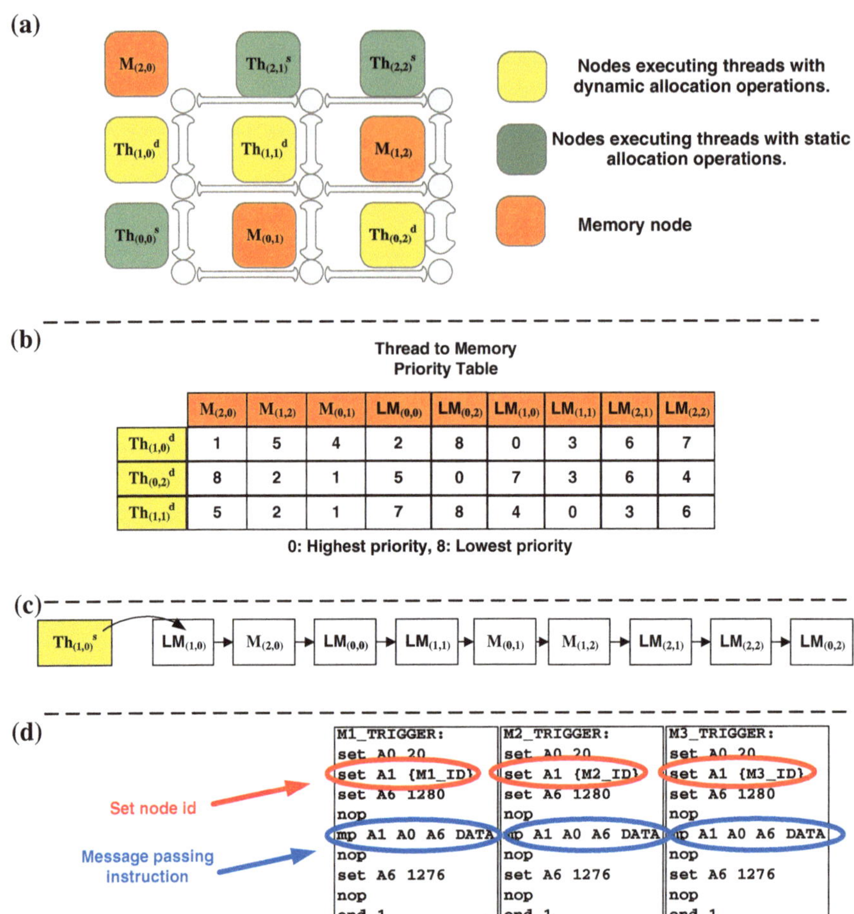


**Fig. 8.6** NoC memory distribution-aware DMM customization example. **a** Selected topology and mapped cores. **b** Thread to memory priority table. **c** SLL structure. **d** Microcoded topology aware function templates

threads with dynamic data. According to Equation 8.1 the priority access table is built and presented in Fig. 8.6b (0 = highest priority, 8 = lowest priority).

Looking at the priority table (Fig. 8.6b) for each $\mathrm{Th}^{d}_{(i,j)}$ a Single Linked List (SLL) structure is built containing microcoded memory distribution-aware functions responsible for triggering the correct physical memory when needed. The SLL structure is presented in Fig. 8.6c. The microcode templates responsible for triggering remote memory nodes are presented in Fig. 8.6d. Microcoded message passing policy has been selected to propagate information to neighboring nodes. The microcoded functions are totally independent and transparent to DMM's code. They are placed at the end of the code and they are automatically triggered when the local node asks for a remote (de)allocation request. With these functions, the execution of microcode to a different node is allowed even if the remote node hasn't received any signal from its own local core.

## 8.3.2 Evaluation of Microcoded DMM

The application used in order to evaluate and compare the microcoded DMM, is a combination of several real-life kernels (performing packet processing, encryption, scheduling, etc.) that are present in network applications [5] and it is presented in Fig. 8.7. The application consists of 5 kernels which are triggered by wireless streams.

The system was triggered with a set of traces from a real wireless network. The representativeness of these traces is assured by the fact that they were obtained from different network "sniffers" in different buildings of the University Campus. The software application is fully multi-threaded as it is increasingly common in computing systems: each kernel is executed in its own independent thread and communicates asynchronously with the other kernels through asynchronous FIFO queues.

- Network traffic corresponding to activities like VoIP, FTP and web browsing, and which have been studied in T. Henderson [11]. This execution thread feeds the entire system with data packets containing the network traffic. The information available for each data packet are: time, IP addresses of the source and destination, source and destination ports and the size of the package. This thread engages the memory needed for the data packet (without including the information in the header).
- Creation of a TCP / IP packet. This thread is responsible for assembling a complete TCP/IP packet completing the information needed in the packet header. We can liken the operation of this thread by calling the `write()` (a system call), and

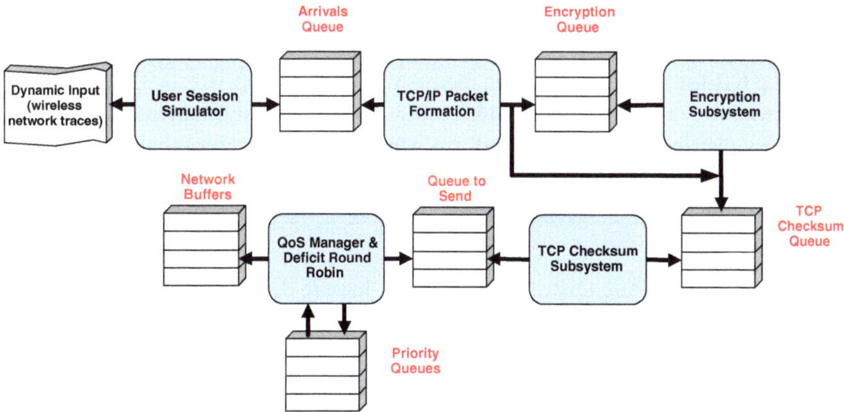


**Fig. 8.7** The used multi-threaded application. Squares define the different threads which communicate asynchronously through asynchronous FIFO queues [5]

thereby built in the entire package. The total packet size is increased by 40 Byte (as is the size of the title). The completed packet is inserted in the corresponding queues to be encrypted or for error checking (TCP checksum) depending on whether the connection is encrypted or not.
- Encryption (packages that are part of an encrypted connection are encrypted according to the DES algorithm). This thread reads data (payload) in packet block sizes of 8 Byte, and after the encryption it transfers the packet in the queue where packages wait for error checking.
- Creation of the TCP checksum. This thread calculates the checksum by applying the procedure described in Information Sciences Institute [13]. The package contents are read by 16 bit and the thread applies to them the corresponding process. Once the checksum is created it is written in the CRC field header and then the package is driven to the next queue and execution thread.
- The quality service manager (QoS manager) builds a list of the destinations of different packets and uses priorities for managing them. Whenever a packet enters the system, it is placed in one of the queues based on its priority type. The packets are extracted from these queues and they are forwarded to the network output according to the Deficit Round Robin (DRR) algorithm. When a packet is forwarded to the output then the weight of the particular queue is decreased according to the packet size.

Authors in I. Anagnostopoulos et al. [2] tested the network application with two microcoded DMMs generated by their framework under four different platform configurations. The description of the microcode DMM is depicted in Table 8.2. Column 2 depicts the application-specific characteristics (Number and type of fixed size freelists) of each of the selected DMMs.

The topology used for the evaluation of the middleware acceleration is a 2×2 NoC. Nodes $(0, 0)$, $(0, 1)$, $(1, 0)$ are processing nodes with their own local memory. Specially each of $(0, 1)$, $(1, 0)$ execute 2 threads, one that handles dynamic data and another that handles only static data. Node $(0, 0)$ executes only one thread that handles dynamic data. Node $(1, 1)$ is a memory node that serves all requests that can't be handled by local memory. For local memories the Heap size is 4 KB(2 KB for fixed lists and 2 KB for free lists). For the memory node the Heap size is 32 KB. For the selected topology four different DMM configurations were implemented depending on memory distribution over the platform. Directed edges present that an allocation request is possible to the destination from the source node while weights, based on $PT_{s,d}$, show the priority of choosing the destination node (0 = highest priority, 3 = lowest priority).

- **Configuration 1:** Pure Distributed Memory. In pure distributed memory configuration (Fig. 8.8), each node sends allocation requests for dynamic data to its Local Heap. There is no Global Heap.
- **Configuration 2:** Centralized single Heap. In centralized single Heap configuration (Fig. 8.9), each node sends allocation requests for dynamic data only to Global Heap $(1, 1)$. There are no Local Heaps.

**Table 8.2** Description of the Selected DMM Configurations [2]

| | | Code size (# microcode instructions) | | | |
|---|---|---|---|---|---|
| DMM | Description | Configuration 1 | Configuration 2 | Configuration 3 | Configuration 4 |
| DMM 1 | $\text{FixList}_0(block = 40B)$<br>$\text{FixList}_1(block = 1460B)$<br>$\text{FixList}_2(block = 1500B)$<br>Generic heap | 1407 | 485 | 1736 | 1859 |
| DMM 2 | $\text{FixList}_0(block \in [0B, 40B])$<br>$\text{FixList}_1(block \in [1280B, 1460B])$<br>$\text{FixList}_2(block \in (1460B, 1500B])$<br>$\text{FixList}_3(block = 92B)$<br>Generic heap | 1467 | 485 | 1796 | 1919 |

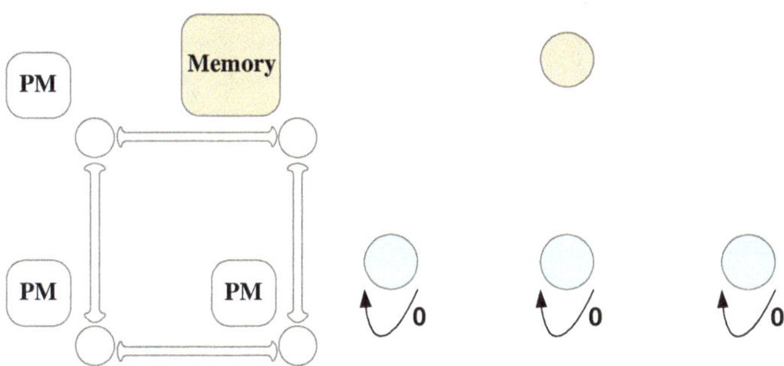


**Fig. 8.8** Pure Distributed Memory [2]

- **Configuration 3:** Distributed multiple-Heap with global Heap. In distributed multiple-Heap with global Heap configuration (Fig. 8.10), each node first sends allocation requests to its Local Heap. If Local Heap is not able (due to lack of space) to serve any more allocation requests, the request then is sent to Global Heap (1, 1).
- **Configuration 4:** Memory distribution-aware multiple-Heap with global Heap. In memory distribution-aware multiple-Heap with global Heap configuration (Fig. 8.11), each node first sends allocation requests to its Local Heap. If Local Heap is not able (due to lack of space) to serve any more allocation requests, then, *according to priorities*, the Global Heap or the Local Heap of another node is selected in order to serve the allocation request.

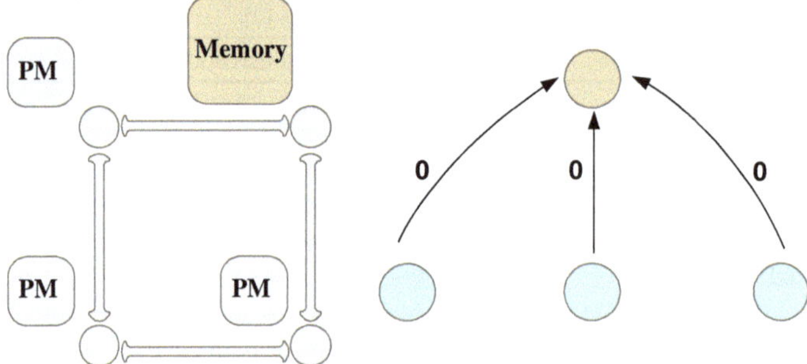


Fig. 8.9 Centralized single Heap [2]

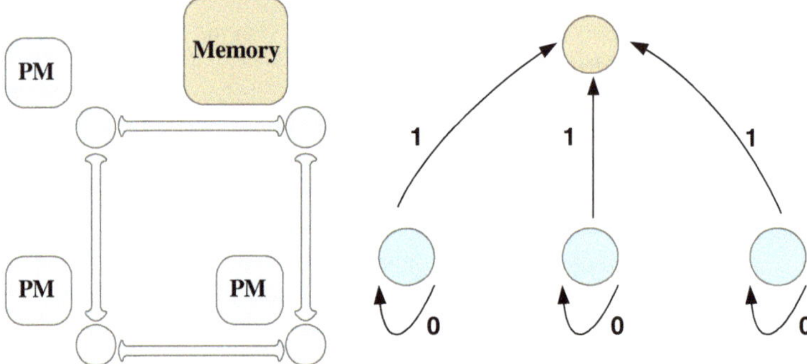


Fig. 8.10 Distributed multiple-Heap with global Heap [2]

For the two selected DMMs and for each of the four aforementioned configurations, authors in I. Anagnostopoulos et al. [2], compared the microcode implementation against the equivalent C implementation on the LEON3 processor in terms of: (i) the cycles performed until a Heap memory overflow event appears, (ii) the DMM event distribution and (iii) the microcode performance. According to Fig. 8.12, when DMM is aware of the memory distribution, the time in which Heap overflow appears, increases. Specifically, configuration 4 performs 7× more cycles for DMM 2 compared to configuration 1. Above each bar the actual count of served DMM functions ***(Local Heap/ Global Heap)*** is presented until Heap memory overflow appears. Additionally, microcode implementation serves the same number of DMM events in fewer cycles, performing faster, up to 25%, than its corresponding C implementation.

Figure 8.13 shows: (i) the average accelerator cycles, (ii) the cycles spent due to memory stall and (iii) the average energy consumption (pJoule) consumed per DMM function for DMM 1 and 2.

Configuration 1 even though it is the fastest one, it is the first one that consumes all the available memory for handling dynamic data, and since the system is not available to provide any additional memory, and hence the system fails. Configuration 2 is the slowest since most of time is spent on synchronization and lock memory issues.

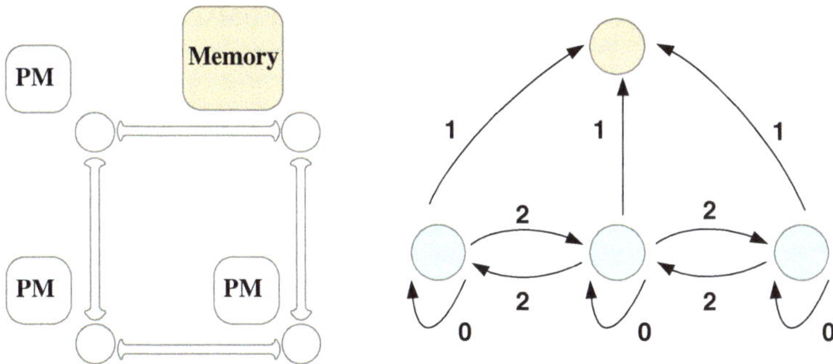


Fig. 8.11 Memory distribution-aware multiple-Heap with global Heap [2]

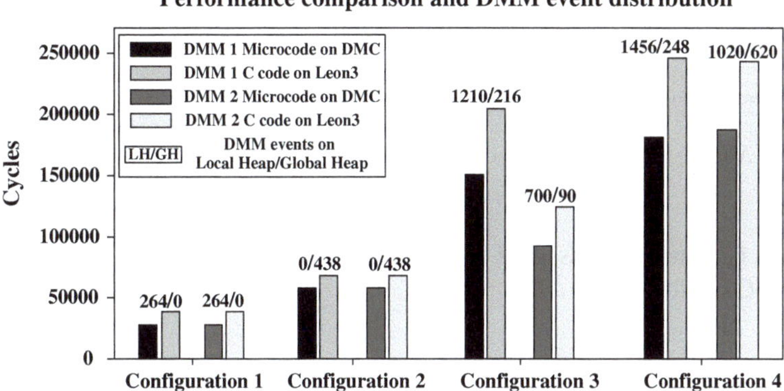


Fig. 8.12 Performance comparison and DMM event distribution [2]

Configuration 3 offers good performance but Configuration 4 is the best among all due to the bigger variety of available heaps. However, Configuration 4, consumes approximately 25% more energy in comparison to Configuration 1 due to the often communication for detecting the most available Heap to use.

## Questions

**8.1** What is the difference between software and middleware DMM implementations?
**8.2** Why hardware DMM cannot be considered as a wide method for serving allocation requests?
**8.3** Name the basic decision blocks for building a dynamic memory manager.
**8.4** Name some middleware services provided in NoCs.
**8.5** What is the message passing technique and how can it be used for microcoded DMM?

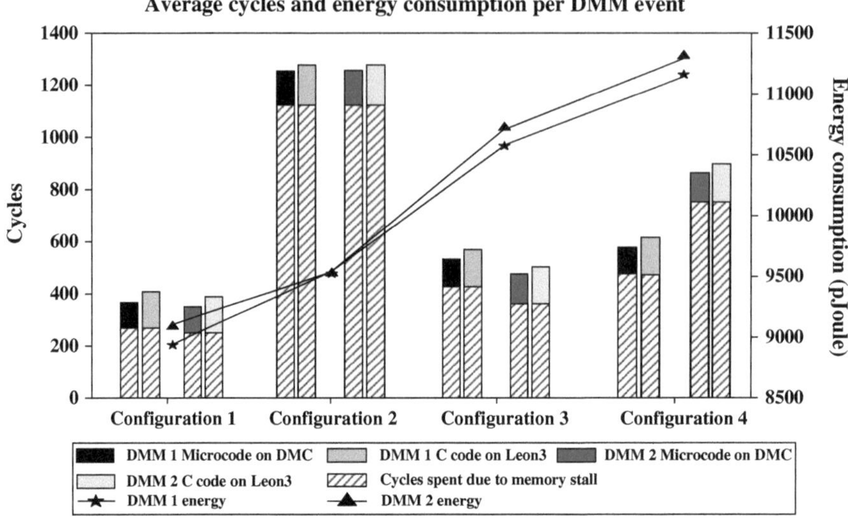


**Fig. 8.13** Average cycles and energy consumption per DMM event [2]

**8.6** Is cache coherence essential for microcoded dynamic memory management in DSM NoC platforms?

## Problems

**8.7** Transform the BEST, LIFO and EXACT fitting policies from C to microcode pseudocode.
**8.8** Assume a 3×3 torus NoC. Build the appropriate priority tables from the example depicted in Fig. 8.6.
**8.9** Explore the computational effort and cost for handling priority tables in an 8×8 and 10×10 NoC. Will there be a delay problem? Provide an appropriate solution.

## Projects and Lab Exercises

**8.10** Design and simulate a cache coherence protocol for DSM NoC platforms.
**8.11** Explore the power consumption of different fitting policies while allocating memory in a 3×3 NoC.
**8.12** Explore the power consumption of different fitting policies while allocating memory in a 3×3 NoC.

## References


1. A. Agarwal et al., The MIT alewife machine: architecture and performance, in *Proceeding of ISCA*, 1995, 2–13
2. I. Anagnostopoulos et al., Custom microcoded dynamic memory management for distributed on-chip memory organizations. IEEE Embed. Syst. Lett. **3**, 66–69 (2011)
3. D. Atienza et al., Systematic dynamic memory management design methodology for reduced memory footprint. ACM Trans. Design Autom. Electr. Syst. **11**, 465–489 (2006)
4. K. Bai, A. Shrivastava, Heap data management for limited local memory (LLM) multi-core processors, in *CODES + ISSS*, 2010, pp. 317–326
5. A. Bartzas et al., Software metadata: systematic characterization of the memory behaviour of dynamic applications. J. Syst. Softw. **83**, 1051–1075, (2010)
6. E.D. Berger et al., Hoard: a scalable memory allocator for multithreaded applications, in *Proceeding of ASPLOS*, (ACM, Cambridge, MA, 2000), PP. 117–128, 0362–1340
7. S. Borkar, Thousand core chips: a technology perspective, in *Proceeding of DAC*, 2007, pp. 746–749
8. F. Catthoor et al., *Data Access and Storage Management for Embedded Programmable Processors* ( Kluwer Academic Publishers, Boston, 2002)
9. M. Chaudhuri, M. Heinrich, SMTp: an architecture for next-generation scalable multithreading, in *Proceeding of ISCA*, 2004, 124–135
10. X. Chen et al., Supporting distributed shared memory on multi-core network-on-chips using a dual microcoded controller, in *Proceeding of DATE*, Dresden, Germany, 2010, 39–44
11. T. Henderson, D. Kotz, I. Abyzov, The changing usage of a mature campus-wide wireless network, in *Proceeding of MobiCom*, 2004
12. J.L. Hennessy, D.A. Patterson, *Computer Architecture: A Quantitative Approach* (Morgan Kaufmann, San Francisco, 2007)
13. Information Sciences Institute, RFC 793: Transmission control protocol, http://tools.ietf.org/html/rfc793, 1981
14. A. Iyengar, Parallel and distributed processing, in *Proceedings of the Fifth IEEE Symposium on Parallel Dynamic Storage Allocation Algorithms*, 1993, pp. 82–91
15. V. Kiem-Phong, Vmalloc: a general and efficient memory allocator. Soft. Practice Experience **26**, 1–18 (1996)
16. J. Kuskin et al., The stanford FLASH multiprocessor, in *Proceeding of ISCA*, 1994, 302–313
17. P. Larson, M. Krishnan, Memory allocation for long-running server applications, in *Proceedings of the 1st International Symposium on Memory Management*, 1998, PP. 176–185
18. D. Lea, A memory allocator (1996), http://g.oswego.edu/dl/html/malloc.html,
19. M. Monchiero et al. Exploration of distributed shared memory architectures for NoC-based multiprocessors. JSA, **53**, 719–732, (2007) (Elsevier North-Holland, Inc)
20. S.K. Reinhardt et al., Tempest and typhoon: user-level shared memory, in *Proceeding of ISCA*, 1994, 325–336
21. S. Schneider, Scalable locality-conscious multithreaded memory allocation, in *Proceeding of the ACM SIGPLAN International Symposium on Memory Management*, 2006
22. M. Shalan, V.J. Mooney, Hardware support for real-time embedded multiprocessor system-on-a-chip memory management, in *Proceeding of CODES*, (ACM, Estes Park, Colorado, USA), 2002, PP. 79–84
23. SIA, Semiconductor Industry Association, International Technology Roadmap for Semiconductors (2011), http://www.itrs.net/Links/2009ITRS/Home2011.htm
24. S. Vassiliadis et al., Microcode processing: positioning and directions. IEEE MICRO. **23**, 21–30 (2003)
25. V. Voon-Yee, H. Wen-Jing, A scalable and efficient storage allocator on shared-memory multiprocessors, in *International Symposium on Parallel Architectures, Algorithms, and Networks*, I-SPAN, 1999, pp. 230–235
26. P. R. Wilson et al., Dynamic storage allocation: a survey and critical review, in *Proceeding of IWMM* Kinross, Scotland, UK, Springer-Verlag (1995), pp. 1–116, 3–540-60368-9

27. S. Xydis et al., Custom mutli-threaded dynamic memory management for multiprocessor system-on-chip platforms, in *Proceeding of ICSAMOS*, Samos Island, Greece, pp. 102–109, 2010

# Chapter 9
# On Designing 3-D Platforms

**Abstract** Several new topologies for on-chip interconnect networks are supported by vertical integration. These three-dimensional (3-D) topologies improve the performance of an on-chip network primarily in two ways. The length of the physical links connecting the switches of the network is shorter. Additionally, the data can be routed across the on-chip network through a smaller number of switches. 3-D NoC topologies include two different types of physical links implemented with horizontal and vertical interconnects. Among others, these links exhibit differentiations in terms both of physical, as well as electrical characteristics. Though a number of topology exploration frameworks for quantifying the potential improvements from this new design paradigm, the assumptions made from the majority of them usually leads to results with considerable variation as compared to the actual 3-D platforms. On the other hand, there are only a few CAD tools for designing 3-D chips (e.g., R3Logic [1]). Throughout this chapter we introduce a framework for quantifying the potential gains of employing this new design technology onto digital designs. In contrast to relevant approaches, which are mainly based on models from academic tools, the solution discussed here is based on Cadence toolflow [2].

## 9.1 3-D Integration

Traditional technology scaling has been the engine fueling the ever increasing trend toward system miniaturization and functionality integration. It is unfortunately running out of steam, though, as requirements are becoming more stringent and physical/manufacturing limitations are being encountered. Hence, integrating more functionality in a smaller form factor with lower power consumption pushes traditional semiconductor technology scaling to its limits. This has generated many

This chapter was contributed by Dionysios Diamantopoulos, Kostas Siozios, George Economakos, and Dimitrios Soudris of the School of ECE, National Technical University of Athens.

K. Tatas et al., *Designing 2D and 3D Network-on-Chip Architectures*,

DOI: 10.1007/978-1-4614-4274-5_9, 

discussions concerning the end of device scaling as we know it, and has hastened the search for solutions beyond the perceived limits of current 2-D architectures.

Stacking multiple die in the vertical axis and interconnecting them using very fine-pitch Through-Silicon Vias (TSVs) enables the creation of chips that combine different process technologies and exhibit higher performance. 3-D chip stacking is touted as the silver bullet technology that can keep Moore's momentum and fuel the next wave of consumer electronic products. This chapter introduces a framework that enables rapid evaluation of 3-D SoCs with existing physical design tools.

Along with the technology updates, there are several published works dealing with the 3-D physical design problem [3, 33]. Among others, tools for partitioning, floor-plan [4], placement, and routing for 3-D architectures, have been proposed. Most of them apart from wire-length reduction also address issues related to reliability and thermal [5] that become important for the implementation of 3-D stacked ICs. Moreover, from modeling electrical characteristics, thermal and cost modeling is also an important design issue [6].

A methodology for evaluating the impact of 3-D microarchitectural designs on overall system performance is presented in [7]. The developed automated physical design and architecture performance estimation flow for 3-D systems includes tools for 3-D floor-planning, routing, and thermal via insertion.

Existing approaches for physical design pay effort to improve the performance metrics of 3-D architectures by letting systems components to be swapped among dies. However, this stems from the assumption that all stacked dies are made in the same process technology. Hence, heterogeneous integration is not supported by the majority of the algorithms/tools presented in literature.

Modeling of 3-D stacks is also receiving attention. Apart from studying the electrical characteristics, thermal and cost modeling are becoming significant issues. An industrially-oriented cost analysis tool for the manufacturing of 3-D ICs is presented in [8]. It evaluates the cost of a given TSV process flow by taking into account the capital investment needed for 3-D enabled foundries, as well as operating costs and process yield to evaluate the overall cost of 3-D stacks. A similar work for evaluating a number of performance and cost metrics regarding 3-D stacks can be found in [9]. Thermal modeling [10–12] is also gaining a lot of attention mainly due to the increased power density that comes with 3-D integration.

These approaches are based almost exclusively on academic tools. On the other hand, the only known commercial framework for supporting the design of 3-D SoCs is provided by R3Logic Corporation [1].

The last major track is case studies that illustrate the advantages offered by 3-D integration at system-level performance. The majority of these works are focused on optimizing memory organizations, where the benefits of going to a 3-D system become clearer (due to higher bandwidth).

In this chapter, we introduce a novel framework for supporting rapid evaluation of 3-D SoCs with the usage of existing 2-D CAD tools. Such a framework is crucial even before physical design tools for the 3-D domain become commercially available, since it provides a good estimation about the potential benefits from designing 3-D chips.

For demonstration purposes we depict how this framework is applicable on deriving 3-D instances of two test-cases. Specifically, the first of them affects an IP core, namely the LEON3 processor [13], whereas the second one provides results for a NoC-based interconnection scheme based on 16 routers.

## 9.2 Problem Formulation

In this section, we formulate the problem of implementing a complex digital system with 3-D process technology. More specifically, the problems we tackle are: (*i*) the 3-D stack generation, and (*ii*) the physical implementation of 3-D Stack.

**Definition:** *Application Graph*
We consider as application graph a directed graph App$G(F, N)$, where each vertex $f_i \in F$ represents an application's functionality, while the directed edge $n_{(i,j)} \in N$ corresponds to the communication between logic functions $f_i$ and $f_j$. The weight of the edge $n_{(i,j)}$ denoted as $\text{com}_{\text{weight}(i,j)}$, represents the communication load/bandwidth between vertexes $f_i$ and $f_j$.

**Definition:** *Platform Graph*
We consider as platform graph a directed graph Plat$G(C, W)$ where each vertex $c_i \in C$ represents an element of the target architecture (e.g., logic block, processor, memory, etc.), while the directed edge $w_{(i,j)} \in W$ represents a communication path between the hardware elements $c_i$ and $c_j$. The weight of the edge $w_{(i,j)}$, denoted as $\text{inter}_{\text{weight}(i,j)}$, denotes the fabricated interconnection hardware resources among these logic blocks.

**Problem1:** *3-D Stack Generation*
Given the application App$G(F, N)$ and the design requirements (in terms of maximum number of layers, acceptable interlayer connections, power/delay constraints, etc), find candidate 3-D stacks under different design criteria. During the 3-D stack generation, the application's components $(l_i)$ have to be assigned at architecture layers and then to appropriately order them to meet design constraints (i.e., minimize power/delay/number of interlayer connections, etc). The outcome of this problem is a directed graph, named 3-DStack$G(F, L)$, which represents the layer $(l_i \in L)$ where each of the application's functionalities is assigned to.

**Problem2:**
Physical Implementation of 3-D Stack Given the application App$G(F, N)$, the platform Plat$G(C, W)$, as well as the 3-D stack 3-DStack$G(F, L)$ graphs, find a physical implementation that meets the system's requirements.

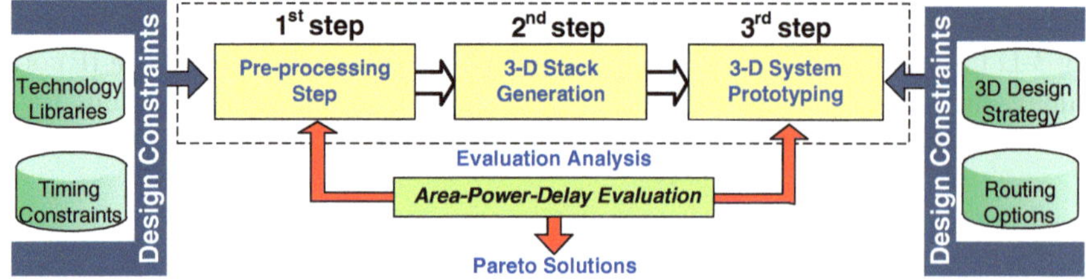


**Fig. 9.1** Proposed framework for supporting rapid evaluation of 3-D SoCs

## 9.3 Employed Framework for Quantifying 3-D Chips

Figure 9.1 depicts the proposed framework, which consists of three modular steps in order to enable interaction with tools from similar and/or complementary flows. More specifically, the steps of our framework are summarized as follows:

- *Pre-processing Step*: Verification of functional integrity for the design and extraction of its XML description.
- *3-D Stack Generation*: Generation of the 3-D stack and determination of the communication (routing paths) among layers.
- *3-D System Prototyping*: Physical implementation of 3-D SoC and evaluation of the derived solution.

## 9.4 Pre-processing Step

The first step in our methodology is depicted in Fig. 9.2. Initially, the architecture's HDL description (i.e., VHDL, Verilog) is simulated under various parameters and constraints (e.g., clock period, on-chip memories organization) in order to verify the system's functionality. For this purpose, we employ the *Cadence NC-sim* RTL simulator.

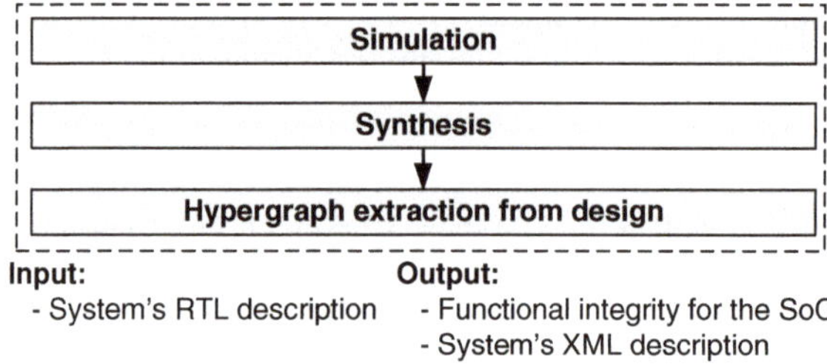


**Fig. 9.2** Tasks for the pre-processing step

Then, we determine the desired hierarchy for target 3-D architecture. Our framework can handle different levels of hierarchy, each of which exhibits advantages and disadvantages. For instance, a block-based system's description leads to a coarse-grain solution, whereas a gate-level netlist comes with a finer system implementation. In other words, the fine-grain approach imposes the highest performance enhancement for the 3-D architecture, but it also introduces the maximum computational complexity for performing architecture-level exploration. For the scopes of this chapter, we choose (without affecting the generality of proposed framework) to maintain the system's hierarchy among heterogeneous modules (e.g., logic, memory), while each of them is flatten in order to improve timing constraint.

After defining the SoC's hierarchy, the HDL description is synthesized with *Synopsys Design Compiler*. As long as design constraints (e.g., timing slacks, DRC's, etc) are met, the output from synthesis is translated to an equivalent XML description, which corresponds to the system's hypergraph representation. The derived XML description is fed as input to the second step of our proposed framework, which deals with the 3-D stack generation under the selected design constraints.

## 9.5 3-D Stack Generation

The 3-D stack generation takes as input the XML description derived from previous step, which represents the SoC's netlist after technology synthesis, and derives the parts of application's functionality that are assigned to each of the layers of 3-D stack. Figure 9.3 gives the tasks for performing 3-D stack generation under the selected design criteria. Since the decisions of this step are based both on architectural issues (e.g., number of layers, distribution of TSVs, etc), as well as designer selections (e.g., timing and power specifications for target product), careful study and analysis should

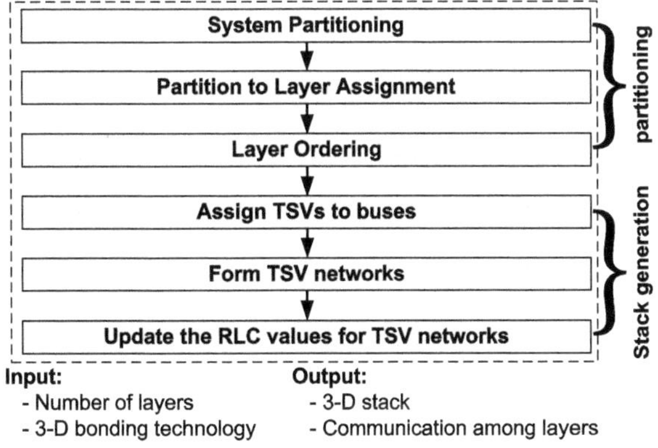


**Fig. 9.3** Tasks for 3-D stack generation

be applied at each layer of stack generation because a non-optimal selection of a former task might lead to a non-acceptable final solution.

Initially, application is partitioned into a number of subsets. During this task, both fabrication and cost parameters are taken into consideration. Though one might think that it is preferable that the number of subsets to be equal to the number of layers of the target 3-D architecture, previous studies shown that such a selection leads to sub-optimal solutions. On the other hand, if the number of subsets is higher than the corresponding number of layers, the upcoming tool, dealing with the assignment of application's partitions to layers, exhibits higher flexibility to derive a solution more closer to the optimal one. Consequently, a careful balance between the number of derived partitions and the availability of layers should be performed, whereas the absolute value of this parameter is application- and 3-D-chip specific. Different partitioning algorithms, which are able to handle designs described in XML-based format, may be employed for this purpose, such as the *hmetis* [14] and *Tabu* [15] algorithms.

Given the number of application's subsets, in conjunction to the availability of layers, we can proceed with the partition to layer assignment task. A number of technology oriented selections are taken into consideration during this task. More specifically, for a given layer, only technology combatable components can be assigned to, while the layers have to exhibit sufficient area utilization.

Then, the layers are appropriately ordered to maximize the performance of derived 3-D stack. This is feasible by assigning to adjacent spatial locations over the $z$-axis layers with increased interlayer signal activity. For this purpose, we pay effort to minimize the number of interlayer communication, if this is possible. In order to highlight this, Fig. 9.4b, c shows two alternative partition to layer assignments regarding the application's hypergraph depicted in Fig. 9.4a. The vertical lines in this figure correspond to TSVs. As we can conclude, the assignment depicted in Fig. 9.4b needs 80 TSVs (45 TSVs between $Layer_1$ and $Layer_2$ and 35 TSVs between $Layer_2$ and $Layer_3$). On the other hand, the solution depicted in Fig. 9.4c requires

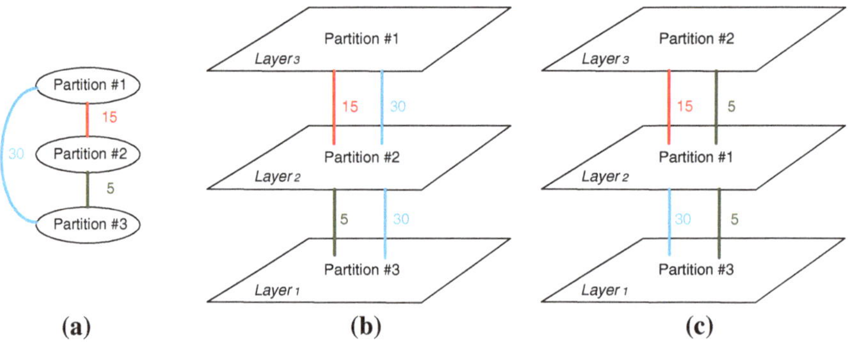


**Fig. 9.4** Two alternative solutions derived (**b** and **c**) from partitioning to layer assignment for the same application (depicted in (a))

only 55 TSVs (20 TSVs between $Layer_1$ and $Layer_2$ and 35 TSVs between $Layer_2$ and $Layer_3$), without affecting either the functionality of target application, or its performance.

The application's partitioning and layer ordering is software supported by *TABU* algorithm [15]. Rather than similar approaches, which mainly perform min-cut partitioning [14], our solution provides additional flexibility, since it is aware also about the selected bonding technology (e.g., TSV, Face-to-Face, etc), the desired density of TSVs (per layer) and the shape of 3-D stack (e.g., cube, pyramid, etc). Finally, the derived solutions are evaluated with models for wire-length [16], delay [17] and power consumption [18].

Next step involves the assignment of a TSV array to each bus that connects application's functionalities in different layers (i.e., the colored lines in Fig. 9.4b, c). Note that whenever a bus needs to be routed in layers $i$ and $j$, silicon area equals to the area occupied by the TSV array has to be reserved in both layers. Although our framework can also handle distinct TSVs, throughout this study we select to employ arrays of TSVs because they introduce fewer constraints to routing algorithm [19]. Then, the TSV arrays that provide bus connectivity between adjacent layers $i$ and $j$ are connected through special purpose routing paths, named *TSV networks*. As we will discuss later, these networks are actually implemented with additional metal layers with tunable *RLC* characteristics.

## 9.6 3-D System Prototyping

The last step in our framework, as it was depicted in Fig. 9.1, deals with the system prototyping in order to derive the final 3-D stack. Figure 9.5 gives the tasks applied during this step. More specifically, during this step we perform floor-planning, power and ground network generation, placement of physical library cells, clock tree synthesis, and global/detail signal routing with the *Cadence SoC Encounter* tool.

The main differentiation of the flow discussed throughout this chapter, as compared to approaches found in relevant references affects the employed toolset. Specifically,

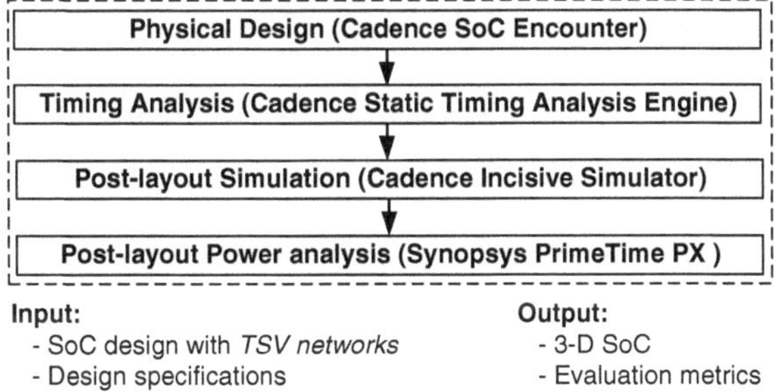


**Fig. 9.5** Tasks for 3-D system prototyping

our framework is software supported by a well-established 2-D commercial flow provided by Cadence [2]. However, since it is not possible to modify the functionality of these tools (their source code is not available), we have to make them aware about the additional flexibility imposed by the third dimension through appropriate design encoding. For this purpose we introduce:

- *TSV networks*: These networks correspond to routing paths that provide connectivity between TSV arrays assigned to adjacent layers. Note that during physical implementation, our framework preserves that TSV arrays assigned at consecutive layers are aligned. This is possible by forcing the placement of TSVs to the same relative $(x, y)$ co-ordinates between adjacent layers. The *TSV networks* are actually implemented through additional metal layers inserted to the technology library file, while their total resistance $(R)$, capacitance $(C)$, and inductance $(L)$ values correspond to the TSV's *RLC* parameters.
- *Virtual layers*: Our framework assumes that target 3-D SoC consists of a number of *virtual layers*, each of which contains hardware resources assigned to different physical layers of the 3-D stack. This enables the usage of existing 2-D physical design tools in order to evaluate the performance enhancement of 3-D chips.

After 3-D physical prototyping, we evaluate the efficiency of derived solution by applying timing analysis. For this purpose we employ *Cadence Static Timing Analysis Engine*, while for shake of completeness the analysis is performed both in advanced, as well as after clock tree synthesis and architecture's routing. In case, the derived 3-D stack does not meet system's constraints/specifications, a number of design optimization could be applied for additional improvements.

Then, we verify the functional integrity of physical design by applying a post-layout simulation with *Cadence Incisive Simulator*. For this reason, the delay for all the architecture's routing paths is extracted in SDF format (Standard Delay Format), and then we appropriately annotate the delay values for *TSV networks*. Note that the annotation of delay values for *TSV networks* is an important task, since these routing paths exhibit the *RLC* characteristics of the selected TSV technology. For the scopes of this letter, the electrical characteristics for *TSV networks* are retrieved from commercially available TSV models [20].

Furthermore, our framework supports the evaluation of 3-D SoCs in term of power consumption by applying a post-layout analysis with *Synopsys PrimeTime PX* tool. The inputs to this analysis are a trace file that contains signal activities in VCD (Value Change Dump) format, as well as the annotated SPEF (Standard Parasitic Exchange Format) file with extracted parasitic values for all the design's resources.

Another important aspect of the 3-D system prototyping step is the capability of going back and changing already taken decisions. Flexibility is very important, because this methodology is meant to be used for fast search-space exploration. If the resulting 3-D chip does not meet the system specifications, there is a feedback loop back to the partitioning step to allow designers to modify some of the decisions already made, like using different IP blocks or different die assignment options. A high-level estimation of performance metrics (i.e., area, yield, power, delay, etc) is needed to assess compliance to the specification.

## 9.7 Case Study 1: Implementation of a 3-D Instance for LEON3 Processor

This section depicts how it is possible to employ the proposed framework for designing a 3-D instantiation (with two layers) of LEON3 processor [13]. The core of LEON3 (shown at Fig. 9.6) is a synthesizable 32-bit processor compliant with the SPARC V8 architecture [21] and configurable through VHDL generics. We instantiated a single-core processor attached as a master to the AMBA Advanced High Performance bus (AHB). The processor has 7 pipeline stages, while the internal instruction and data cache include 1 set of 4 KB each (Harvard architecture).

Since the target architecture is an embedded system, our methodology was tuned to derive a low-power solution. Figure 9.6 depicts the block diagram of LEON3 processor, as it is retrieved after the 3-D stack generation step. Different colors in this figure denote blocks assigned to different (*virtual*) layers. Though additional 3-D stacks can be derived from *TABU* algorithm [15], the selected one corresponds to Pareto optimal solution (maximum performance enhancement with the minimum fabrication cost in terms of number of layers and TSVs).

The synthesis of LEON3 processor is performed with *Synopsys Design Compiler* at 130 nm CMOS technology under a timing constraint of 4.35 ns (or 230 MHz). The derived netlist consists of 38,988 standard cells, 42,626 nets, and 110 I/O ports. Figure 9.7 gives the output from floor-planning, assuming a 3-D device consisted of two layers. In this figure, red and green color dots denote arrays of TSV and their landing blocks assigned to *virtual* $\text{layer}_1$ and $\text{layer}_2$, respectively, whereas the *TSV networks* are depicted with blue color lines. Similarly, red and green color lines

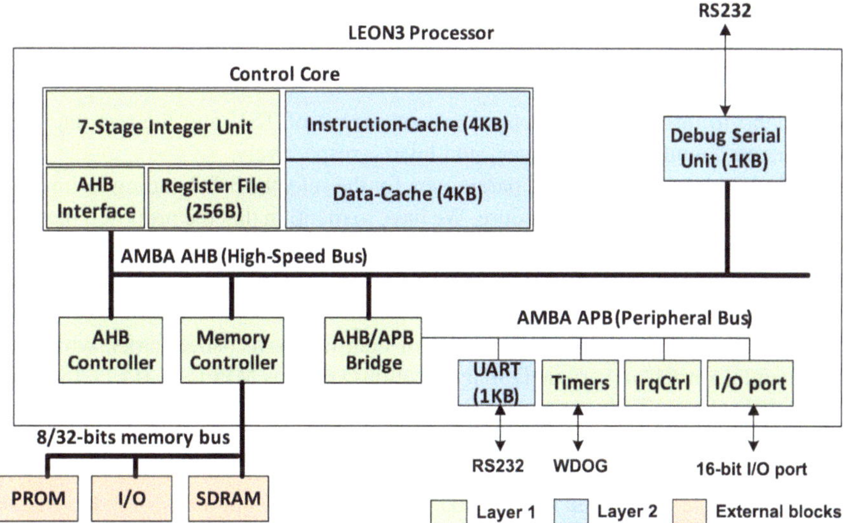


**Fig. 9.6** Block diagram for the Leon3 processor

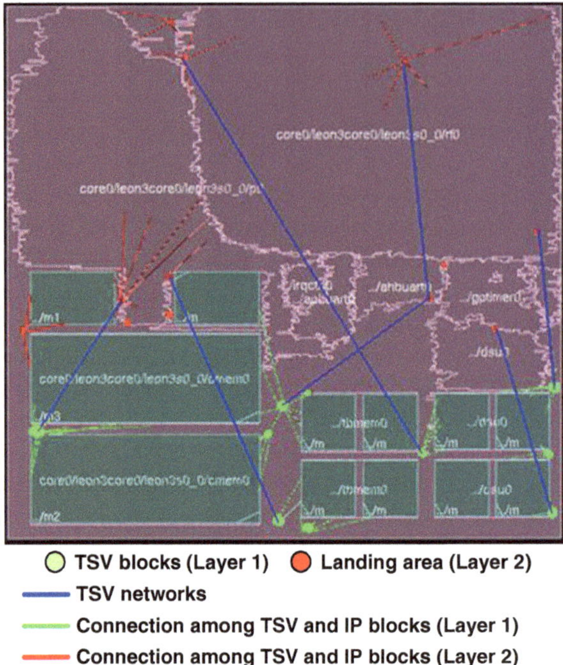


**Fig. 9.7** Example of designing a 3-D instantiation of LEON3 processor

**Table 9.1** Electrical characteristics for TSVs [20]

| Diameter | Min. Pitch | Resistance | Capacitance | Height |
|---|---|---|---|---|
| 1.2 μm | 4 μm | 0.35 Ω | 2.5 fF | 3–9 μm |

correspond to intralayer connections among arrays of TSV and the rest hardware components found in *Virtual* $\text{layer}_1$ and $\text{layer}_2$, respectively.

Table 9.1 summarizes the RC parameters for the selected 3-D bonding technology used during the evaluation procedure. We have to mention that the proposed methodology, as well as the tools that support the evaluation of 3-D SoCs supports also more advanced technologies both for each layer (e.g., 90 nm, 65 nm, 45 nm, etc), as well as for TSVs.

For evaluation purposes we use a set of data intensive benchmarks, which are fundamental kernels in various DSP applications (such as MPEG-4, JPEG, filtering, and H.263). In particular, we used five motion estimation algorithms: full search (FS), hierarchical search (HS), three-step logarithmic step (3SLOG), parallel hierarchical one-dimensional search (PHODS), and spiral search (SS). It has been noted that their complexity ranged from 60 to 80 % of the total complexity of video encoding (MPEG-4) [22]. In addition, we used the 1-D wavelet transformation, cavity

**Table 9.2** Employed DSP kernels

| DSP kernel | Clock cycles | Simulated time (ms) | Computational complexity | Algorithm input size |
|---|---|---|---|---|
| 3SLOG | 19,748,213 | 157 | $M \times N(8\log_2(p)+1)$ | $144 \times 176$ |
| FS | 7,368,117 | 71 | $(2p+1)^2 \times N \times M$ | $144 \times 176$ |
| PHODS | 13,334,375 | 120 | $6 \times M \times N \times \log_2(p)$ | $144 \times 176$ |
| SS | 7,439,388 | 72 | $(2p+1)^2 \times N \times M - S$ | $144 \times 176$ |
| Cavity | 23,777,500 | 216 | $2N_{RUNS}[\mathrm{GB}+N\times M+(N-2\,\mathrm{GB})\times(M-2\,\mathrm{GB})\times(4\,\mathrm{GB}+\mathrm{NB}+1)]$ | $64 \times 64$ |
| HS | 9,603,875 | 93 | $N \times M \times [\frac{(2z+1)^2}{16+45/4}]$ | $144 \times 176$ |
| MMUL | 5,894,285 | 57 | $O(n^3)$ | $50 \times 50$ |
| FFT | 8,428,398 | 91 | $\Omega(N\log(N))$ | 512 points |
| Wavelet | 6,035,932 | 56 | $15hor_{l1} \times ver_{l1} + 0.5 \times ver_{l2} \times (57hor_{l2}+15) + 15 \times ver_{l3} \times (hor_{l3}+2) + 75\mathrm{ver}_{l4} \times \mathrm{hor}_{l4}$ | $256 \times 256$ |
| Bubbles | 5,595,728 | 46 | $O(n_2)$ | 1,000 elements |

detector, and Fast Fourier transformation (FFT) algorithm. We also incorporated basic benchmarks such as Matrix Multiplication and Bubble-sort sorting algorithm.

Table 9.2 summarizes the employed DSP kernels used during the evaluation procedure. At this table, $\mathrm{hor}_{li}$ and $\mathrm{ver}_{li}$ corresponds to the $i$-layer horizontal and vertical space, respectively. $W$ corresponds to the width, $N \times M$ is the image size, $p$ defines the search space parameter. $N_{\mathrm{RUNS}}$, $GB$, and $NB$ give the number of loops, the shift parameter, and the block size, respectively.

These benchmarks were implemented in C language and compiled to LEON3 binaries with BCC cross compiler. The binaries were firstly fed to TSIM LEON3 instruction-level simulator [22] for functional verification and resource utilization at host-machine. For instance, Cavity detector algorithm for an input of $64 \times 64$ pixels image takes 69 hours simulation time, whereas it requires 58 GB of hard disk space for VCD storage and up to 800 MB physical RAM, on a host machine with Intel Core2 Duo processor and 4 GB RAM. On the other hand, this simulation in TSIM environment is completed in less than 3 minutes.

Table 9.3 gives some technical details about the physical implementation of LEON3 processor. Based on these results we can conclude that the derived 3-D architecture reduces total wire-length by 36 %, as compared to the corresponding 2-D system implementation. Since throughout this study, we focus on designing a low-power instantiation of LEON3 processor, the wire-length reduction is expected to come with considerable power savings, without compromising the performance of derived architecture.

For evaluation purposes, the efficiency of derived 3-D architecture is quantified with a number of DSP kernels. Table 9.3 provides also results about the performance

**Table 9.3** Metrics about the physical implementation

| Characteristics | 2-D system | 3-D system | |
|---|---|---|---|
| | | Layer 1 (Logic) | Layer 2 (Memory) |
| Wire-length (μm) | 855,637 | 530,443 | 100,440 |
| Half-perimeter (μm) | 823,033 | 495,010 | 97,527 |
| Number of TSVs | 0.00 | 817 | 817 |
| Area for TSVs ($\mu m^2$) | 0.00 | $24.7 \times 24.7$ | $24.7 \times 24.7$ |
| Aspect ratio | 1.00 | 1.00 | 1.00 |
| Area per layer ($mm^2$) | 2.89 | 1.30 | 1.53 |
| Operation frequency (MHz) | 230 | 230 | |
| Power consumption (mW) | 53.96 | 43.30 | |

and average power dissipation of LEON3 processor. Note that the performance between 2-D and 3-D architectures is constant, since we assumed same throughput. However, the wire-length reduction imposed by the usage of 3-D integration leads to average power savings 20 %, as compared to the corresponding 2-D implementation.

## 9.8 Case Study 2: Implementation of a 3-D NoC

The existing approach on designing 3-D NoCs, as it is discussed in the majority of relevant approaches, affects uniform networks consisted solely of 3-D routers. Assuming a 3-D mesh topology, these routers, apart from the direct connection to their four neighbors assigned to the same layer, also provide connectivity to vertically aligned routers (upper and lower layers).

Although such a selection leads to a "uniform" underlying hardware, however, it is not always the most efficient solution. Specifically, since a NoC is usually an application-oriented communication infrastructure, it is strongly recommended to study carefully the demand for interlayer communication.

Toward this direction, we performed application mapping onto a homogeneous mesh 3-D NoC. The optimization goal during application mapping for this analysis is the minimization of routing traffic, while we are also preserving the proper application's functionality (in order to avoid packet conflicts). For this study, we employ 4 different applications from the multimedia domain:

- Video Object Plane Decoder (VOPD) is a digital signal processing application that has been proposed for use on NoC and studied before [23, 24]. This application offers quality video transition with decent bandwidth performance. The tested VOPD decoder includes 16 nodes, such as two length decoders, an AC-DC prediction, an ARM processor, two memory components, and a VOP reconstructor.

- Multi-Window Display (MWD) is another digital signal processing application [25], which consists of 12 nodes and it is also suitable for a NoC architecture.
- MPEG-4 is a broadly used protocol for audio and video encoding [26]. A hardware encoder and decoder consist of many components, so a NoC approach is suitable. The employed implementation of MPEG-4 has 12 nodes, including various processing elements, such as a video unit, an audio unit, a RISC processor, a med CPU, a binary alpha block, and three SRAMs.
- The last case study is a multimedia system (MMS) [27] consisting of 25 nodes, including several memories and DSP processors.

As we will discuss more thoroughly in the experimental results section (please refer to Figs. 9.13, 9.14, 9.15 and 9.16), none of these applications requires vertical connectivity across all the routers, since such an approach. Thus, it is preferable to employ an architecture, where in case there is no demand for vertical connectivity inside a router, then this 3-D router could be replaced with a 2-D one, without affecting the final application mapping (since there is no change in the data traffic). On contrast, due to the simpler architecture of a 2-D router, as compared to a 3-D one, such a selection is expected to impose mentionable gains in number of transistors for the entire system, which in turn introduces performance and power consumption enhancements.

In addition to that, connections between routers assigned to different layers are actually implemented through TSVs. Each of these TSVs occupy a significant amount of silicon area, as compared to a gate [20].

Though there are some prior works on discussing issues related to design of heterogeneous 3-D NoCs [3, 28–30], these approaches exhibit a number of limitations, due to the fact that are based solely on abstract models. Specifically, both the performance of retrieved NoC, as well as its power/energy dissipation usually is retrieved by counting the hops (connections between adjacent routers) that a packet has to traverse in order to be delivered from source to destination nodes. Also, these approaches do not take into consideration constraints posed by the selected 3-D technology, and consequently they usually lead to architectural solutions that are not possible to be fabricated (e.g., due to the excessive amount of TSVs). On the other hand, the experimentation throughout this work is performed with the usage of Cadence SoC Encounter tool, which was appropriately tuned (as we discuss in upcoming section) in order to be aware about the vertical connections. Such a selection can guarantee that the derived results are closer to those foe 3-D chips.

The usage of introduced framework enables the design of efficient heterogeneous NoCs, consisted of a mixture of 2-D and 3-D routers. This section introduces the architectural organization of these routers, while it also provides an overview of the entire communication scheme. Two different types of routers are considered in the target 3-D NoC. A 2-D router can be used where an incoming routing track is connected to wires on the same layer ($F_s = 3$). Specifically, the router's flexibility, denoted as $F_S$, gives the number of directions to which each incoming wire can be connected. Alternatively, a 3-D router supports connections to the third dimension (upper and lower layers, $F_s = 5$).

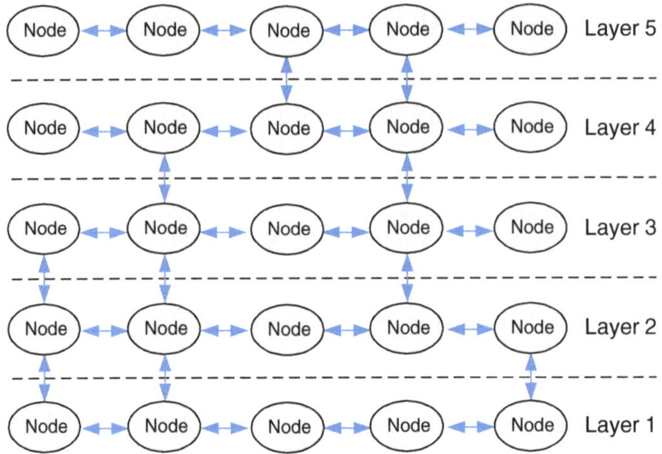


**Fig. 9.8** Example for an application's communication graph

In order to depict the differences between these two routers, we assume an application's task graph (depicted in Fig. 9.8) mapped onto a 3-D chip consisted of five layers, whereas the communication between nodes is provided through a NoC infrastructure. The arrows in this figure denote links between adjacent routers, either across the horizontal, or vertical, direction. As we can conclude from this diagram, none all of the routers exhibit requirement for data transfer between layers assigned to adjacent layers, and hence this is possible to employ a heterogeneous 3-D NoC.

Figure 9.9 shows two alternative assignments of 2-D and 3-D routers for this NoC. Specifically, the solution depicted in Fig. 9.9a corresponds to the approach which is widely accepted up to now to relevant literature, whereas the second solution (Fig. 9.9b) highlights the concept of incorporating different flavors of 3-D routers, as it is proposed by the heterogeneous NoC. More specifically, the main differentiation between these two approaches affects the usage of three different flavors for 3-D routers:

- *3-D Router*$_1$: It supports connectivity only for links from lower to higher layer (marked as "1" in Fig. 9.9b). Since an incoming packet can be routed to four different directions (note that the input and output ports could not be the same in order to avoid packet deadlocks), the router's flexibility is $F_s = 4$.
- *3-D Router*$_2$: It supports connectivity only for links that realize connections from higher to lower layers (marked as "2" in Fig. 9.9b). Similar to previous case, this router has $F_s = 4$.
- *3-D Router*$_3$: It supports connectivity both for bidirectional links with $F_s = 5$ (from lower to higher layer and vice versa).

Though these scenarios lead to valid communication schemes for the employed example, i.e., they provide exactly the same data transfer between source and

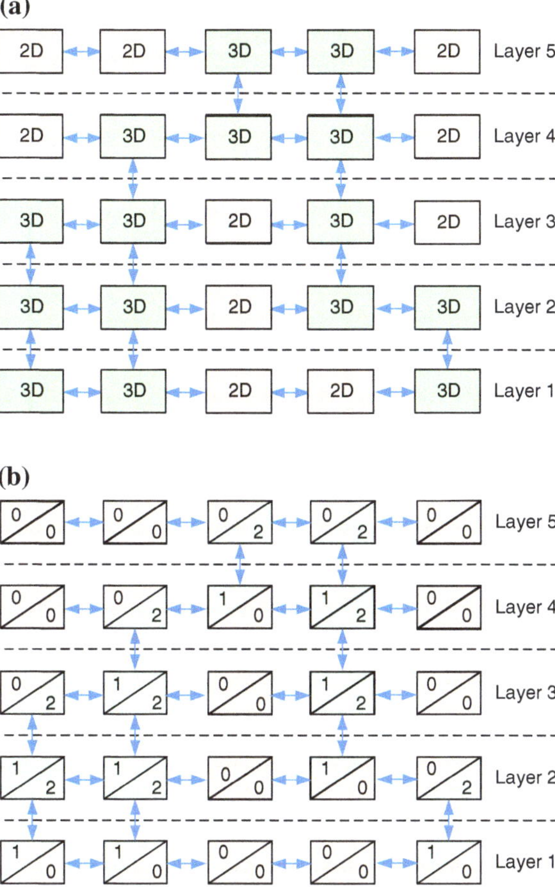


**Fig. 9.9** Two different NoC flavors for the example discussed in Fig. 9.8: **a** consisted only of 3-D routers for interlayer communication and **b** the proposed solution

destination nodes, however, the two solutions depicted at Fig. 9.9 exhibit remarkable variations both in performance (maximum operation frequency and power/energy consumption), as well as the occupied silicon area. Moreover, since the scenario depicted in Fig. 9.9a incorporates additional TSVs, it introduces constraints regarding the yield and fabrication cost.

The basic component of the proposed interconnection architecture is the NoC router. This architectural concepts of this router were already discussed in earlier chapter (please refer to Chap. 2), however, for the scopes of the designing 3-D NoCs, they should be appropriately extended in order to handle also connections between adjacent layers. Specifically, for regular 2-D NoCs, a typical router has five ports (one port is assigned to the attached node, whereas the remaining ports are used for connecting the neighbors across the four directions). Similarly, as we

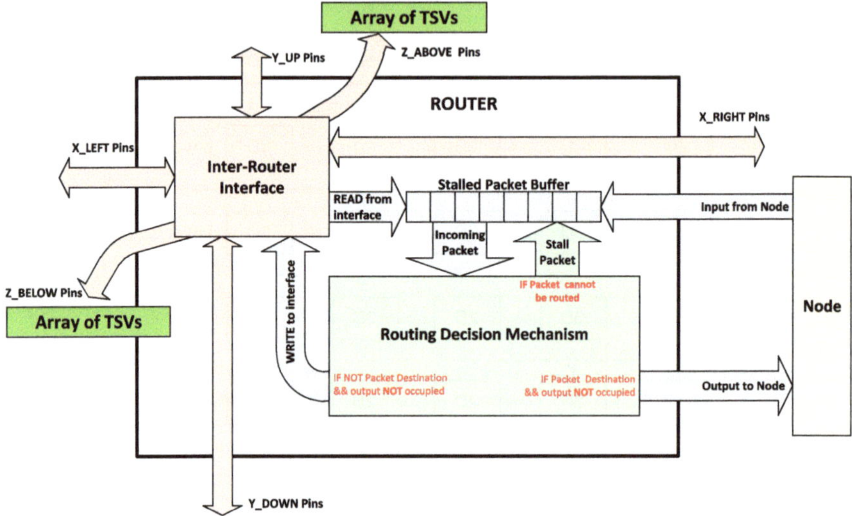


**Fig. 9.10** Architectural template for a 3-D router

have already mentioned, in case we have a 3-D NoC, then a typical router has seven ports (the two additional ports are employed to provide connectivity across the vertical direction).

Figure 9.10 gives the simplified block diagram for a 3-D router. The inputs from the interface (as they were retrieved after a Read operation), as well as the corresponding inputs from the attached node, are stored to a buffer of stalled packets. This buffer contains those packets which could not be routed to any output of the router, because these outputs were occupied by other packets. Furthermore, as only one packet is routed in a clock cycle, this buffer also contains the remaining incoming packets. The selection of output direction for an incoming packet/flit is defined from the Routing Decision Mechanism (RDM) by applying the employed routing algorithm.

Algorithms 9.1 and 9.2 give more details about the procedure for deciding the optimum output port for an incoming packet under to constraint of minimization on-chip traffic. Apart from this modified *ZXY* algorithm, the heterogeneous NoC derived from our framework could also support more flexible routing algorithms if their functionality is appropriately encoded in the router's description.

**Algorithm 9.1:** Employed routing algorithm for the Routing Decision Mechanism (RDM).

```
Algorithm RouterMain Data: NewPacketsArray
Data: StalledPacketsArray
Result: Packet routing
Begin
NewPacketsArray ← Read packets from neighbor routers;
NewPacketsArray ← NewPacketsArray + Read packets from nodes;
foreach StalledPacket in StalledPacketsArray do
  | PacketRoute (StalledPacket);
  | if (StalledPacket is routed) then
  |   | StalledPacketsArray.Remove(StalledPacket);
  | end
end
foreach NewPacket in NewPacketsArray do
  | StalledPacketsArray.Add(NewPacket);
end
End
```

**Algorithm 9.2:** Pseudocode for the "PacketRoute" function.

```
Function PacketRoute Data: Packet
Data: Total number of ports per Router
Data: Router port which received Packet
Result: Router port for sending Packet
Begin
if (Packet reached Destination Node) then
  | if (Destination Node's input ports are NOT occupied) then
  |   | Send Packet to Destination Node;
  | end
end
else
  | // Excludes the Packet from return to the sender's router.
  | // This prevents deadlocks.
  | ValidDirections ← Total Router Sides - Router side which received Packet;
  | foreach Direction in ValidDirections do
  |   | // Priority is given to Z axis
  |   | if (neighbor router not occupied) then
  |   |   | BestDir ← min(Cost(BestDir), Cost(Direction));
  |   | end
  |   | if (BestDir is found) then
  |   |   | Send packet to neighbor router denoted by BestDir; Mark the
  |   |   | neighbor router as occupied;
  |   | end
  | end
end
End
```

### *9.8.1 Evaluation of 3-D NoC*

This subsection discusses a number of experimental results that prove the efficiency of the proposed solution. For demonstration purposes, four applications were employed, as they were described in Sect. 9.8. Different evaluation metrics are employed towards this goal, including both fabrication-oriented, as well as functionality-oriented parameters. All the results summarized in this section were retrieved with the usage of TMSC 45 nm process technology.

Initially, we quantify the efficiency of introduced 3-D router, as compared to the corresponding 2-D implementation, as we modify the number of local ports. The number of these local ports correspond to the nodes attached to each router. Note that the case Node = 0 corresponds to a scenario, where the router is only connected to the four (or six) adjacent routers, depending if it is 2-D, or 3-D, respectively. The results of this analysis are summarized in upcoming figures.

More specifically, Figs. 9.11 and 9.12 quantify the efficiency of each router in terms of latency and power consumption, as we modify the number of local ports. Based on Fig. 9.11, we can conclude that on average the 3-D router exhibits 32 % lower latency, as compared to the corresponding 2-D implementation. Similarly, even though it seems that the power profile differs between 2-D and 3-D routers, however, the average power consumption for the studied scenarios is the almost the same (29 mW).

After quantifying the building blocks of the NoC, this subsection studies also the performance metrics for implementing the four different applications onto NoC-based platforms. Figs. 9.13, 9.14, 9.15, and 9.16 plot the application mapping onto the minimum 2-D, as well as the proposed heterogeneous 3-D NoCs, consisted of

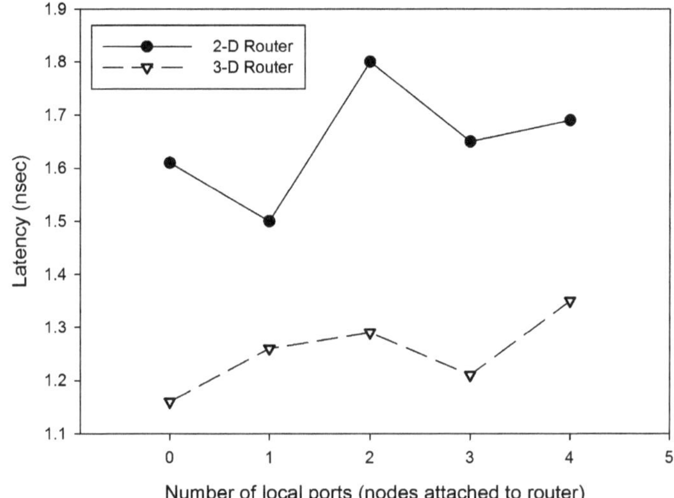


**Fig. 9.11** Evaluation of 2-D and 3-D router in term of latency

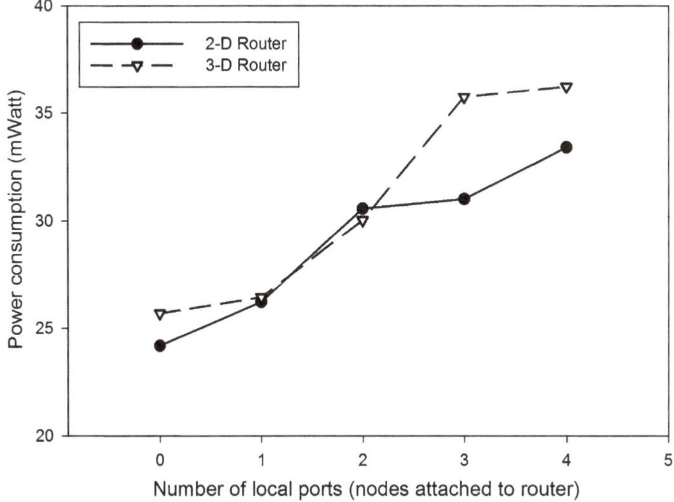


**Fig. 9.12** Evaluation of 2-D and 3-D router in term of power consumption

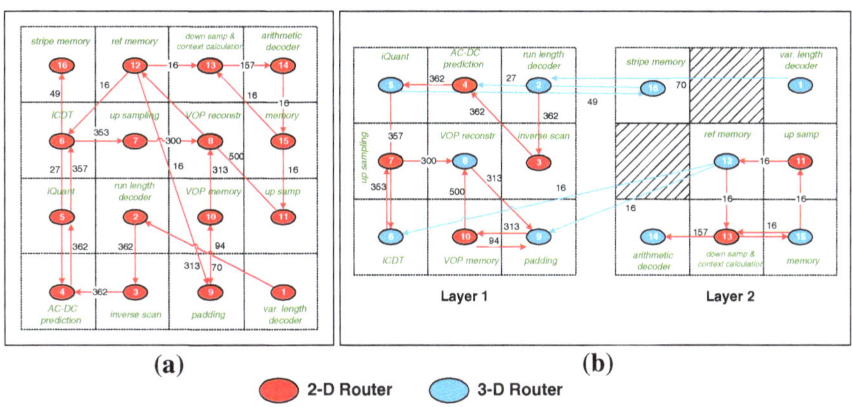


**Fig. 9.13** Mapping of VOPD application onto: **a** 2-D NoC and **b** the proposed heterogeneous 3-D NoC platform

two layers. The term *minimum* corresponds to the smaller number of routers that are required for performing application mapping. For both studies mesh topologies are assumed, whereas our framework for designing heterogeneous 3-D NoCs is also applicable to any other topology (either regular or irregular). Though the analysis discussed in this section affects routers with one local port (for attaching the processing, or storage node), this is not a prerequest for our solution, as we can also handle routers with multiple nodes.

The red and blue colored circles at these figures correspond to 2-D and 3-D routers, respectively. Similarly, the directed red colored arrows denote packet

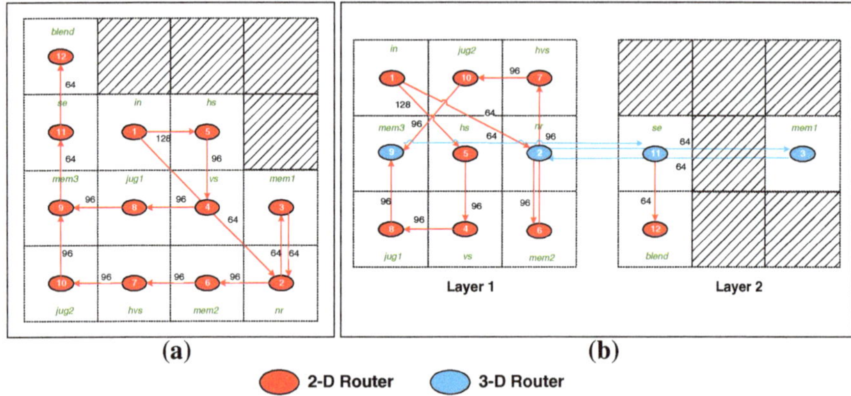


**Fig. 9.14** Mapping of MWD application onto: **a** 2-D NoC and **b** the proposed heterogeneous 3-D NoC platform

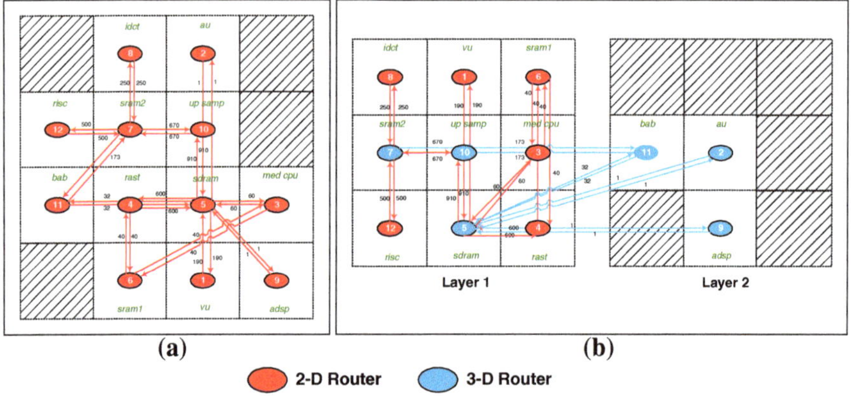


**Fig. 9.15** Mapping of MPEG-4 application onto: **a** 2-D NoC and **b** the proposed heterogeneous 3-D NoC platform

transfers between source and destination routers assigned to the same layer, whereas the blue color arrows correspond to data transfer between routers assigned to different layers, and hence the data transfer is realized through TSVs. The weights at these arrows correspond to the number of packets from source to destination nodes. Moreover, at these figures, some locations at the grid are marked with shadow color. In contrast to relevant approaches dealing with mesh topologies, throughout this chapter we eliminate the number of routers assigned to these locations (marked with dash color), where no IP node is assigned. Such a selection results to a more efficient architectural solution, since it achieves higher operation frequencies and lower requirements for silicon area. As we discuss later, the additional complexity for performing packet routing at heterogeneous 3-D NoCs does not affect the performance of the whole architecture.

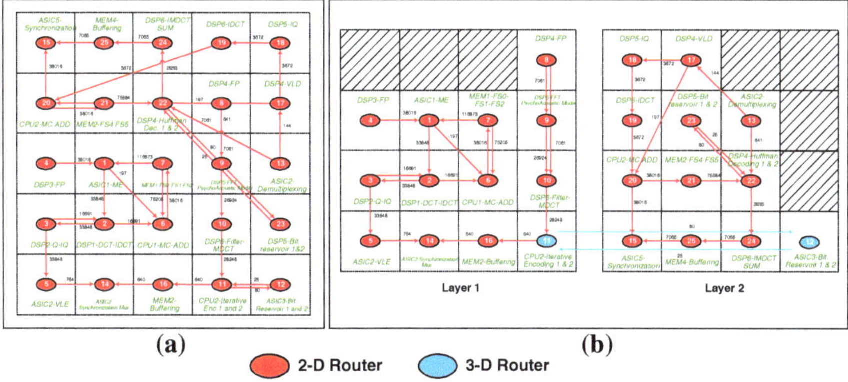


**Fig. 9.16** Mapping of MMS application onto: **a** 2-D NoC and **b** the proposed heterogeneous 3-D NoC platform

Next figure plots the outcome after physical implementation for one of the studied applications (MPEG-4), as it was derived with our methodology for designing 3-D platforms. At this figure, we highlight the two different *Virutal Layers*, as well as the connectivity between them (plotted with yellow color at Fig. 9.17b), which is actually implemented through "Virtual networks" that model TSVs.

The optimization goal during application mapping onto different platforms was the minimization of hops for packets routing. Towards to this direction, 3 alternative scenarios are evaluated. More specifically, we quantify the introduced heterogeneous approach, where 2-D and 3-D routers exist on the same NoC depending on the requirement for interlayer connectivity. As reference to this analysis, two more approaches are discussed: (i) a 2-D NoC with the minimum number of routers and (ii) the corresponding homogeneous 3-D NoC (similar to the one found in relevant references). Note that the homogeneous 3-D NoC consists of exactly the same number of routers compared to the corresponding 2-D system implementation.

The results of this analysis are summarized in Fig. 9.18. Based on this analysis we can conclude that the existing way for designing 3-D NoCs achieves to reduce the number of packet hops, as compared to the corresponding 2-D NoC implementation by 29 %, on average. Such a reduction denotes the significant importance of employing the 3-D integration paradigm as a viable solution for designing products with increased demand for connectivity. This conclusion was also derived from the majority of relevant publications dealing with the topology exploration problem. However, these approaches rarely take into account constraints and limitations posed by the fabrication process. Specifically, the usage of 3-D routers impose a TSV assigned to each wire that provide signal connectivity between adjacent layers. Due to the large diameter and pitch of these TSVs, their excessive usage at 3-D designs usually introduce area and yield overheads.

On the other hand, our proposed architectural solution has the minimum possible number of 3-D routers in order to alleviate the consequences of these problems.

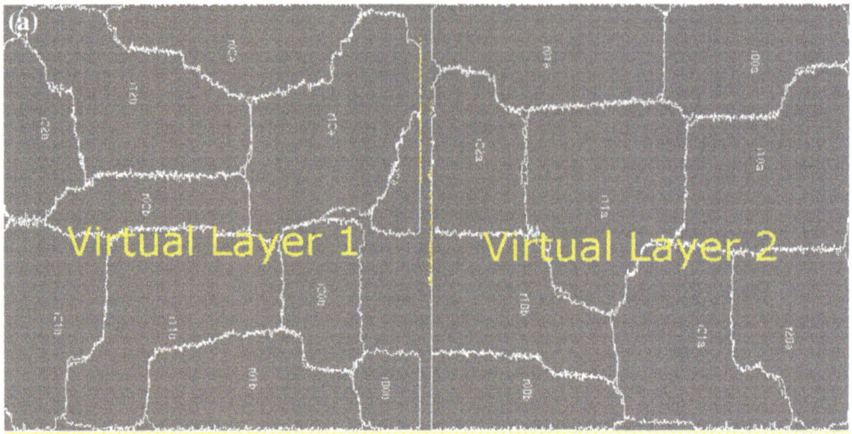


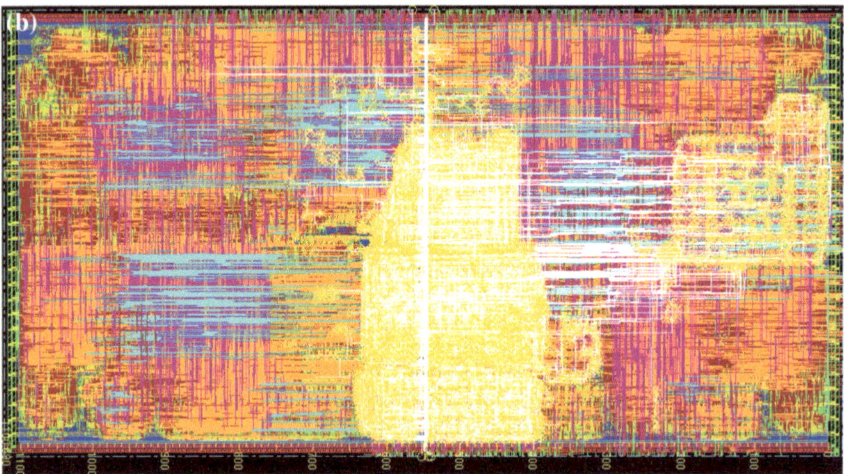


**Fig. 9.17** Physical layouts for the introduced heterogeneous 3-D NoC regarding the MPEG-4 application

Since our approach provides fewer connections to rest layers, a slight increase to the number of hops (due to routing restrictions) could be though as affordable. This penalty is highlighted in Fig. 9.18 for the four applications and it is about 8.5 % on average compared to the uniform 3-D NoC approach.

Table 9.4 provides a number of implementation-oriented parameters as they are retrieved after synthesis and physical implementation. Based on this table, we can conclude that both the 3-D, as well as the introduced solution has aspect ratio 2 (in order to realize the two virtual layers shown previously at Fig. 9.17). Two more parameters are worth in mentioning at this table. More specifically, since the proposed solution has a combination of 2-D and 3-D routers, as compared to the 3-D approach,

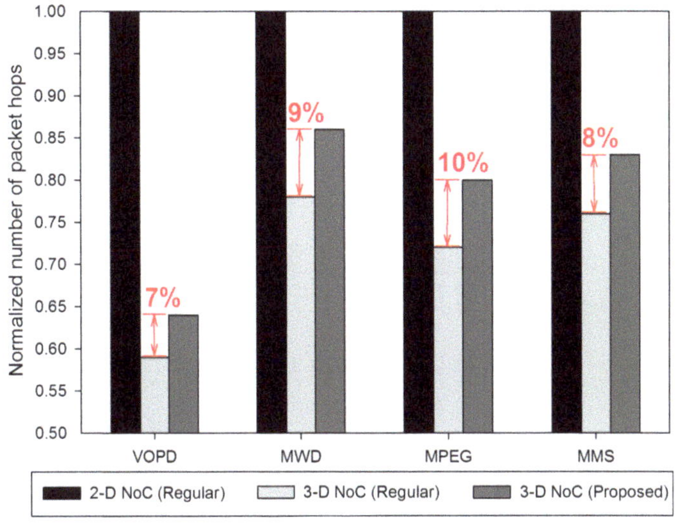


**Fig. 9.18** Number of packet hops for different architectural solutions: (i) a homogeneous 2-D NoC, (ii) a homogeneous 3-D NoC, and (iii) the proposed heterogeneous 3-D NoC

we need fewer TSVs to be fabricated in order to provide signal connectivity between adjacent layers. However, such a selection does not impose any wirelength degradation, as the proposed heterogeneous approach leads almost to an average 57 % wirelength reduction compared to the case where all the routers are 3-D.

Next, we provide some quantification results regarding the different instantiations of NoCs. More specifically, Fig. 9.19 shows the maximum operation frequency, whereas the results about power consumption are summarized in Fig. 9.20. From these values we can conclude that on average the proposed solution, consisted of a mixture of 2-D and 3-D routers, exhibits higher maximum operation frequency compared to 2-D and homogeneous 3-D by 28 % and 25 %, respectively. This performance enhancement occurs mainly due to better manipulation of fabricated routing wires, since we replaced 3-D with 2-D router in case there is no demand for vertical connectivity.

Additionally, such a selective reduction of 3-D routers is expected to lead to lower power consumption, as it is depicted in Fig. 9.20. More specifically, based on this figure we conclude that the proposed solution achieves on average power reduction against to uniform 2-D and 3-D NoCs 48 % and 39 %, respectively. These savings becomes much more important, if we take into consideration the higher operation frequencies retrieved with the usage of our solution.

Another critical issue that quantifies the efficiency of a 3-D chip affects the number of fabricated TSVs, which is mainly tackled by the partitioning algorithm. With the usage of our introduced design methodology, we lead to a reduced number of interlayer connection, and hence TSVs that have to be fabricated. Of course, this number guarantees the proper application's functionality in terms of data throughput.

**Table 9.4** Implementation properties for the target applications

| Benchmark | | Properties | | | | | | | | |
|---|---|---|---|---|---|---|---|---|---|---|
| | | Area ($\mu m^2$) | Ratio | Utilization | Wirelength | # of TSVs | # of vias | # of cells | # of nets | # of IOs |
| MMS | 2-D | 991.5 × 990.3 | 0.99 | 0.58 | 10,261,064 | – | 4,429,472 | 458,823 | 462,825 | 3,251 |
| | 3-D | 599.3 × 1,198.2 | 1.99 | 0.58 | 11,480,859 | 704 | 2,926,287 | 324,689 | 326,317 | 3,251 |
| | Proposed | 762.8 × 1,523.3 | 1.99 | 0.58 | 7,864,344 | 130 | 4,342,636 | 534,731 | 539,864 | 3,251 |
| MPEG-4 | 2-D | 754.74 × 753.48 | 0.99 | 0.58 | 5,915,764 | – | 2,517,520 | 256,980 | 259,307 | 1,561 |
| | 3-D | 432.03 × 861.84 | 1.99 | 0.58 | 6,373,170 | 576 | 1,659,742 | 167,911 | 168,709 | 1,561 |
| | Proposed | 561.7 × 1,122.66 | 1.99 | 0.56 | 3,882,465 | 192 | 2,576,354 | 271,828 | 275,552 | 1,561 |
| MWD | 2-D | 758.9 × 756.0 | 0.99 | 0.58 | 5,566,222 | – | 2,170,216 | 262,609 | 265,219 | 1,561 |
| | 3-D | 435.4 × 870.6 | 1.99 | 0.58 | 5,794,055 | 1,170 | 1,477,280 | 171,979 | 172,762 | 1,561 |
| | Proposed | 551.5 × 1,101.24 | 1.99 | 0.58 | 3,673,350 | 260 | 2,436,398 | 272,823 | 275,523 | 1,561 |
| VOPD | 2-D | 781.3 × 777.4 | 0.99 | 0.58 | 6,156,379 | – | 2,506,653 | 283,098 | 285,961 | 2,081 |
| | 3-D | 441.3 × 885.7 | 2.00 | 0.58 | 7,081,633 | 1,170 | 1,565,790 | 178,034 | 179,182 | 2,081 |
| | Proposed | 585.8 × 1,170.5 | 1.99 | 0.58 | 4,121,908 | 650 | 2,852,871 | 318,237 | 320,404 | 2,081 |

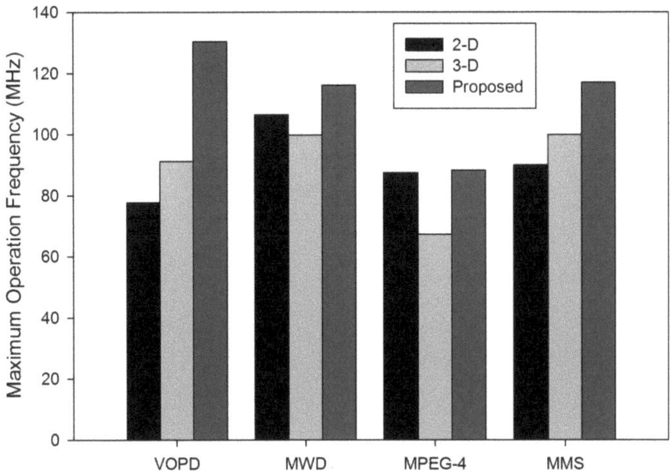


**Fig. 9.19** Maximum operation frequency for different instantiations of NoC

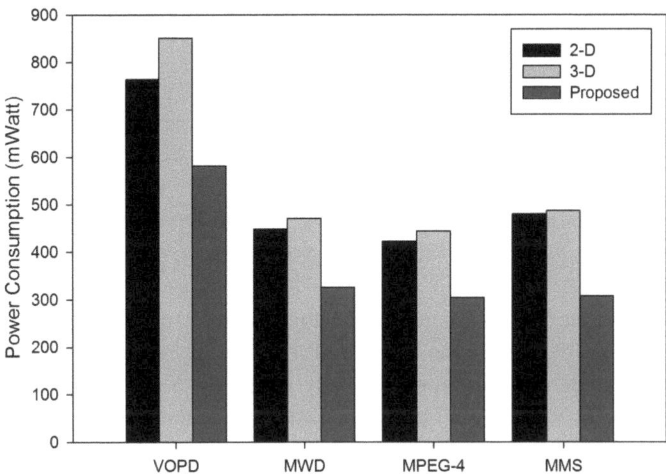


**Fig. 9.20** Power consumption for different instantiations of NoC

Such a reduced number of TSVs in turn leads to architectures that are characterized by higher yield and lower fabrication cost, as compared to existing approaches. The yield of a physical layer within a homogeneous, or heterogeneous 3-D NoC can be, respectively, described as follows [31]:

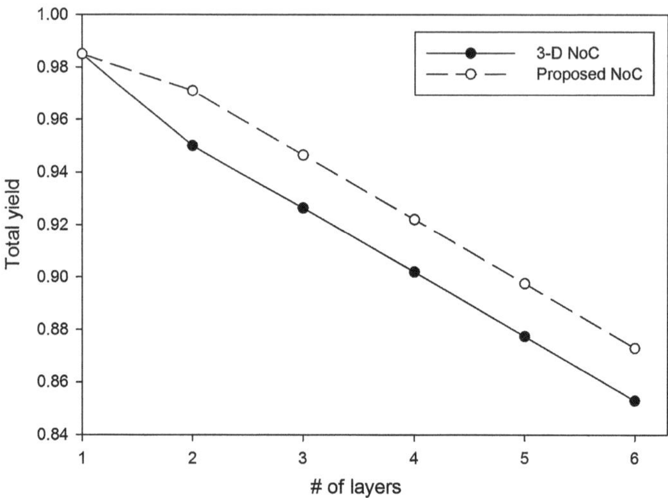


**Fig. 9.21** Process yield of homogeneous and heterogeneous (proposed) 3-D NoCs for different number of layers

$$Y_{\text{fully 3-D}} = \frac{1}{\left(1 + S \times D_0 \times A_{\text{fully 3-D router}}\right)^{N/S}} \tag{9.1}$$

$$Y_{\text{partially 3-D}} = \frac{1}{\left(1 + S \times D_0 \times A_{\text{partially 3-D router}}\right)^{N/S}} \tag{9.2}$$

where the area of the NoC is assumed to be dominated by the area of the routers. The parameters $N$ and $D_0$ denote the number of the mask layers for each layer and the average electrical defect density, respectively. Similarly, $S$ corresponds to the shape factor related to $D_0$ [31]. Based on these parameters, the total yield of a 3-D NoC consisting of $n$ layers can be written as:

$$Y_{t\text{ fully 3-D}} = Y^{n}_{\text{fully 3-D}} \times Y^{n-1}_{\text{bond}} \tag{9.3}$$

$$Y_{t\text{ partially 3-D}} = Y^{n}_{\text{partially 3-D}} \times Y^{n-1}_{\text{bond}} \tag{9.4}$$

where $Y_{\text{bond}}$ is the yield of the stacking process. Based on (9.1)–(9.4), the yield of a homogeneous 3-D NoC and an NoC with the proposed heterogeneous interconnect fabric are plotted in Fig. 9.21. The proposed heterogeneous NoC exhibits an improved yield due to the significant savings in area which results by the limited use of 3-D routers. Note that these yield expressions do not incorporate the effect of using fewer TSVs, which will further increase the yield for the heterogeneous NoC [32].

## 9.9 Conclusion

A novel framework for supporting rapid evaluation of 3-D SoCs was introduced. For the scopes of this chapter, the proposed methodology was applied to design a 3-D instantiation of LEON3 processor under low-power constraints, as well as a heterogeneous 3-D NoC. Experimental results with various DSP kernels prove the effectiveness of proposed solution, since it leads to significant performance and power gains.

## References


1. http://www.r3logic.com/
2. http://www.cadence.com/us/pages/default.aspx
3. V. Pavlidis, E. Friedman, *Three-dimensional Integrated Circuit Design* (Morgan Kaufmann, San Francisco, 2008)
4. L. Zhuoyuan Li, H. Xianlong, Z. Qiang, C. Yici, B. Jinian, H. Yang, V. Pitchumani, C. Chung-Kuan, Hierarchical 3-D floorplanning algorithm for wirelength optimization. IEEE Trans. Circuits Syst. I Regul. Pap. **53**(12), 2637–2646 (2006)
5. J. Cong, Z. Yan, Thermal-driven multilevel routing for 3D ICs, in *Asia and South Pacific Design Automation Conference (ASP-DAC)*, pp. 121–126 (2005)
6. G. Katti, M. Stucchi, K. De Meyer, W. Dehaene, Electrical modeling and characterization of through silicon via for three-dimensional ICs. IEEE Trans. Electr. Dev. **57**(1), 256–262 (2010)
7. B. Black, D. Nelson, C. Webb, N. Samra, 3D processing technology and its impact on iA32 microprocessors, in *International Conference on Computer Design: VLSI in Computers and Processors (ICCD)*, pp. 316–318 (2004)
8. Y. Deng, W. Maly, 2.5D system integration: a design driven system implementation schema, in *Asia and South Pacific Design Automation Conference (ASP-DAC)*, pp. 450–455 (2004)
9. Academic tools aiming at 3-D ICs, http://proteas.microlab.ntua.gr/ksiop/software.html
10. Y. Yonghong, G. Zhenyu, Z. Changyun, R. Dick, S. Li, ISAC: integrated space-and-time-adaptive chip-package thermal analysis. IEEE Trans. Comput. Aided Des. Integr. Circuits Syst. **26**(1), 86–99 (2007)
11. C. Ting-Yen, S. Souri, Chi On Chui, K. Saraswat, Thermal analysis of heterogeneous 3D ICs with various integration scenarios, in *International Electron Devices Meeting (IEDM)*, pp. 31.2.1–31.2.4 (2001)
12. G. Link, M. Vijaykrishnan, Thermal trends in emerging technologies, in *International Symposium on Quality Electronic Design (ISQED)*, 8 pp. (2006)
13. LEON3, http://www.gaisler.com
14. N. Selvakkumaran, G. Karypis, Multiobjective hypergraph-partitioning algorithms for cut and maximum subdomain-degree minimization. IEEE Trans. Comput. Aided Des. Integr. Circuits Syst. **25**(3), 504–517 (2006)
15. K. Siozios, D. Soudris, A Tabu-based partitioning and layer assignment algorithm for 3-D FPGAs. IEEE Embed. Syst. Lett. **3**(3), 97–100 (2011)
16. S. Das, A. Chandrakasan, R. Reif, Calibration of Rent's rule models for three-dimensional integrated circuits, in *IEEE Transactions on Very Large Scale Integration (VLSI) Systems*, vol. 12, No. 4, pp. 359–366, April 2004
17. T. Okamoto, J. Cong, Buffered Steiner tree construction with wire sizing for interconnect layout optimization, in *International Conference on Computer-Aided Design (ICCAD)*, pp. 44–49 (1996)

18. J. Rabaey, A. Chandrakasan, B. Nikolic, *Digital Integrated Circuits*, 2nd edn. (Prentice-Hall, Prentice Hall, 2003)
19. A. Sheibanyrad, F. Petrot, A. Jantsch, *3D Integration for NoC-Based SoC Architectures* (Springer Editions, Berlin, 2011)
20. S. Gupta, M. Hilbert, S. Hong, R. Patti, Techniques for producing 3-D ICs with high-density interconnect, in *International VLSI Multi-Level Interconnection Conference* (2004)
21. SPARC V8, http://www.sparc.org/standards/V8.pdf
22. V. Bhaskaran, K. Konstantinides, *Image and Video Compression Standards: Algorithms and Architectures*, 2nd edn. (Kluwer Academic, Dordrecht, 1998)
23. V. Ngo, H. Nguyen, H. Choi, The optimum network on chip architectures for video object plane decoder design, in *International Conference on Parallel and Distributed Processing and Applications (ISPA)*, pp. 75–85 (2006)
24. S. Murali, G. De Micheli, Bandwidth-constrained mapping of cores onto NoC architectures, in *Proceedings of Design, Automation and Test in Europe Conference and Exhibition*, pp. 896–901 (2004)
25. D. Bertozzi, A. Jalabert, S. Murali, R. Tamhankar, S. Stergiou, L. Benini, G. De Micheli, NoC synthesis flow for customized domain specific multiprocessor systems-on-chip. IEEE Trans. Parallel Distrib. Syst. **16**(2), 113–129 (2005)
26. I. Richardson, *H.264 and MPEG-4 Video Compression: Video Coding for Next Generation Multimedia*, 1st edn. (Wiley, New York, 2003)
27. J. Hu, R. Marculescu, Energy-and performance-aware mapping for regular NoC architectures. IEEE Trans. Comput. Aided Des. Integr. Circuits Syst. **24**(4), 551–562 (2005)
28. A. Bartzas, K. Siozios, D. Soudris, Three dimensional network-on-chip architectures, in *Networks-on-Chips: Theory and Practice*, ed. by F. Gebali, H. Elmiligi, M.W. El-Kharashi (CRC Press, Boca Raton, 2008)
29. I. Anagnostopoulos, A. Bartzas, D. Soudris, Application-specific temperature reduction systematic methodology for 2D and 3D networks-on-chip, in *International Conference on Integrated Circuit and System Design: Power and Timing Modeling, Optimization and Simulation (PATMOS)*, pp. 86–95 (2009)
30. A. Richard, D. Milojevic, F. Robert, A. Bartzas, A. Papanikolaou, K. Siozios, D. Soudris, Fast design space exploration environment applied on NoCs for 3D-stacked MPSoCs, *Workshop on Parallel Programming and Run-Time Management Techniques for Many-Core Architectures* (2010)
31. R. Weerasekera, D. Pamunuwa, L. Zheng, H. Tenhynnen, Two-dimensional and three-dimensional integration of heterogeneous electronic systems under cost, performance, and technological constraints. IEEE Trans. Comput. Aided Des. Integr. Circuits Syst. **28**(8), 1237–1250 (2009)
32. D. Velenis, M. Stucchi, E. Marinissen, E. Beyne, Impact of design choices on 3D SIC manufacturing cost, in *Proceedings of the Workshop on 3-D Integration, at Design, Automation, and Testing in Europe (DATE)*, pp. 1–5 (2009)
33. http://3d-performance.lancs.ac.uk/

# Chapter 10
# The SYSMANTIC NoC Design and Prototyping Framework

**Abstract** SYSMANTIC is a framework for high-level exploration, Register Transfer-Level (RTL) design and rapid prototyping of Network-on-Chip (NoC) architectures. From the high-level exploration, a selected NoC topology is derived, which is then implemented in RTL using an automated design flow. Furthermore, for verification purposes, appropriate self-checking testbenches for the verification of the RTL and architecture files for the semi-automatic implementation of the system in Xilinx EDK are also generated, significantly reducing design and verification time, and therefore NRE cost. Simulation and FPGA implementation results are given for four case studies of multimedia applications, proving the validity of the SYSMANTIC approach.

## 10.1 Introduction and Previous Work

A number of NoC architectures have been implemented and evaluated in both FPGA and ASIC platforms, among them [1]. Furthermore, frameworks and tools for high-level exploration for NoC architectures exist [2].

In Leary [3] a holistic algorithm for NoC synthesis able to address all these requirements together in an integrated manner was presented. However, the synthesis methodology provided does not provide Register Transfer-Level descriptions that can readily be used for system implementation. In Strano [4] a more complete framework for the design of NoC is presented, but it does not include FPGA rapid prototyping.

FPGA rapid prototyping has been explored in [5], where using FPGA long links was explored, however, the authors mention having to modify a router manually to

---

This chapter was contributed by Konstantinos Tatas and Costas Kyriacou from the Department of Computer Science and Engineering of Frederick University, Cyprus, and Kostas Siozios, Alexandros Bartzas and Dimitrios Soudris of the School of ECE, National Technical University of Athens.

K. Tatas et al., *Designing 2D and 3D Network-on-Chip Architectures*, 
DOI: 10.1007/978-1-4614-4274-5_10, 

insert long links, requiring less than a day. In the proposed framework, this modification would have been trivial, due to the automatic RTL generation tool. Similarly, a $3 \times 3$ mesh NoC for an image processing benchmark was implemented in [6]. In Krasteva [7] a more flexible prototyping framework was presented, which includes RTL NoC models and also supports dynamic reconfiguration. However, our framework also includes design space exploration starting from the target application requirements. In short, all existing approaches either include exploration and synthesis to RTL but not rapid prototyping, or they are based on ad-hoc NoC implementations in FPGAs in order to prove the feasibility of NoC architectures. While certainly valid, this isolates the prototyping from the exploration phase in the design flow.

Therefore, existing approaches, either provide exploration and design solutions without taking rapid prototyping into account, or provide rapid prototyping without exploration. Our work bridges this gap by introducing a novel EDA tool flow which performs high-level exploration, automatic RTL implementations of the interconnection network and rapid prototyping using off-the-shelf FPGAs. The flow performs high-level design exploration and selects the optimal NoC topology and application mapping, given a set of design constraints and goals. From the selected topology, the proposed flow automatically generates the corresponding NoC RTL code, testbenches for RTL verification, as well as the appropriate files for rapid prototyping in Xilinx FPGA devices, thus greatly speeding up the design, verification and prototyping process, and therefore NRE cost.

Furthermore, the automation of the RTL design and rapid prototyping hides many low-level implementation details from the designer, while still allowing him to control important NoC parameters such as topology, router buffer size, wordlength etc. or optimize them for specific metrics such as performance, power consumption and area.

## 10.2 Exploration Methodology

The SYSMANTIC NoC exploration and implementation framework described in [8], is depicted in Fig. 10.1. The inputs to this methodology are a set of target applications to be mapped on a NoC and a number of design constraints and goals, such as the desired throughput, the power/energy budget and the affordable silicon area overhead. More specifically, the target application is represented as a communication graph, where we edges denote the communication link between computational kernels and/or memory blocks, whereas their weight gives the amount of data that has to be exchanged between them.

The first design phase involves the application modeling and optimization. A careful and precise analysis of application characteristics, mainly the communication bandwidth needed is essential in order to make a rough plan about the design space and components needed by it. In order to perform the exploration for alternative topologies of NoC architectures, we have used as a basis the Worm_Sim NoC

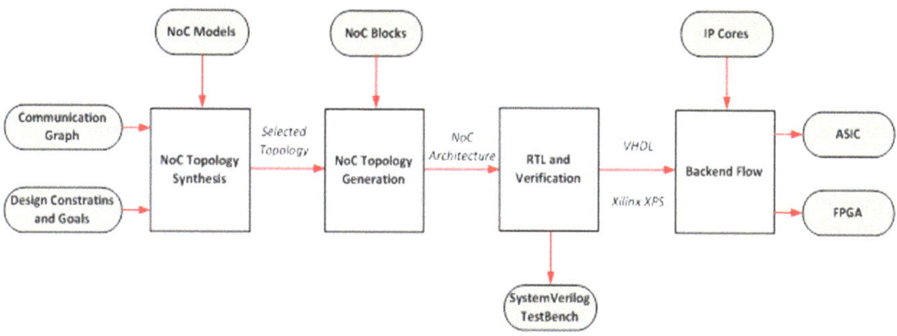


**Fig. 10.1** SYSMANTIC exploration and implementation flow

Simulator [9] that utilizes wormhole Routing. The main objective of the methodology is to derive an architecture, either homogeneous or heterogeneous that performs best to the generated traffic based on the application(s) activity. For this purpose, a number of different traffic models might be employed (e.g. random, bit reversal, butterfly, etc) [10]. The goal during this phase is to stress the NoC and find an efficiency topology, without conjunction points and with an acceptable trade-off between minimum delay and energy consumption. The outcome from this step is an abstract instantiation of the target topology.

Then, application mapping onto the target NoC-based platform is performed. During the second step of the methodology, we also deal with the optimization of the available communication resources. The goal is to choose an efficient routing algorithm and a flow control mechanism, including specifications of the routers and their number of connections with PEs that will ensure that the on-chip traffic will be handled, according to the desired specifications. The key performance metrics here are the average and maximum packet latency, the throughput of the network and the communication bandwidth, while important cost metrics are power consumption of communication on network and its overhead in the overall consumption.

We employ a bandwidth-constrained mapping algorithm [11]. The mapping decisions are evaluated using a high-level NoC Simulator [12]. The simulation results (average packet latency and energy consumption) reveal which is the NoC architecture that accommodates best the requirements of the application(s), deriving the optimal NoC topology. We have to notice that the employed mapping algorithm is orthogonal to any other available solution. The derived NoC can then be constructed by appropriately configuring and connected a number of NoC Routers. The results of this early stage high-level exploration phase are the directives that are passed to next step and used to create the RTL description of the interconnection network. Constraints to this procedure are the timing, power and area specifications, in respect to meet the desired communication (traffic) among IP cores. For the sake of completeness, the derived results were also evaluated with the usage of commercial flows.

Finally, the last step in our framework deals with the validation and synthesis of the NoC platform which result to a prototype system. The NoC components are based on a highly configurable at compile-time NoC Router, which can be instantiated with different number of ports, according to the derived topology. The SYSMANTIC framework, automatically VHDL descriptions of all NoC routers and a top VHDL description providing the appropriate connection of the NoC routers. Furthermore, a set of self-checking SystemVerilog testbenches that correspond to the selected topology is also generated. The testbenches create random packets at random intervals at random nodes, and check if they are received at the nodes they were sent to.

Moreover, in order to facilitate FPGA emulation for further validation, the architecture .mhs file of the Xilinx XPS is automatically generated based on a base template that was developed. The automatically generated .mhs file together with the corresponding NoC component RTL and the IP cores provided by XPS or third parties, can then be implemented in an off-the-shelf FPGA with minimal designer interaction.

## 10.3 NoC RTL Components

The basic component of the NoC architecture is the NoC Router, designed in reusable VHDL [13] in order to be highly configurable at compile time. For regular NoCs, it is a typical five-port Router (one port for the processing element, and the remaining four for connections in four directions). However, the number of ports, phit size (word-length), flit size, packet size and buffer size are configurable, through a package file. The Router simplified block diagram is shown in Fig. 10.2. For simplicity, the processing element link is not shown.

The Router communication schema is shown in Fig. 10.3. As shown, a Router is composed of a series of buffers to store incoming packets, a switch matrix implementing the input/output connections and a control logic, composed of two state machines for each port, one controlling the buffer write (RX control logic) and the other one responsible for scheduling/synchronizing the output port (TX control logic). The output port selector, allocates the appropriate output buffer, depending on the incoming packet destination address. The router uses output and not input queuing [14], with the modification that there is a different buffer for each path, which results in maximum throughput for an additional buffer area penalty, a clear performance/area trade-off. Each Router in the NoC has an ID register that contains the address of the Router. Finally, the prioritizer block selects between buffered packets that request the same output port in a round-robin fashion.

Furthermore, besides the generic router described above, a novel router architecture based on a fuzzy-logic adaptive routing algorithm was developed and presented in [15]. The proposed routing algorithm decides on the output port of an incoming flit by taking into account the dynamic traffic and power consumption on neighboring router links.

We are going to briefly describe the function of each of these components and how they communicate with each other.

NoC RTL generation is done with a novel software tool we developed, called *NoCGen*. The input to *NoCGen* is the output of the topology synthesis tool. The

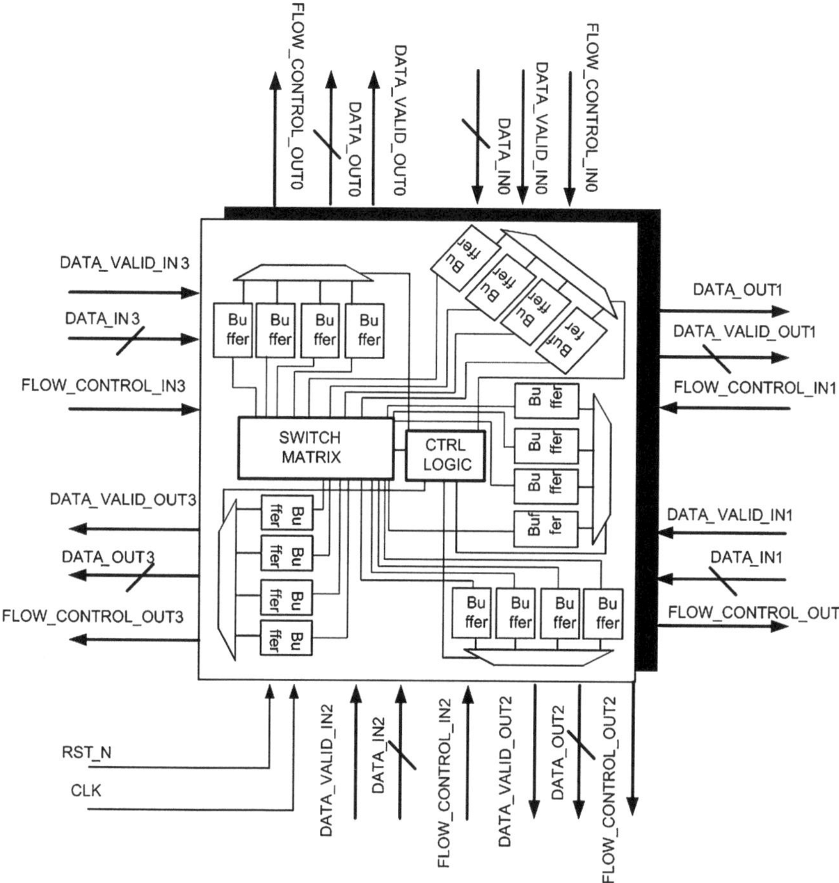


**Fig. 10.2** SYSMANTIC 5-port router simplified block diagram (local node link not shown)

input is composed of three configuration files written in XML as shown in Fig. 10.4. The first one named "elements.xml" contains information about the buffer sizes and number of output ports of each of the routers. These characteristics are determined by system-level exploration as well and consideration of the feedback coming from the network function. The second file, "netlist.xml", contains information about the topology of the design, namely the way the cores are connected with each other. The last file, "traffic.xml", contains some samples of the traffic generated during the execution of the above applications. The above XML document describes the topology of Fig. 10.5.

*NocGen* is an XML-based tool, which generates automatically an RTL description of a Network on Chip with the properties described in the above mentioned configuration files. It parses the aforementioned xml files and writes the VHDL RTL description of the NoC including the appropriate package files, based on the router

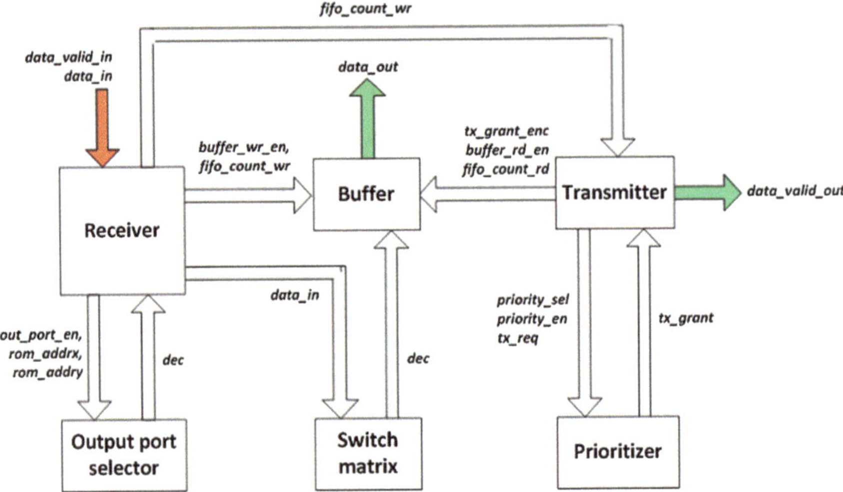


**Fig. 10.3** Communication schema of SYSMANTIC router

```
<netlist>
<link ID="link1">
<sourceRouter> 0 </sourceRouter>
<sourcePort> 1 </sourcePort>
<destinationRouter> 1 </destinationRouter>
<destinationPort> 3 </destinationPort>
</link>
<link ID="link2">
<sourceRouter> 1 </sourceRouter>
<sourcePort> 2 </sourcePort>
<destinationRouter> 3 </destinationRouter>
<destinationPort> 0 </destinationPort>
</link>
<link ID="link3">
<sourceRouter> 0 </sourceRouter>
<sourcePort> 2 </sourcePort>
<destinationRouter> 2 </destinationRouter>
<destinationPort> 0 </destinationPort>
</link>
</netlist>
```

**Fig. 10.4** Example XML topology description

architecture template described in the previous section. These VHDL files have some standard and some reconfigurable parts. The tool writes the reconfigurable part, which has to do with the topology, the traffic and some of the sizes of the router according to the specifications. Essentially, the tool instantiates routers such as the one described in the previous section, configured and connected together according to the XML topology description provided by the exploration tool.

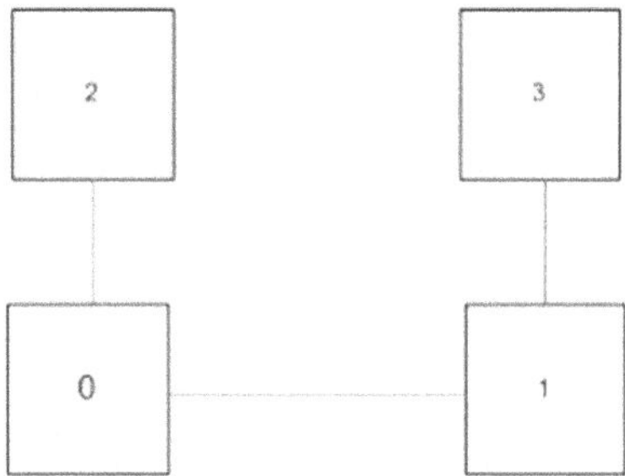


**Fig. 10.5** The topology of the NoC described in the XML document

## 10.4 Verification Components

Similarly, a set of self-checking SystemVerilog testbenches for the generated RTL is also produced by *NoCGen*. This facilitates the verification of the RTL code which can be a very time-consuming process. The generated testbenches are based on an OVM-based NoC testbench template that was developed and is configured by *NoCGen* by writing appropriate package files, similarly to the RTL.

The basic components of the testbench are the generator, the driver, the DUV interface, the monitor and the checker. The generator creates random NoC traffic, according to the specified NoC topology and feeds it to the driver that converts them to transactions according to the NoC flow control protocol. The DUV interface is a wrapper that facilitates the connection of the SystemVerilog testbench to the VHDL NoC DUV. The monitor reads the DUV response and feeds it to the checker which compares it with the expected response. Using this template a high coverage was achieved in short simulation time.

## 10.5 Rapid Prototyping Components

Finally, the architecture file (.mhs) [16] required for the rapid prototyping of the system in Xilinx devices using XPS [17] that corresponds to the selected architecture is automatically generated. Together with the automatically generated RTL of the NoC and the processing element IPs, this significantly reduces prototyping and, therefore, verification time.

*NoCGen* is implemented in C++. In order to parse the XML files, we used the library *xerces-c-3.0.1*. Xerces is a validating XML parser written in a portable subset of C++. It gives the ability to read and write XML data. A shared library is provided for parsing, generating, manipulating, and validating XML documents using the DOM, SAX, and SAX2 APIs. Xerces-C++ is faithful to the XML 1.0 recommendation and many associated standards. The parser provides high performance, modularity, and scalability.

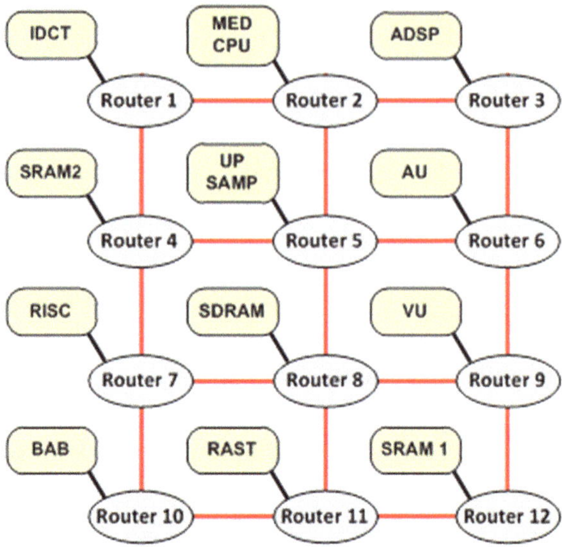


**Fig. 10.6** Block diagram of MPEG-4 (12 cores)

## 10.6 Case Studies and Experimental Results

In order to validate the functionality of the tool flow, we consider four case studies, one with a regular and one with an irregular topology. Prototyping was done on a Xilinx Virtex-4 device [18].

The traffic trace as well as the position of every router on chip is derived from the simulation results of previous step, whose task was the system-level exploration of NoCs. The buffer depth of each router was determined by the network synthesis tool according to the sample traffic needs but we include experimental results for various buffer sizes in order to investigate the effect of buffer size on performance, area and power. Measurements include power dissipated, the maximum operating frequency, the slices and utilization for each implementation for different buffer sizes.

### *10.6.1 Case Study 1: MPEG-4*

For the first case study, as target application we used MPEG-4, a broadly used protocol for audio and video encoding [19]. A hardware encoder and decoder consist of many components, so a NoC approach is suitable. The tested MPEG-4 includes various processing elements, such as a video unit, an audio unit, a RISC processor, a med CPU, a binary alpha block and three SRAMs. The total number of cores needed for

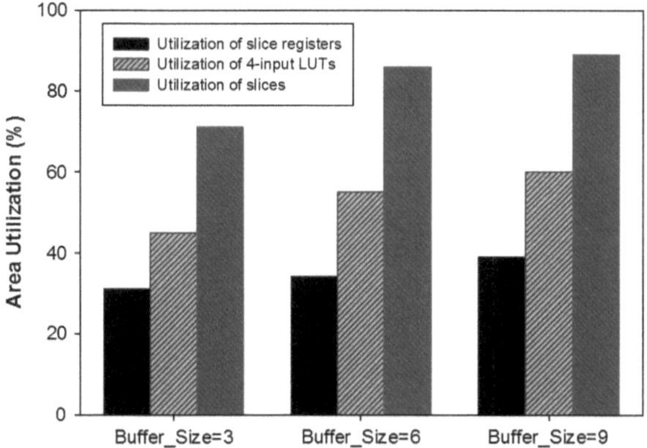


**Fig. 10.7** MPEG-4 area utilization

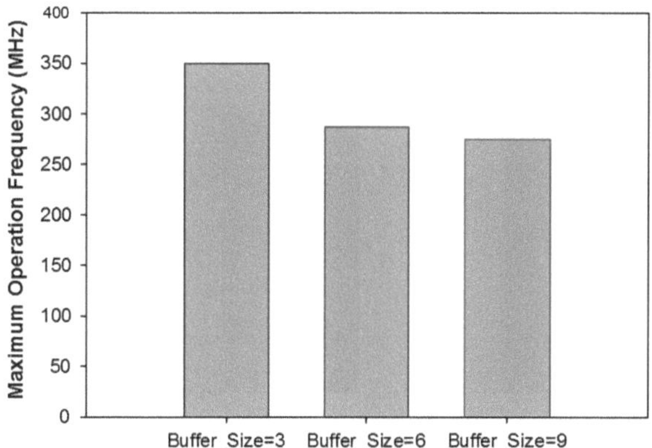


**Fig. 10.8** MPEG-4 maximum frequency

the application is 12, with an equal number of routers attached to them as shown in Fig. 10.6.

The maximum frequency at which the chip can function is 276 MHz in case of buffer size 9. This frequency is well greater than our clock frequency (100 MHz). Furthermore, we notice that the ratio in performance degradation in term of maximum operation frequency is greater from buffer size 3 to 6 than buffer size 6 to 9. This is due to the saturation, which takes place when the buffer size is 6.

Figure 10.7 shows the area utilization, in terms of registers, slices and LUTs. As we can see in Fig. 10.7 the required resources of the chip increase as the buffer size increases. In the case of buffer size 6 and 9 a significant percentage of the chip slices are used (almost 90 %). A larger buffer size would take up all the chip resources.

Figure 10.8 shows the maximum operating frequency for various buffer sizes, while Figs. 10.9 and 10.10 show the total and leakage power dissipation respectively. It can be seen that the total power dissipation increases as the buffer size increases.

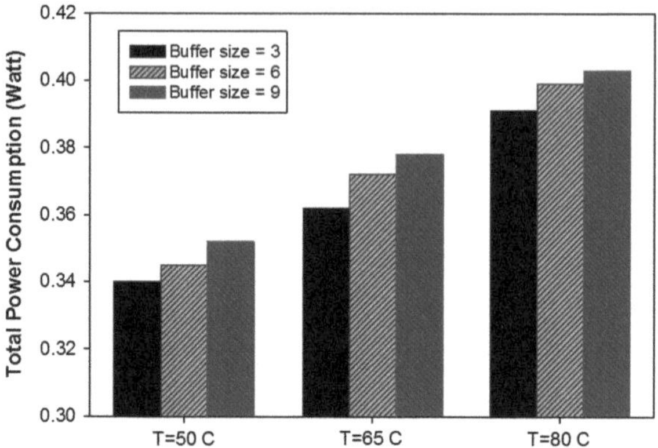


**Fig. 10.9** MPEG-4 total power dissipation

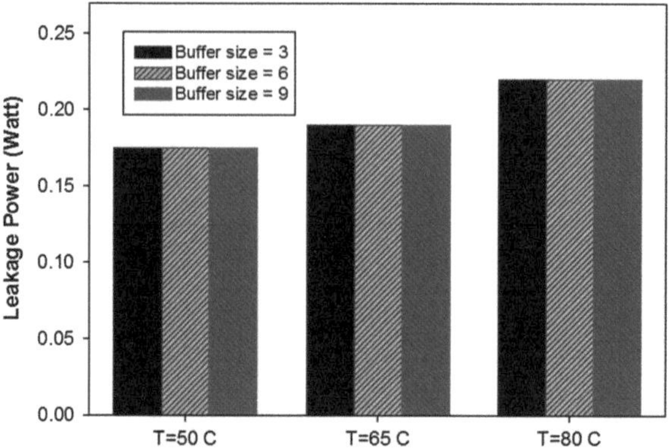


**Fig. 10.10** MPEG-4 leakage power dissipation

Furthermore, as expected increase of chip temperature influences significantly the power. For this application leakage power (Fig. 10.10) is 0.168 at 50 °C, 0.191 at 65 °C, 0.219 at 80 °C independent of the buffer size. The leakage power for given temperature is constant among the different buffer sizes, because it depends mainly on temperature.

### *10.6.2 Case Study 2: VOPD*

Video object plane decoder is another digital signal processing application that has been SYSMANTIC for use on NoC and studied before [20]. VOPD offers quality video transition with decent bandwidth performance. The tested VOPD decoder includes twelve processing elements, such as two length decoders, an AC-DC

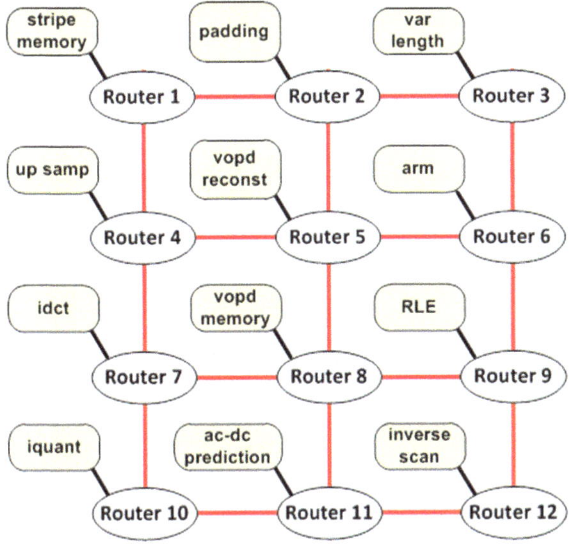


**Fig. 10.11** Block diagram of VOPD (12 cores)

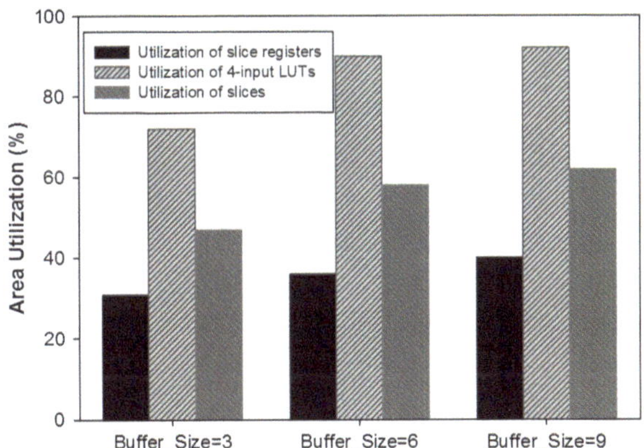


**Fig. 10.12** VOPD area utilization

prediction, an ARM processor, two memory components and a VOP reconstructor. The cores needed for the application are 12 and so many are the routers as well (Fig. 10.11). Area utilization, frequency and power dissipation are depicted in Figs. 10.12, 10.13, 10.14 and 10.15.

As we can see the resources of the chip required increase as the buffer size increases. In the case of buffer size 6 and 9 a significant percentage of the chip slices are used (almost 90 %). A larger buffer size would take up all the chip resources.

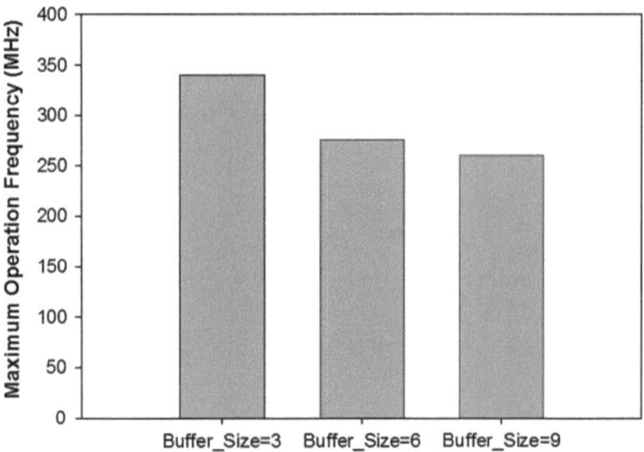


**Fig. 10.13** VOPD maximum frequency

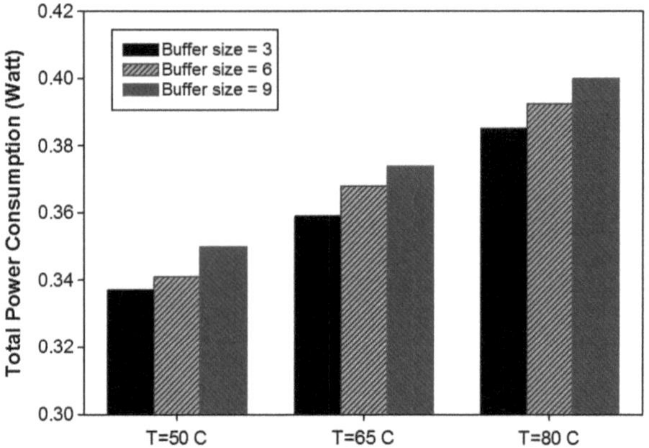


**Fig. 10.14** VOPD total power dissipation

The maximum frequency at which the chip can function is 260 MHz in case of buffer size 9. This frequency is well greater than our clock frequency (100 MHz). Furthermore, we notice that the ratio in performance degradation in term of maximum operation frequency is greater from buffer size 3 to 6 than buffer size 6 to 9. This is due to the saturation, which takes place when the buffer size is 6.

The total power dissipation increases as the buffer size increases. Furthermore, as expected increase of chip temperature influences significantly the power. For this application leakage power is again 0.168 at 50 °C, 0.191 at 65 °C, 0.219 at 80 °C independent of the buffer size. The leakage power for given temperature is constant among the different buffer sizes, because it depends mainly on temperature.

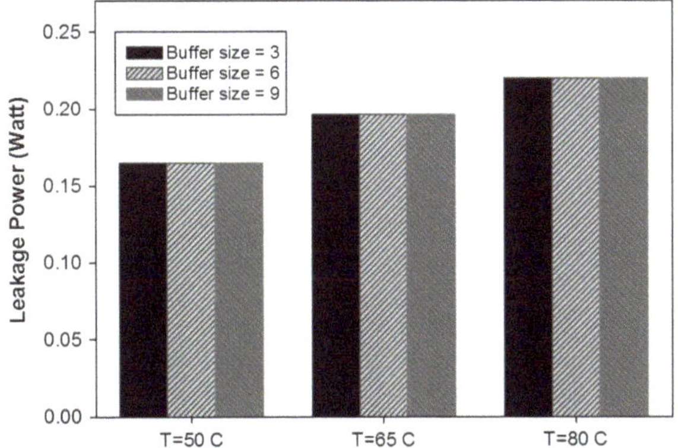


**Fig. 10.15** VOPD leakage power dissipation

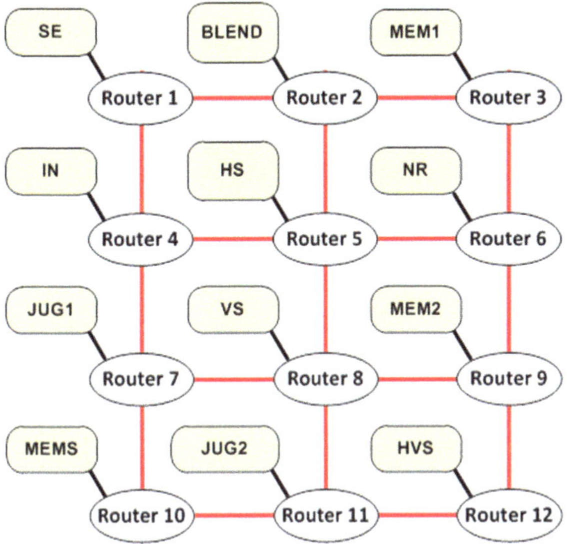


**Fig. 10.16** Block diagram of MWD (12 cores)

### *10.6.3 Case Study 3: MWD*

Multi-window display is another digital signal processing application which is also suitable for NoC architectures and also uses twelve processing elements. The cores needed for the application are 12 and so many are the routers as well (Fig. 10.16).

As we can see the resources of the chip required increase as the buffer size increases (Fig. 10.17). In the case of buffer size 6 and 9 a significant percentage of the chip slices are used (almost 86 %). A larger buffer size would take up all the chip resources.

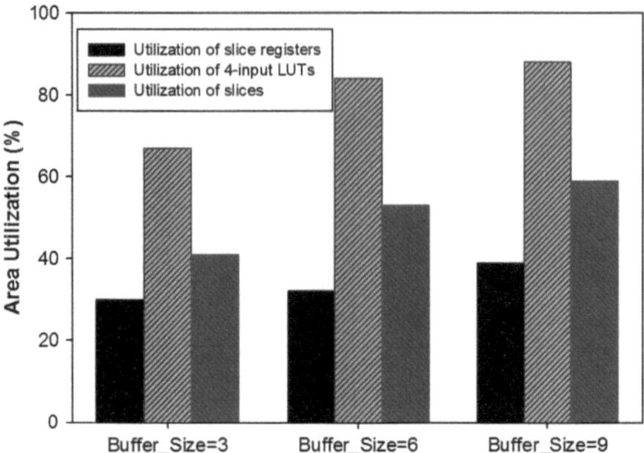


**Fig. 10.17** MWD area utilization

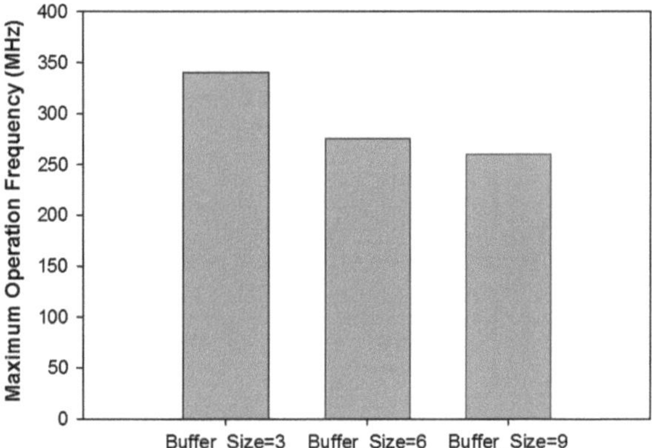


**Fig. 10.18** MWD maximum frequency

The maximum frequency at which the chip can function is 260 MHz in case of buffer size 9 (Fig. 10.18). This frequency is well greater than our clock frequency (100 MHz). Furthermore, we notice that the ratio in performance degradation in term of maximum operation frequency is greater from buffer size 3 to 6 than buffer size 6 to 9. This is due to the saturation, which takes place when the buffer size is 6.

The total power dissipation increases as the buffer size increases. Furthermore, as expected increase of chip temperature influences significantly the power (Fig. 10.19). For this application leakage power is again 0.168 at 50 °C, 0.191 at 65 °C, 0.219 at 80 °C independent of the buffer size. The leakage power for given temperature is constant among the different buffer sizes, because it depends mainly on temperature (Fig. 10.20).

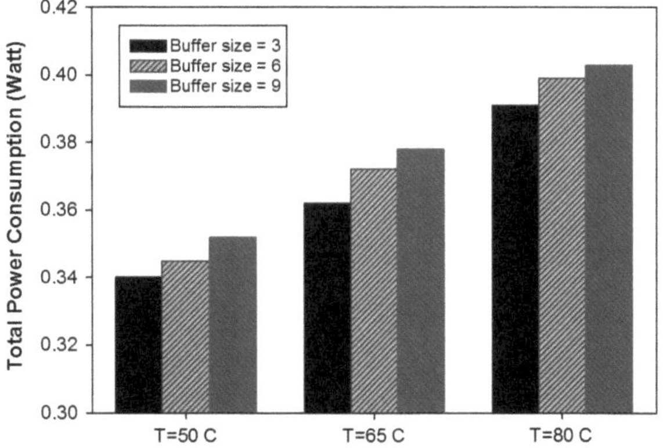


**Fig. 10.19** MWD total power dissipation

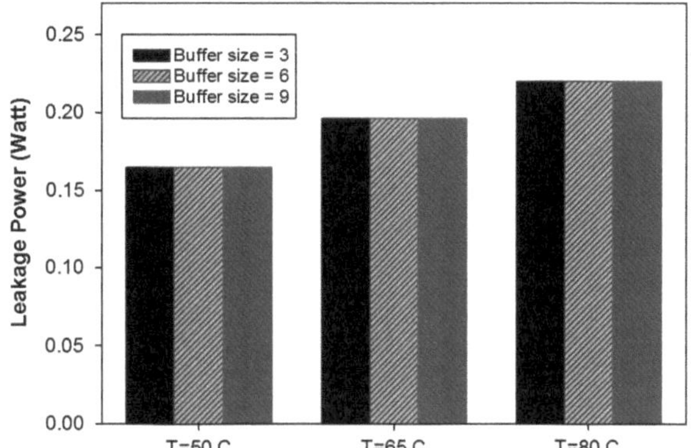


**Fig. 10.20** MWD leakage power dissipation

## *10.6.4 Case Study 4: MMS*

The last case study is a multimedia system (MMS). The system contains 25 cores, including several memories and DSP processors as shown in Fig. 10.21. In this case, we use three routers, since the strategy according to which a router attached to each core requires far too many resources. The clustering of routers was selected in order to achieve traffic minimization. The existence of communication links between routers ensures the lack of deadlock and livelock. This case study was selected because, the NoC topology is very different from the mesh of case studies 1, 2 and 3, and it is important to validate that the flow can support both regular and irregular topologies.

Similar to the previous case studies, Fig. 10.22 shows the area utilization in terms of registers, slices and LUTs. As we can see the resources of the chip required increase as the buffer size increases. But even in the case of buffer 9, only a small percentage

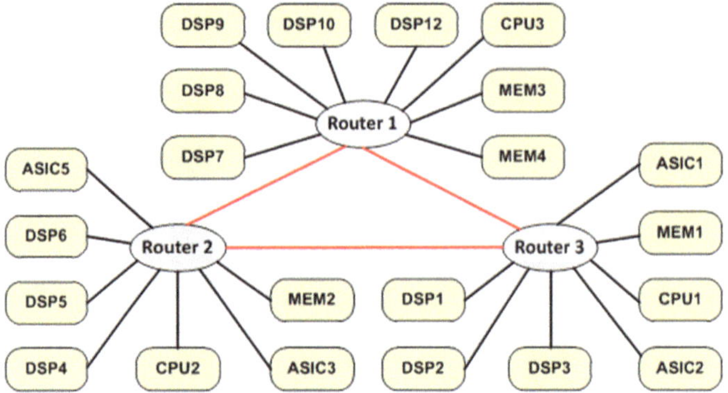


**Fig. 10.21** Block diagram of MMS

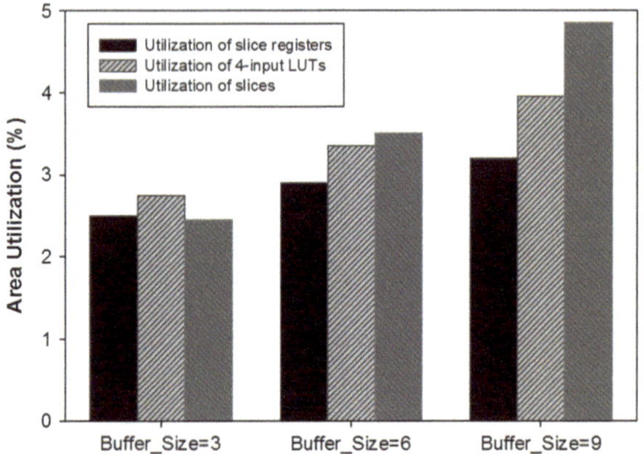


**Fig. 10.22** MMS area utilization

of the chip slices are used (about 6 %). A larger buffer size could be used in this case, if the traffic is to be serviced more efficiently.

Figure 10.23 shows the maximum operating frequency. The maximum frequency at which the chip can function is 312 MHz in case of buffer size 9. This frequency is well greater than our clock frequency (100 MHz).

Figures 10.24 and 10.25 show the total and leakage power consumption respectively. The total power dissipation increases as the buffer size increases. Furthermore, as expected increase of chip temperature influences significantly the power. For this application leakage power is 0.165 at 50 °C, 0.187 at 65 °C, 0.216 at 80 °C independent of the buffer size. Although the design is significantly smaller than MPEG-4, the leakage values are close.

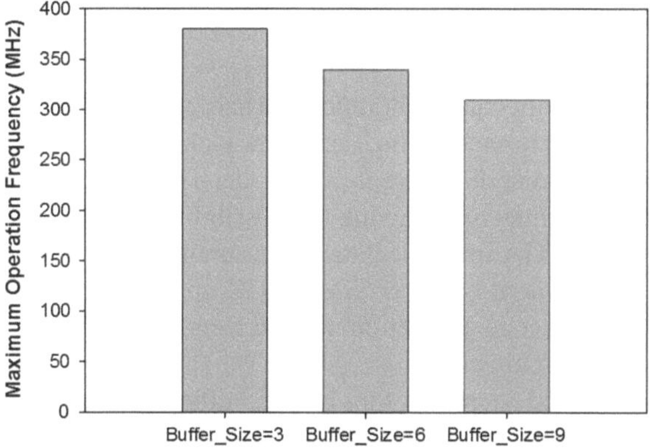


**Fig. 10.23** MMS maximum frequency

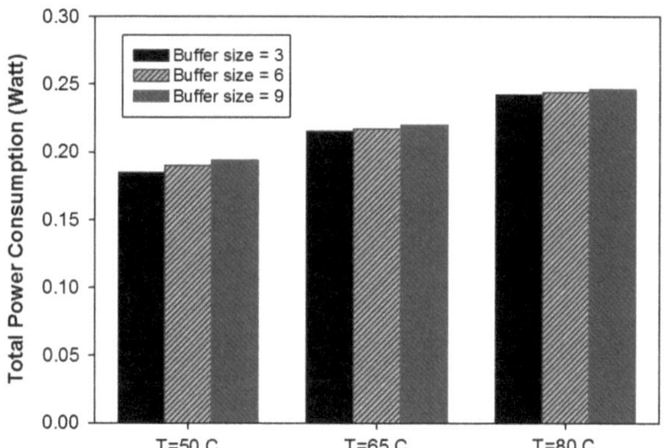


**Fig. 10.24** MMS total power dissipation

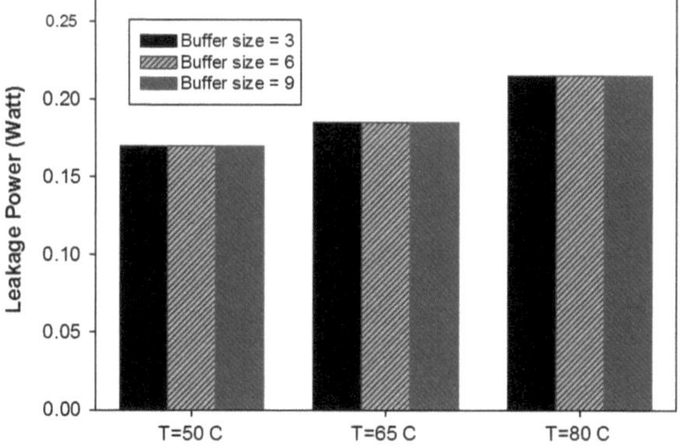


**Fig. 10.25** MMS leakage power dissipation

## 10.7 Conclusion

A framework for the architecture exploration and hardware implementation of custom NoC architectures was presented. The framework includes the automated translation of the derived optimized architecture to RTL code. The flow contributes to the fast prototyping of NoC architectures, giving the possibility for efficient generation and debugging of NoCs. FPGA implementation measurements for regular and irregular topologies proved the merit of the SYSMANTIC approach. Future work includes full automation of RTL code generation, as well as providing hardware prototyping results for NoC mesh architecture.

The power dissipated increases too, as the buffer size increases. This can be explained by the fact that the additional buffer space needs more chip area, which increases the overall power (dynamic and through leakage) produced. As the buffer size increases, the maximum frequency at which the design can function decreases. That happens because a design with many slices requires more interconnects, which results in delays and need for a slower clock. Future plans include evaluating the SYSMANTIC framework with more topologies and prototyping 3-D NoC architectures using state-of-the-art Xilinx Virtex-7 HT devices which feature 3-D stacking.

## References

1. A. Ehliar, D. Liu, An FPGA Based Open Source Network-on-Chip Architecture, in *International Conference on Field Programmable Logic and Applications (FPL)*, pp. 800–803 (2007)
2. A. Kumar, A. Hansson, J. Huisken, H. Corporaal, An FPGA Design Flow for Reconfigurable Network-Based Multi-Processor Systems on Chip, in *Design, Automation & Test in Europe Conference & Exhibition (DATE)*, pp. 1–6 (2007)
3. G. Leary, K. Chatha, A Holistic Approach to Network-on-Chip Synthesis, in *International Conference on Hardware/Software Codesign and System, Synthesis (CODES+ISSS)*, pp. 213–222, (2010)
4. A. Strano, D. Bertozzi, F. Angiolini, L. Di Gregorio, F. Sem-Jacobsen, V. Todorov, J. Flich, F. Silla, T. Bjerregaard, Quest for the Ultimate Network-on-Chip: the NaNoC Project, in *Interconnection Network Architecture: On-Chip, Multi-Chip Workshop (INA-OCMC)*, pp. 43–46 (2012)
5. Y. Ogras, R. Marcillescu, H. Lee, P. Choudhary, D. Marculescu, M. Kaufman, P. Nelson, Challenges and Promising Results in NoC Prototyping Using FPGAs, in *IEEE Micro*, vol. 27, No. 5, pp. 86–95, Sep–Oct 2007
6. T. Le, M. Khalid, NoC Prototyping on FPGAs: A Case Study Using an Image Processing Benchmark, in *International Conference on Electro/Information Technology (EIT)*, pp. 441–445 (2009)
7. Y. Krasteva, F. Criado, E. de la Torre, T. Riesgo, A Fast Emulation-Based NoC Prototyping Framework, in *International Conference on Reconfigurable Computing and FPGAs (ReConFig)*, pp. 211–216 (2008)
8. K. Tatas, K. Siozios, A. Bartzas, C. Kyriacou, D. Soudris, A Novel Prototyping and Evaluation Framework for NoC-based MSoC, accepted for publication, in *Networked Embedded Systems: Special Issue International Journal of Adaptive IGI-Global, Resilient and Autonomic Systems (IJARAS)*, 2012

9. H. Lee, N. Chang, U. Ogras, R. Marculescu, On-chip communication architecture exploration: A quantitative evaluation of point-to-point, bus, and network-on-chip approaches, in *ACM Transactions on Design Automation of Electronic Systems (TODAES)*, vol. 12, No. 3, Article No. 23, 2007
10. W. Dally, B. Towles, *Principles and Practices of Interconnection Networks*, (Morgan Kaufmann Publishers Inc., San Francisco, 2003)
11. S. Murali, G. De Micheli, Bandwidth-Constrained Mapping of Cores onto NoC Architectures, in *Design, Automation and Test in Europe Conference and Exhibition (DATE)*, pp. 896–901, (2004)
12. A. Bartzas, N. Skalis, K. Siozios, D. Soudris, Exploration of Alternative Topologies for Application-Specic 3D Networks-on-Chip, in *Workshop on Application Specific Processors (WASP)*, 2007
13. M. Keating, P. Bricaud, *Reuse Methodology Manual for System-On-A-Chip Designs*, (Springer, Berlin, 2006)
14. E. Rijpkema, K. Goossens, A. Radulescu, J. Dielissen, J. van Meerbergen, P. Wielage, E. Waterlander, Trade-Offs in the Design of a Router With both Guaranteed and Best-Effort Services for Networks on Chip, in *IEE Proceedings on Computers and Digital, Techniques*, vol. 150, No. 5, pp. 294–302, Sept 2003
15. C. Chrysostomou, K. Tatas, A. R. Runcan, A Dynamic Fuzzy Logic Based Routing Scheme for Bufferless NoCs, in *IEEE 15th International Conference on Computational Science and Engineering (CSE)*, pp. 295–302 (2012)
16. Platform Specification Reference Manual (available at http://www.xilinx.com/support/documentation/sw_manuals/xilinx13_2/psf_rm.pdf)
17. Xilinx Platform Studio and the Embedded Development Kit (EDK) (available at http://www.xilinx.com/tools/platform.htm)
18. Virtex-4 FPGA Data Sheets (available at http://www.xilinx.com/support/documentation/virtex-4_data_sheets.htm)
19. I. Richardson, *H.264 and MPEG-4 Video Compression: Video Coding for Next Generation Multimedia*, (Wiley, Chichester, 2003)
20. V. Ngo, H. Nguyen, H. Choi, The Optimum Network on Chip Architectures for Video Object Plane Decoder Design, in *International Conference on Parallel and Distributed Processing and Applications (ISPA)*, pp. 75–85 (2006)

# Part II
# Suggested Projects

# Chapter 11
# Projects on Network-on-Chip

## 11.1 Develop a NoC Architecture in SystemC

Throughout this project, we aim to develop a regular NoC topology in System C. More specifically:

- Develop a router in SystemC. This router should be parametric in order for it's architectural parameters to be easily extended. Initiate the derived solution by setting one local port per router and each router to be connected with up to four neighbor routers (2-D mesh topology). The packet routing in this NoC has to be based on XY algorithm. Simulate both the router, as well as the NoC in order to verify proper functionality.
- Extend the router of previous step in order to be 3-D aware. For this purpose, both the architectural, as well as the routing algorithm should be enhanced. Could you derive a solution that minimizes delay more than the XYZ algorithm? Evaluate the derived results.
- Simulate both the 2-D and 3-D routers in order to verify their proper functionality.
- Finally, develop a $3\times3$, as well as a $3\times3\times2$ (3-D with two layers) NoC architecture consisted of routers developed in former steps.

**Additional practice:** Apply the concept of virtual layers and virtual networks discussed in Chap. 9 for evaluating the performance and power metrics of the 3-D NoC.

## 11.2 Fairness in a Packet Router with VCs

For this project, first download SystemC (http://www.systemc.org) and extend the SystemC Helix multicast 2x2 router in the examples directory (developed by Synopsys) to a $5\times5$ unicast router with virtual channels [1]. More specifically, add a 2-bit flag to the packet structure (called VC_ID) to select one of four different virtual channels which form distinct physical paths through the switch but share the links; distinct paths are supported by four sets of independent input and output buffers.

K. Tatas et al., *Designing 2D and 3D Network-on-Chip Architectures*, 
DOI: 10.1007/978-1-4614-4274-5_11, 

Within the extended 5×5 switch, both VC allocation and link arbitration (port scheduling) must be tackled [1]. Moreover, in respect to a particular VC allocation, network fairness refers to whether requests from different VCs and/or different incoming ports are receiving a fair share of system resources. Focusing on mathematical and conceptual definitions of different fairness metrics (see bibliography [2–5]) and additional papers from ATM networking domain (provided later if necessary), propose and evaluate the performance and fairness of one or more combinations of (VC allocation, link arbitration) algorithms able to drive packets from different incoming VCs/ports to different outgoing VCs/ports, ideally in an equally balanced way.

In order to evaluate your routing algorithm, consider a random destination scenario (with/without hot spots) and compute performance, as well as corresponding fairness metrics. For your performance computations you must apply the following definitions and CNF normalizations.

The performance of an on-chip network under dynamic load is usually assessed by two quantitative parameters, the accepted bandwidth or throughput and the latency.

- Accepted bandwidth is defined as the sustained data delivery rate given some offered bandwidth at the network input. Two important characteristics are the saturation point and the sustained rate after saturation. Saturation is defined as the minimum offered bandwidth where the accepted bandwidth is lower than the global packet creation rate at the source nodes. It is worth noting that, before saturation, offered and accepted bandwidth are the same. The behavior above saturation is important (and complex to analyze mathematically) because the network and/or the routing (VC allocation, link arbitration) algorithms can become unstable, leading to sharp performance degradation. We usually expect the accepted bandwidth to remain stable after saturation, both in the presence of bursty applications that require peak performance for a short period of time and applications that operate after saturation in normal conditions, e.g., when executing a global permutation pattern.
- Network latency is the average delay spent by a packet in the network, from the insertion of the header flit in the injection lane till the reception of the tail flit at the destination. It does not include the source queuing delay. The end-to-end latency rises to infinity above saturation and is impossible to gain any information in this case. For this reason, the network latency is often preferred to analyze the network performance near saturation.

The experimental results of your traffic patterns must be presented according to the Chaos Normal Form (CNF) [2]. The CNF uses two graphs, one to display the accepted bandwidth and the other to display the network latency. In both graphs, the x-axis corresponds to the offered bandwidth normalized with the unidirectional maximum bandwidth of the links connecting the processing nodes to the network switches. This makes the analysis independent of the link bandwidth and the packet (or flit) size. In a more general setting with interconnection networks that are bisection-bandwidth limited (e.g., if you connect several routers together in a mesh), CNF throughput is normalized with the bisection bandwidth, the upper bound on the throughput for uniform traffic.

The following literature is helpful. The last paper refers to VC scheduling for deadlock avoidance and several similar papers/patents exist that relate to improved performance or QoS [1–6].

**Additional practice:** extend your SystemC version to RTL and co-simulate these two versions, e.g., using ModelSim.

## 11.3 SoC Simulation Using GEM5

- Install GEM5 (http://www.m5sim.org) and build build/ARM/gem5.opt (ARM architecture) in gem5 simulator.
- Run gem5.opt, using the simulation script configs/example/ruby_network_test.py, 4 CPUS, 4 DIRS (memories) and Point-to-Point topology (*–topology=Pt2Pt*).

– Enable tracing using flags *–trace-start=0 –trace-file=output.txt*
– Enable the following debug flags: *RubyPort*, *RubySequencer*, *NetworkTest*, *RubyNetwork*, *RubyTest* and *RubyGenerated*.

- Check the tracefile of the simulation in file *m5out/output.txt*. Extract information about the Message (NetworkMessage) as printed by the *PerfectSwitch* class. Describe the various fields that are printed.
- Create a new chordal topology (named *xnet*) based on the Pt2Pt topology described in configs/topologies/Pt2Pt.py. In this xnet topology, each node $n$ is connected a) to the previous node $(n + max\_nodes - 1)\,\%max\_nodes$, b) to the next node $(n+1)\,\%max\_nodes$ and c) to the node across $(n+\frac{max\_nodes}{2})\,\%max\_nodes$. For example in an 8-node xnet, $n$=0, 1, ..., 7 network node 0 is connected respectively to nodes 7, 1, and 4. All other links from Pt2Pt network topology should be eliminated. (In Pt2Pt topology each node is directly connected to all other nodes of the network).
- Which python file should be modified in order to change the number of Virtual Networks (VNETs) used in your Network?
  - What is the minimum number of VNETs you can use? Why it cannot be reduced beyond that number?
  - What types of VNETs are used? *Hint: Check method MessageBuffer* getToNetQueue(NodeID id, bool ordered, int network_num, std::string vnet_type) which is defined in src/mem/ruby/network/simple/SimpleNetwork.hh in order to get more information.*

## 11.4 NoC Simulation Using OMNeT++

- Install OMNeT++ [7] and compile the hypercube example found in folder samples.
- In this example, the hypercube router implemented in file HCRouter.cc uses deflection routing. This means that there is no buffer in the router module: if more than one packet needs to be sent out through the same gate, then one is deflected: it is sent out through another gate.
- Change this implementation in order to add buffers before the output ports in order to avoid packet deflection.
- Examine the use of routing tables in order to make the hypercube routing algorithm adaptive.

## 11.5 Mapping Communication to NoC Topology

Many available commercial and freely available software packages perform partitioning and mapping [13]. Common non-commercial packages include Chaco [8], Jostle [9], Metis [10], and Scotch [11], with the latter two considered faster. They all require a "dual graph" format used for representation of the application graph and architecture graphs.

Download and apply some of these packages in mapping communication graphs to architecture models. Concentrate on mapping common communication patterns (e.g., mesh, butterfly, trees, finite element graphs) onto different NoC topologies (e.g. mesh, torus, chordal rings, and fat trees). Measure both the execution time for different number of edges/vertices in the source and target graphs and quality of your embedding results by examining the following metrics.

- Edge dilation of an edge of GT is defined as the length of the path in GS onto which an edge of GT is mapped. The dilation of the embedding is defined as the maximum edge dilation of GT. Similarly, we define average and minimum dilation metrics. These metrics measure latency overhead during point-to-point communication in the target graph GT.
- Edge expansion refers to a weighted-edge graph GS. At first, it multiplies each edge dilation with its corresponding edge weight. The edge expansion of the embedding is defined as the maximum edge expansion of GT. Similarly, we define average and minimum edge expansion metrics.
- Edge congestion represents the maximum number of paths containing any edge in GT where every path represents an edge in GS. This metric measures edge contention in global intensive communication.
- Node congestion is the maximum number of paths that contain any node in GT where every path represents an edge in GS. This metric is a measure of node contention during global intensive communication.
- Node expansion of the embedding (called load factor or compression ratio) is the ratio of the number of nodes in GT to the number of nodes in GS. Similarly, max-

imum node expansion represents the maximum number of nodes of GS assigned to any node of GT.
- Number of cut edges, incident to vertices of different partitions; this metric is used for comparing target graphs with the same number of edges, the smaller the metric the better.

**Additional practice:** Consider the use of TGFF [12] to automatically generate samples of communication graphs with different types of properties. The following literature is helpful [8–13].

## 11.6 Multicore Processing Using FPGA

*Prerequisite* You are expected to have a basic experience of using major EDA development tools (e.g., by Xilinx, Altera, Cadence, Mentor, etc.) to design hardware modules in RTL using a hardware description language (HDL), or simple embedded reconfigurable systems. This project will not be providing details on how to build the basic system. A development board (Xilinx's ML403, ML505, or any lower cost Spartan-based board) is recommended for a complete lab experience of the project.

*Description* The purpose of this project is to investigate a complete hardware and software processor system design using the Xilinx EDK/SDK tools. In this project, you will use the BSB of the XPS system to automatically create a processor system and then add a second processor to that processor system connected to the PLB bus and directly to the first processor.

Create an XPS Project by using Base System Builder (BSB). In the next dialog boxes that you choose, I would like to create a new design to select the available board of your choice. The MicroBlaze processor can be connected to many different cores which are system memory-mapped. Some of these will be useful when you are implementing your own project.

From the System Assembly View, you can see that a MicroBlaze-based system with the specified peripherals has been generated, and all the connections between components are created accordingly. The local memory bus (LMB) is used by the MicroBlaze core to control the dlmb_cntlr and ilmb_cntlr units. The microblaze_0 is the master processor while the controlled components are the slaves.

Under Implement Flow, Generate Netlist implements the design using the Xilinx backend, going through synthesis and place and route to create the final netlist. Generate BitStream generates the actual bitstream using bitgen. Export Design is used to create a hardware platform description, which can be used for software development.

Fast Simplex Link (FSL) provides a point to point communication between any two components on the FPGA. Add a second MicroBlaze processor connected on the PLB bus and, two LMB controllers with an additional BRAM from the IP catalog tab. As the FSL is unidirectional FIFO-based point-to-point communication bus, add two FSL links between them. Each should support transfer of eight 32-bit words.

At the ports tab carefully connect the system clock and reset to the new-added cores. At the Addresses tab allocate at least 16 KB of instruction and data memory to the new processor.

After generating the bitstream export the design to the Software Development Kit (SDK) and launch Xilinx SDK.

1. Initially test the system with the processor_0 to send a series of numbers to processor_1. This second processor should print them (using the *putnum()*, or the *xil_printf()* function) through the UART to the terminal of your host system connected to the prototype platform.
2. Use a publicly available algorithm (e.g., [14, 15]) to perform data compression on the data that arrive to processor_1. Thus, processor_0 can operate as a job scheduler processor assigning data to co-processors.
3. Through a shared memory, either the off-chip SRAM or DDR, perform the same scenario with Step 2, only that now processor_0 can send a pointer to the data chunk that reside at the shared memory.

### *11.6.1 Background Overview*

Xilinx Inc. MicroBlaze is a FPGA-based soft microprocessor IP core. It uses the Harvard RISC architecture and the structure of the 32-bit instruction and data bus, there are 32 general-purpose 32-bit width register; in the 150 MHz clock frequency, up to 125 DMIPS of processing performance. Xilinx offers the use of the EDK (Embedded System Development Kit), so that the provided library IP cores can be parameterized by the graphical interface to facilitate completion of the embedded soft-processor system design. MicroBlaze can be heavily customized to the needs of the target application by configuring its properties such as instruction and data cache sizes, use of a memory management unit, use of a floating point unit, etc.

MicroBlaze soft processor core provides rich interfaces.

- Processor Local Bus (PLB) interface, a traditional system memory-mapped transaction bus with master/slave capability.
- AXI interface as of 6-, 7-series devices.
- High-speed Local Memory Bus (LMB) interface.
- FSL master and slave device interface.
- Xilinx Cache Link (XCL) interface.
- With MDM (Microprocessor Debug Module) interface to connect the debugger.

MicroBlaze soft-core's FSL is a unique one-way communication method based on the FIFO links and can be user-defined; XCL is for the realization of the memory chip high-speed access. There is also a dedicated debug interface. Through parameter setting, developers can use only the required application-specific processor features.

# References


1. J.D. William, Virtual Channel Flow Control. IEEE Trans Parallel Distrib. Syst. pp. 194–205. http://cs472.pbworks.com/f/virtual. Accessed March 1992
2. http://www.cs.washington.edu/research/projects/lis/chaos/www/presentation.html (details on the presentation of network simulation results)
3. http://en.wikipedia.org/wiki/Fairness_measure
4. http://en.wikipedia.org/wiki/Max-min_fairness
5. http://www.ece.rutgers.edu/ marsic/Teaching/CCN/minmax-fairsh.html
6. https://cug.org/5-publications/proceedings_attendee_lists/2007CD/S07_Proceedings/pages/Authors/Abts/Abts_paper.pdf
7. http://www.omnetpp.org
8. Chaco, http://www.cs.sandia.gov/CRF/chac.html
9. Jostle, http://www.gre.ac.uk/_c.walshaw/jostle/
10. Metis, http://www-users.cs.umn.edu/_karypis/metis/
11. Scotch, http://www.labri.fr/Perso/_pelegrin/scotch/
12. TGFF, http://ziyang.eecs.umich.edu/ dickrp/tgff
13. Wikipedia - graph partitioning, http://en.wikipedia.org/wiki/Graph_partition
14. http://en.wikipedia.org/wiki/Run-length_encoding
15. https://code.google.com/p/lz4

MIX
Papier aus verantwortungsvollen Quellen
Paper from responsible sources
FSC® C105338

If you have any concerns about our products,
you can contact us on
**ProductSafety@springernature.com**

In case Publisher is established outside the EU,
the EU authorized representative is:
**Springer Nature Customer Service Center GmbH**
**Europaplatz 3, 69115 Heidelberg, Germany**

Printed by Libri Plureos GmbH
in Hamburg, Germany